AF127305

Biofuels Production and Processing Technology

Biofuels Production and Processing Technology

Editor

Alessia Tropea

MDPI • Basel • Beijing • Wuhan • Barcelona • Belgrade • Manchester • Tokyo • Cluj • Tianjin

Editor
Alessia Tropea
Department of Research and
Internationalization
University of Messina
Messina
Italy

Editorial Office
MDPI
St. Alban-Anlage 66
4052 Basel, Switzerland

This is a reprint of articles from the Special Issue published online in the open access journal *Fermentation* (ISSN 2311-5637) (available at: www.mdpi.com/journal/fermentation/special_issues/biofuels_production_processing).

For citation purposes, cite each article independently as indicated on the article page online and as indicated below:

LastName, A.A.; LastName, B.B.; LastName, C.C. Article Title. *Journal Name* **Year**, *Volume Number*, Page Range.

ISBN 978-3-0365-4824-1 (Hbk)
ISBN 978-3-0365-4823-4 (PDF)

Contents

About the Editor

Alessia Tropea

Dr Alessia Tropea works at the University of Messina, Italy. She holds an Europaeus Doctorate in Chemistry Sciences, achieved through work in both Italy and the United Kingdom. Her research focuses on the implementation of fermentation processes, mainly focused on food waste valorization for obtaining value-added products such as single-cell protein for feed, biofuel, and edible pigments. She has many years of experience in research, evaluation, and teaching in education institutions. She has been published in high-quality journals in the field, serves as a Topical Advisory Panel Member of the journal *Fermentation*, and she is a Guest Editor and Topical Collection Editor for various MDPI journals. Moreover, she has been a reviewer for several international journals.

Preface to "Biofuels Production and Processing Technology"

The negative impacts of global warming and global environmental pollution due to fossil fuels mean that the main challenge of modern society is finding alternatives to conventional fuels. In this scenario, biofuels derived from renewable biomass represent the most promising renewable energy sources. Depending on the biomass used by the fermentation technologies, it is possible to obtain first-generation biofuels produced from food crops, second-generation biofuels produced from non-food feedstock, mainly starting from renewable lignocellulosic biomasses, and third-generation biofuels, represented by algae or food waste biomass.

Although biofuels appear to be the closest alternative to fossil fuels, it is necessary for them to be produced in competitive quantities and costs, requiring both improvements to production technologies and the diversification of feedstock. This Special Issue is focused on technological innovations, including the utilization of different feedstocks, with a particular focus on biethanol production from food waste; different biomass pretreatments; fermentation strategies, such as simultaneous saccharification and fermentation (SSF) or separate hydrolysis and fermentation (SHF); different applied microorganisms used as a monoculture or co-culture; and different setups for biofuel fermentation processes.

The manuscripts collected represent a great opportunity for adding new knowledge to the scientific community as well as industry.

Alessia Tropea
Editor

fermentation

MDPI

Editorial

Biofuels Production and Processing Technology

Alessia Tropea

Department of Research and Internationalization, University of Messina, Via Consolato del Mare, 41, 98100 Messina, Italy; atropea@unime.it

Abstract: The negative global warming impact and global environmental pollution due to fossil fuels mean that the main challenge of modern society is finding alternatives to conventional fuels. In this scenario, biofuels derived from renewable biomass represent the most promising renewable energy sources. Depending on the biomass used by the fermentation technologies, it is possible obtain first-generation biofuels produced from food crops, second-generation biofuels produced from non-food feedstock, mainly starting from renewable lignocellulosic biomasses, and third-generation biofuels, represented by algae or food waste biomass. Although biofuels appear to be the closest alternative to fossil fuels, it is necessary for them to be produced in competitive quantities and costs, requiring both improvements to production technologies and diversification of feedstock. This Special Issue is focused on technological innovations, which include but are not limited to the utilization of different feedstock; different biomass pretreatments; fermentation strategies, such as simultaneous saccharification and fermentation (SSF) or separate hydrolysis and fermentation (SHF); different applied microorganisms used as monoculture or co-culture; and different setups for biofuel fermentation processes.

Keywords: biofuel production technologies; downstream processing; biorefinery; energy; bioethanol production; agroforest and industrial waste feedstock valorization; microorganisms for biofuel; sustainability

Citation: Tropea, A. Biofuels Production and Processing Technology. *Fermentation* **2022**, *8*, 319. https://doi.org/10.3390/fermentation8070319

Received: 24 June 2022
Accepted: 5 July 2022
Published: 7 July 2022

1. Biofuel Production Overview

The world's energy consumption has reached 14 billion tons of oil equivalent (TOE) [1,2], and in 2018 fossil fuels consisted of more than the 80% of the world's energy demand [3]. The main cause of the huge greenhouse gas (GHG) emissions in the atmosphere is ascribable to the continuous utilization of fossil fuels for energy generation [4]. Today, environmental policies are pushing for the reduction GHG emissions, and thanks to the support of innovative advances in crop engineering and fermentation processes, bioethanol, biodiesel, and biogas production represent viable and sustainable surrogates for petroleum-based fuels [5]. In this regard, Lee and Tsai [6] reported a study presenting a trend analysis of the motor gasoline supply/consumption, the bioethanol supply, and the regulatory system relevant to bioethanol production and gasohol use since 2007 in Taiwan.

Additionally, new incoming technologies are focused on the CO_2 capture and conversion into carbon-neutral value-added products, for instance, via microbial electrosynthesis (MES) [7], which has been reviewed by Quraishi et al. [8] in a comprehensive analysis, including original research and patents of numerous products obtained by the use of MES, including downstream processing and its potential commercialization and limitations. Moreover, it further discusses the recent trends, emphasizing MES and the role of electroactive microbes for their various applications, including electricity production and wastewater treatment [8].

Generally, the main feedstock use for bioethanol production is represented by sugar-containing edible crops, such as sugar cane, sugar beet, and sweet sorghum, while those used for biodiesel production are oil-bearing edible crops, such as soybean, rapeseed, sunflower, and palm tree, due to their high sugar and lipid contents with economic feasibility for upgrading processes [9,10]. The use of edible crops for the production of biofuels give rise to many concerns for their potential competition with food and feed supplies. In addition, the insecure supply chain of biomass feedstock due to regional and seasonal variations is considered as one of the critical constraints for hindering the commercialization of biofuels in many countries [11–13]. Hence, alternatives, such as the production on fallow fields of crops and grasses to produce bioethanol, have recently attracted attention and much effort has been made to discover new feedstock from various lignocellulosic waste materials [4].

Moreover, in order to mitigate GHG emissions and meet the global fuel demand, biofuel technology advancements need to be focused on the optimization of current biofuel-production processes to obtain higher productivity and efficiency of lignocellulosic feedstock bio-conversion; on the diversification of the biomass in order to guarantee the feasibility of biofuel production within existing ecological and economic constraints; and on the increase of the chemical scenario toward designer molecules able to improve fuel performance and economy, reducing in the meantime the carbon emissions. More efforts need to be addressed not only to overcome technological barriers but also to integrate social, economic, and environmental factors in order to provide long-term and cost-effective production systems for biofuel industries [5].

Rai et al. [14] reported an interesting review on the developments in lignocellulosic biofuels as a renewable source of bioenergy, where the impact of environmental factors on biofuel production and the approaches for enhanced biofuel production are well investigated, as is the production of second-generation lignocellulosic biofuels from non-edible plant biomass (i.e., cellulose, lignin, hemi-celluloses, non-food material) in a more sustainable manner.

Another original research has been carried out by Lin and Ma [15] regarding biofuel-production technology. The study was focused on the water-removal process using molecular sieves vibrated using a rotary shaker, since it can be considered a competitive method during the biofuel production reaction to achieve a superior quality of biofuels, starting from feedstock oil. In particular, the aim of the study was to evaluate the effects of vibration modes and operating time of molecular sieves on the fraction of water removal from palm oil and ethanol and to investigate the structural damage of the water-absorbing material after the process. Molecular sieves accompanied by two different kinds of vibrating motions, including electromagnetic stirring and rotary shaking, were used to absorb water from the reactant mixture of trans-esterification. The study pointed out that the rotary shaking motion represents an adequate agitation method for increasing the contact frequency and the area among the reactant mixtures of feedstock oil, water, and alcohol, resulting in a higher reaction rate and faster water-removal efficiency; moreover, the vibrating motion could facilitate the fluidity and mixing extents of the reactant mixture and thus accelerate the chemical reaction [15].

2. Bioethanol Production from Food Waste

Bioethanol production from waste, such as municipal waste and food waste, has consistently been one of the most popular alternative energy production pathways. Bioethanol emits lower greenhouse gases in comparison to fossil fuel. For this reason, it represents a valid alternative as vehicle fuel source. It can be mixed in various proportions with gasoline, obtaining gasohol, to be used immediately in internal combustion engines without requiring further engine modifications [16].

With regard to food waste utilization as bioethanol feedstock, nowadays it represents an interesting solution to the environmental crisis caused by the current amount of food waste, which is steadily increasing as the economy and population grow. Globally, 931 million tons of food waste were produced in 2019, with approximately 30% of food produced being discarded as waste [17]. These wastes show a high potential, due to their micro and macro composition [16,18–23], as a low-cost high-potency second-generation feed-stock that can easily undergo biodegradation by different fermentation approaches, such as direct fermentation (DF), separate hydrolysis and fermentation (SHF), and simultaneous saccharification and fermentation (SSF) [24,25] for bioethanol production. Salafia et al. [26] reported a study focused on the evaluation of pineapple waste cell wall sugars as an alternative source of second-generation bioethanol by *Saccharomyces cerevisiae*, carrying out an SSF process using a supplemented medium, by the addition of a specific nitrogen source, salts, and vitamins, which are required by the yeast in order to improve its ability to use the substrate both for alcohol production and for its own growth [27]. The study pointed out that the amount of cell wall sugars detected in pineapple waste after enzymatic hydrolysis makes this substrate an interesting resource for bioethanol production. The ethanol theoretical yield, calculated according to dry matter lost, reached up to 85% (3.9% EtOH), making pineapple waste an excellent raw material for ethanol production by *S. cerevisiae*. Moreover, the resulting fermentation substrate was enriched in single cell protein (SCP). In fact, the protein content increased from 4.45% up to 20.1% during the process and this allows the final fermented product to be suitable as animal feed, thus replacing expensive conventional sources of protein, like fishmeal and soymeal, and preventing the production of further waste by the end of the fermentation process, with respect to environmental sustainability [26].

Food waste bioconversion has also been reported by Jarunglumlert et al. [16]. The main goal of their research was to evaluate how increase the potential of energy production from food waste by the co-production of bioethanol and biomethane, testing different concentrations of enzymes for food waste hydrolysis. It was pointed out that when increasing the enzyme concentration, the amount of reducing sugar produced were increased as well, reaching a maximum amount of 0.49 g/g food waste. The resulting sugars were used as fermentative substrate by *Saccharomyces cerevisiae*, to be converted to ethanol. After 120 h of fermentation the ethanol yield reached up 0.43–0.50 g ethanol/g reducing sugar, ranging between the 84.3–99.6% of theoretical yield. The solid residue resulting from fermentation process was subsequently subjected to anaerobic digestion, allowing the production of biomethane, which reached a maximum yield of 264.53 ± 2.3 mL/g. This study shown how food waste represents a raw material with high energy production potential [16].

Vucurovic et al. [28] referred to a process and cost model of bioethanol production starting from spent sugar beet pulp, with the aim of applying it in the evaluation of new technologies and products based on lignocellulosic raw materials. The model developed allows the determination of the capital and production costs for a bioethanol-producing plant, processing about 17,000 tons of spent sugar beet pulp per year. Moreover, it can predict the process and economic indicators of the tested biotechnological process, determine the contribute of major components in bioethanol production cost, and compare different model scenarios for processing co-products [28].

KOH-pretreated seed pods of *Bombax ceiba* for ethanol by *S. cerevisiae* in SSF and SHF were used as second-generation feedstock by Ghazanfar et al. [29]. The study shows that the SSF process allows the maximum saccharification (58.6% after 24 h) and highest ethanol yield (57.34 g/L after 96 h) to be obtained. The SSF process was optimized for physical and nutritional parameters by one factor at a time (OFAT) and central composite design (CCD), allowing to set the optimum fermentation parameters for highest ethanol production (72.0 g/L): 0.25 g/L yeast extract, 0.1 g/L K_2HPO_4, 0.25 g/L $(NH_4)_2SO_4$, 0.09 g/L $MgSO_4$, 8% substrate, 40 IU/g commercial cellulase, 1% *Saccharomyces cerevisiae* inoculum, pH 5. This study proposed an inexpensive and novel source as a promising feedstock for pilot-scale second-generation bioethanol production [29].

Usually, research on bioethanol production is lacking in economic information on efficiency and profit at larger scales. This gap has been investigated by Rosentrater and Zhang [30] in their study on the techno-economic analysis of integrating soybean biorefinery products into corn-based ethanol fermentation operations. In order to determine the economic feasibility of this bio-refining, a techno-economic analysis for combining corn and soybean bio-refinery processes was carried out. The aim of the study was to use the techno-economic analysis (TEA) for estimate the costs associated with the construction and the operation of this type of integrated system. Moreover, the research compared an integrated corn and soybean bio-refinery with an original corn-based ethanol process in economic performance, for exploring the effect of new applications on the corn-based ethanol production under 40- and 120-million-gallon ethanol production scales.

Derman et al. [31] reported a study where a microbial consortium of *Saccharomyces cerevisiae* and *Trichoderma harzianum* were used in simultaneous saccharification and fermentation (SSF) process of pretreated empty fruit bunches (EFBs) by employing the central composite design of response surface methodology. According to the authors, this represents an innovative study based on the contemporary utilization of a new combination of enzymes and microbes employed in the fermentation process for bioethanol production from EFBs. In the study, the combination of enzymes and microorganisms for bioethanol production was screened in order to determine the optimum concentration of this combination suitable for SSF. It was pointed out that the enzyme combinations of cellulase and β-glucosidase with the microbial consortium of *S. cerevisiae* and *T. harzianum* allowed the best conversion of the EFBs into bioethanol. Several parameters that could affect the fermentation process, such as the fermentation time, the temperature, the pH, and the inoculums concentration, have been evaluated by the authors in their research. The highest bioethanol yield (9.65 g/L) was obtained after 72 h fermentation, at 30 °C, pH 4.8, and by adding an inoculum concentration of 10% (*v/v*) [31].

3. Processing Technology

As stated above, the use of fossil-based energy has been declining since its use causes climate changes and air pollution [32] and new solution need to be addressed to solve out this issue. An example is represented by the utilization of biochar that, due to its chemical and physical characteristics, can be used as a product itself or as an ingredient, within a mixed product for multiple objectives, including soil improvement, waste management, energy (or fuel) production, water pollution, and mitigation of climate change, as reported by Tsai et al. [33].

Additionally, there is an increasing interest in the production of renewable and carbon-neutral fuels, mainly obtained by fermentation [34], and in the development of new promising technologies such as the indirect fermentation. This technique consists of the conversion of several kind of carbonaceous compounds to synthesis gas, named syngas, through gasification, followed by its fermentation for obtaining desired products by specific biocatalysts [35]. Syngas is mainly composed by carbon monoxide, carbon dioxide, and hydrogen. It can be produced from biomass, coal, animal or municipal solid waste, and industrial CO-rich off-gases [36]. Benevenuti et al. [37] carried out a study, using *Clostridium carboxidovorans* for syngas fermentation, evaluating the effect of different concentrations of Tween® 80 in the culture medium and the best process conditions were validated in a stirred tank bioreactor (STBR). The study pointed out that the supplementation with Tween® 80 to the culture medium was characterized by an increasing in biomass and ethanol production during *Clostridium carboxidivorans* syngas fermentation in serum bottles and validated in a stirred tank bioreactor. In particular, biomass and ethanol production increased by 15% and 200% using Tween® 80 in the culture medium, respectively, compared to pure culture medium. In the bioreactor, 106% more biomass was produced compared to serum bottle fermentation, but the same ethanol concentration was achieved [37].

Syngas fermentation has been evaluated also by de Medeiros et al. [38]. Their work presented a strategy for optimizing the ethanol production process via integrated gasification and syngas fermentation by using two types of waste feedstock, wood residues, and sugarcane bagasse. The energy efficiency was found to be 32% in both cases, and the main critical variables of the process were found to be the gasification zone temperature, the split fraction of the unreformed syngas sent to the combustion chamber, the dilution rate, and the gas residence time in the bioreactor.

Another promising technology and advantageous solution for the treatment and valorization of organic waste and wastewater is represented by the pressurized anaerobic digestion (PDA) as it allows the generation of a high-quality biogas with a low CO_2 content [39]. In pressurized anaerobic digestion the pressure of the biogas is gradually auto-generated during fermentation. Therefore, PAD processes are carried out at pressures greater than atmospheric, which allows the obtainment of a biogas with a high methane fraction and a low carbon dioxide content [40].

The study reported by Siciliano et al. [39] assessed the effects of pressure increase, at different organic load rate (OLR) values, on the pressurized anaerobic digestion of compost leachate process performance. Biogas composition, specific biogas yield (SBY), specific methane yield (SMY), and the main process parameters, such as pH, volatile fatty acids (VFA)/alkalinity ratio, and nutrient concentrations, were evaluated in response to the pressure change. The study pointed out that even if the biogas quality was enhanced by the pressure increasing, the overall amount of methane was lowered. Indeed, the pressured conditions did not cause substantial modification in the characteristics of digestates [39].

A lot of research has been carried out regarding the development of new technologies and the implementation of new feedstock suitable for biofuels production. This topic still represents an interesting challenge for the scientific and industrial world, and many efforts are still needed in this field in order to reduce the negative global warming impact and global environmental pollution due to fossil fuels in accordance with the environmentally sustainable development.

Funding: This research received no external funding.

Institutional Review Board Statement: Not applicable.

Conflicts of Interest: The author declares no conflict of interest.

References

1. Csefalvay, E.; Horvath, I.T. Sustainability assessment of renewable energy in the united states, canada, the european union, china, and the russian federation. *ACS Sustain. Chem. Eng.* **2018**, *6*, 8868–8874. [CrossRef]
2. Bilgen, S. Structure and environmental impact of global energy consumption. *Renew. Sustain. Energy Rev.* **2014**, *38*, 890–902. [CrossRef]
3. Davidson, D.J. Exnovating for a renewable energy transition. *Nat. Energy* **2019**, *4*, 254–256. [CrossRef]
4. Sungyup, J.; Nagaraj, P.S.; Kakarla, R.R.; Mallikarjuna, N.N.; Young-Kwon, P.; Tejraj, M.A.; Eilhann, E.K. Synthesis of different biofuels from livestock waste materials and their potential as sustainable feedstocks—A review. *Energy Convers. Manag.* **2021**, *236*, 114038. [CrossRef]
5. Liu, Y.; Cruz-Morales, P.; Zargar, A.; Belcher, M.S.; Pang, B.; Englund, E.; Dan, Q.; Yin, K.; Keasling, J.D. Biofuels for a sustainable future. *Cell* **2021**, *184*, 1636–1647. [CrossRef]
6. Lee, Y.-R.; Tsai, W.-T. Bottlenecks in the Development of Bioethanol from Lignocellulosic Resources for the Circular Economy in Taiwan. *Fermentation* **2021**, *7*, 131. [CrossRef]
7. Dessì, P.; Rovira-Alsina, L.; Sánchez, C.; Dinesh, G.K.; Tong, W.; Chatterjee, P.; Tedesco, M.; Farràs, P.; Hamelers, H.M.V.; Puig, S. Microbial electrosynthesis: Towards sustainable biorefineries for the production of green chemicals from CO_2 emissions. *Biotechnol. Adv.* **2021**, *46*, 107675. [CrossRef]
8. Quraishi, M.; Wani, K.; Pandit, S.; Gupta, P.K.; Rai, A.K.; Lahiri, D.; Jadhav, D.A.; Ray, R.R.; Jung, S.P.; Thakur, V.K.; et al. Valorisation of CO_2 into Value-Added Products via Microbial Electrosynthesis (MES) and Electro-Fermentation Technology. *Fermentation* **2021**, *7*, 291. [CrossRef]
9. Demichelis, F.; Laghezza, M.; Chiappero, M.; Fiore, S. Technical, economic and environmental assessement of bioethanol biorefinery from waste biomass. *J. Cleaner Prod.* **2020**, *277*, 124111. [CrossRef]
10. Rezania, S.; Oryani, B.; Park, J.; Hashemi, B.; Yadav, K.K.; Kwon, E.E.; Jin, H.; Jinwoo, C. Review on transesterification of non-edible sources for biodiesel production with a focus on economic aspects, fuel properties and by-product applications. *Energy Convers. Manag.* **2019**, *201*, 112155. [CrossRef]
11. Li, Q.; Hu, G. Techno-economic analysis of biofuel production considering logistic configurations. *Bioresour. Technol.* **2016**, *206*, 195–203. [CrossRef] [PubMed]
12. Ghaderi, H.; Pishvaee, M.S.; Moini, A. Biomass supply chain network design: An optimization-oriented review and analysis. *Ind. Crops Prod.* **2016**, *94*, 972–1000. [CrossRef]
13. Varun, I.K.; Bha, R.P. LCA of renewable energy for electricity generation systems—A review. *Renew. Sustain. Energy Rev.* **2009**, *13*, 1067–1073. [CrossRef]
14. Rai, A.K.; Al Makishah, N.H.; Wen, Z.; Gupta, G.; Pandit, S.; Prasad, R. Recent Developments in Lignocellulosic Biofuels, a Renewable Source of Bioenergy. *Fermentation* **2022**, *8*, 161. [CrossRef]

15. Lin, C.-Y.; Ma, L. Comparison of Water-Removal Efficiency of Molecular Sieves Vibrating by Rotary Shaking and Electromagnetic Stirring from Feedstock Oil for Biofuel Production. *Fermentation* **2021**, *7*, 132. [CrossRef]
16. Jarunglumlert, T.; Bampenrat, A.; Sukkathanyawat, H.; Prommuak, C. Enhanced Energy Recovery from Food Waste by Co-Production of Bioethanol and Biomethane Process. *Fermentation* **2021**, *7*, 265. [CrossRef]
17. UNEP. *Food Waste Index Report 2021*; UNEP: Nairobi, Kenya, 2021; ISBN 978-92-807-3868-1. Available online: https://www.unep.org/resources/report/unep-food-wasteindex-report-2021 (accessed on 4 March 2021).
18. Potortí, A.G.; Lo Turco, V.; Saitta, M.; Bua, G.D.; Tropea, A.; Dugo, G.; Di Bella, G. Chemometric analysis of minerals and trace elements in Sicilian wines from two different grape cultivars. *Nat. Prod. Res.* **2017**, *31*, 1000–1005. [CrossRef]
19. Tuttolomondo, T.; Dugo, G.; Leto, C.; Cicero, N.; Tropea, A.; Virga, G.; Leone, R.; Licata, M.; La Bella, S. Agronomical and chemical characterisation of *Thymbra capitata* (L.) Cav. biotypes from Sicily, Italy. *Nat. Prod. Res.* **2015**, *29*, 1289–1299. [CrossRef]
20. La Torre, G.L.; Potortì, A.G.; Saitta, M.; Tropea, A.; Dugo, G. Phenolic profile in selected Sicilian wines produced by different techniques of breeding and cropping methods. *Ital. J. Food Sci.* **2014**, *26*, 41–55.
21. Lo Turco, V.; Potortì, A.G.; Tropea, A.; Dugo, G.; Di Bella, G. Element analysis of dried figs (*Ficus carica* L.) from the Mediterranean areas. *J. Food Compos. Anal.* **2020**, *90*, 103503. [CrossRef]
22. Tropea, A.; Potortì, A.G.; Lo Turco, V.; Russo, E.; Vadalà, R.; Rand, R.; Di Bella, G. Aquafeed production from fermented fish waste and lemon peel. *Fermentation* **2021**, *7*, 272. [CrossRef]
23. Tropea, A. Food Waste Valorization. *Fermentation* **2022**, *8*, 168. [CrossRef]
24. Tropea, A.; Wilson, D.; Lo Curto, R.B.; Dugo, G.; Saugman, P.; Troy-Davies, P.; Waldron, K.W. Simultaneous saccharification and fermentation of lignocellulosic waste material for second generation ethanol production. *J. Biol. Res.* **2015**, *88*, 142–143.
25. Pandit, S.; Savla, N.; Sonawane, J.M.; Sani, A.M.; Gupta, P.K.; Mathuriya, A.S.; Rai, A.K.; Jadhav, D.A.; Jung, S.P.; Prasad, R. Agricultural waste and wastewater as feedstock for bioelectricity generation using microbial fuel cells: Recent advances. *Fermentation* **2021**, *7*, 169. [CrossRef]
26. Salafia, F.; Ferracane, A.; Tropea, A. Pineapple Waste Cell Wall Sugar Fermentation by Saccharomyces cerevisiae for Second Generation Bioethanol Production. *Fermentation* **2022**, *8*, 100. [CrossRef]
27. Tropea, A.; Wilson, D.; Cicero, N.; Potortì, A.G.; La Torre, G.L.; Dugo, G.; Richardson, D.; Waldron, K.W. Development of minimal fermentation media supplementation for ethanol production using two Saccharomyces cerevisiae strains. *Nat. Prod. Res.* **2016**, *30*, 1009–1016. [CrossRef]
28. Vucurovic, D.; Bajic, B.; Vucurovic, V.; Jevtic-Mucibabic, R.; Dodic, S. Bioethanol Production from Spent Sugar Beet Pulp—Process Modeling and Cost Analysis. *Fermentation* **2022**, *8*, 114. [CrossRef]
29. Ghazanfar, M.; Irfan, M.; Nadeem, M.; Shakir, H.A.; Khan, M.; Ahmad, I.; Saeed, S.; Chen, Y.; Chen, L. Bioethanol Production Optimization from KOH-Pretreated Bombax ceiba Using Saccharomyces cerevisiae through Response Surface Methodology. *Fermentation* **2022**, *8*, 148. [CrossRef]
30. Rosentrater, K.A.; Zhang, W. Techno-Economic Analysis of Integrating Soybean Biorefinery Products into Corn-Based Ethanol Fermentation Operations. *Fermentation* **2021**, *7*, 82. [CrossRef]
31. Derman, E.; Abdulla, R.; Marbawi, H.; Sabullah, M.K.; Gansau, J.A.; Ravindra, P. Simultaneous Saccharification and Fermentation of Empty Fruit Bunches of Palm for Bioethanol Production Using a Microbial Consortium of *S. cerevisiae* and *T. harzianum*. *Fermentation* **2022**, *8*, 295. [CrossRef]
32. Gildemyn, S.; Molitor, B.; Usack, J.G.; Nguyen, M.; Rabaey, K.; Angenent, L.T. Upgrading syngas fermentation effluent using Clostridium kluyveri in a continuous fermentation. *Biotechnol. Biofuels* **2017**, *10*, 1–15. [CrossRef] [PubMed]
33. Tsai, W.-T.; Jiang, T.-J.; Lin, Y.-Q.; Chang, H.-L.; Tsai, C.-H. Preparation of Porous Biochar from Soapberry Pericarp at Severe Carbonization Conditions. *Fermentation* **2021**, *7*, 228. [CrossRef]
34. Benevenuti, C.; Botelho, A.; Ribeiro, R.; Branco, M.; Pereira, A.; Vieira, A.C.; Ferreira, T.; Amaral, P. Experimental Design to Improve Cell Growth and Ethanol Production in Syngas Fermentation by *Clostridium carboxidivorans*. *Catalysts* **2020**, *10*, 59. [CrossRef]
35. Datar, R.P.; Shenkman, R.M.; Cateni, B.G.; Huhnke, R.L.; Lewis, R.S. Fermentation of biomass-generated producer gas to ethanol. *Biotechnol. Bioeng.* **2004**, *86*, 587–594. [CrossRef] [PubMed]
36. Sun, X.; Atiyeh, H.K.; Zhang, H.; Tanner, R.S.; Huhnke, R.L. Enhanced ethanol production from syngas by Clostridium ragsdalei in continuous stirred tank reactor using medium with poultry litter biochar. *Appl. Energy* **2019**, *236*, 1269–1279. [CrossRef]
37. Benevenuti, C.; Branco, M.; do Nascimento-Correa, M.; Botelho, A.; Ferreira, T.; Amaral, P. Residual Gas for Ethanol Production by Clostridium carboxidivorans in a Dual Impeller Stirred Tank Bioreactor (STBR). *Fermentation* **2021**, *7*, 199. [CrossRef]
38. De Medeiros, E.M.; Noorman, H.; Maciel Filho, R.; Posada, J.A. Multi-Objective Sustainability Optimization of Biomass Residues to Ethanol via Gasification and Syngas Fermentation: Trade-Offs between Profitability, Energy Efficiency, and Carbon Emissions. *Fermentation* **2021**, *7*, 201. [CrossRef]
39. Siciliano, A.; Limonti, C.; Curcio, G.M. Performance Evaluation of Pressurized Anaerobic Digestion (PDA) of Raw Compost Leachate. *Fermentation* **2022**, *8*, 15. [CrossRef]
40. Scamardella, D.; De Crescenzo, C.; Marzocchella, A.; Molino, A.; Chianese, S.; Savastano, V.; Tralice, R.; Karatza, D.; Musmarra, D. Simulation and Optimization of Pressurized Anaerobic Digestion and Biogas Upgrading Using Aspen Plus. *Chem. Eng. Trans.* **2019**, *74*, 55–60.

Article

Bottlenecks in the Development of Bioethanol from Lignocellulosic Resources for the Circular Economy in Taiwan

Yu-Ru Lee [1] and Wen-Tien Tsai [2,*]

[1] Graduate Institute of Environmental Management, Tajen University, Pingtung 907, Taiwan; yuru@tajen.edu.tw

[2] Graduate Institute of Bioresources, National Pingtung University of Science and Technology, Pingtung 912, Taiwan

* Correspondence: wttsai@mail.npust.edu.tw; Tel.: +886-8-7703202

Abstract: Strategies and actions for mitigating the emissions of greenhouse gas (GHG) and air pollutants in the transportation sector are becoming more important and urgent due to concerns related to public health and climate change. As a result, the Taiwanese government has promulgated a number of regulatory measures and promotion plans (or programs) on bioethanol use, novel fermentation research projects and domestic production since the mid-2000s. The main aim of this paper was to present a trend analysis of the motor gasoline supply/consumption and bioethanol supply, and the regulatory system relevant to bioethanol production and gasohol use since 2007 based on the official database and the statistics. The motor gasoline supply has shown a decreasing trend in the last five years (2016–2020), especially in 2020, corresponding to the impact of the COVID-19 outbreak in 2020. Although the government provided a subsidy of NT$ 1.0–2.0 dollars per liter for refueling E3 gasohol based on the price of 95-unleaded gasoline, the bioethanol supply has shown decreasing demand since 2012. In addition, the plans for domestic bioethanol production from lignocellulosic residues or energy crops were ceased in 2011 due to non-profitability. To examine the obstacles to bioethanol promotion in Taiwan, the bottlenecks to bioethanol production and gasohol use were addressed from the perspectives of the producer (domestic enterprise), the seller (gas station) and the consumer (end user).

Keywords: gasohol; trend analysis; promotion policy; regulatory measure; bottleneck

Citation: Lee, Y.-R.; Tsai, W.-T. Bottlenecks in the Development of Bioethanol from Lignocellulosic Resources for the Circular Economy in Taiwan. *Fermentation* **2021**, *7*, 131. https://doi.org/10.3390/fermentation7030131

Academic Editor: Alessia Tropea

Received: 30 June 2021
Accepted: 22 July 2021
Published: 26 July 2021

1. Introduction

Taiwan, situated in East Asia, features a high population density (i.e., 650 people per km^2 based on a population of 23.4 million and a land area of 36,000 km^2) and dependence on imported energy (i.e., 97.56%) in 2020 [1]. Based on the official book of statistics [2], the total energy consumption in Taiwan increased from 76.84 million kiloliters of oil equivalent (KLOE) in 2005 to 84.91 million KLOE in 2019. When classified by sector, the transportation sector accounted for 15.78% of all energy consumption in 2019. When classified by the form of energy, petroleum-based products provided 46.18 million KLOE, accounting for 52.42%. It was indicated that petroleum-based products (e.g., gasoline) account for a significant proportion of greenhouse gas (GHG) emissions. On the other hand, the net GHG emissions in Taiwan have grown greatly, from 113×10^6 metric tons of carbon dioxide equivalent (CO_{2eq}) in 1990 to 275×10^6 metric tons of CO_{2eq} in 2018 [3]. Since the Kyoto Protocol came into force in 2005, the Taiwan government has been seeking a balance between energy security, the green economy, and environmental sustainability, and for this reason, the "Guidelines on Energy Development in Taiwan" were first issued by the central competent authority (i.e., Ministry of Economic Affairs, MOEA) in 2012, and were recently revised in 2017 [4]. During the energy security phase, the energy supply side must achieve diversification, energy autonomy, and low carbon. One of the guiding principles is to promote the technological development and application of alternatives to fossil fuel energy in order to reduce dependency on fossil fuel energy.

Against the background mentioned above, the Taiwanese government decided to act in accordance with the direction concluded by the National Energy Conference in June 2005. A tentative policy was determined for the promotion of bioethanol [5–7], which was supplied by imports during the early stages, but which was mostly supplied by domestic producers during later stages. In 2007, the MOEA launched the Bioethanol Execution Plan with several development stages. To be able to enforce the policy, the Petroleum Management Act was revised in January 2008 to add new clauses describing fixed ratios for blends of fossil fuels with biofuels (i.e., E3 gasohol). In addition, the Taiwan Sugar Corporation (TSC, a state-owned enterprise) was encouraged to invest in the mass production of bioethanol in Taiwan in 2007, despite the fact that it had already terminated production of bioethanol from sugar-processing by-products (i.e., molasses) in 2003 due to the high costs and the low revenues. The fact that the establishment of a new bioethanol plant by the TSC was concluded to be unfeasible and unprofitable can be attributed to the high domestic production costs, insufficient support by end users, and the rapid increase in imported crude oil prices during the period of 2007–2009.

Although the technologies for the fermentation of bioethanol from various lignocellulosic sources [8–11], as well as its lifecycle analysis, have been studied by many Taiwanese scholars and researchers [12–16], few works have been devoted to a systematic description of the status of bioethanol use and its regulatory promotion, and the bottlenecks to domestic development from the perspectives of producers, sellers, and consumers in Taiwan [7]. Therefore, the main topics of this paper will cover the following subjects:

- Trend analysis of motor gasoline supply/consumption and bioethanol supply.
- Trend analysis of bioethanol supply during the period of 2007–2020.
- Regulatory systems relevant to bioethanol production and use.
- Official plans for bioethanol use and domestic production since 2007.
- Bottlenecks to domestic development of bioethanol.

2. Data Mining and Methodology

In this work, an analytical description of trends related to Taiwan's registered vehicles, motor gasoline supply/consumption, and bioethanol supply during the period of 2007–2020 is carried out using the latest databases of the relevant central government agencies (i.e., Bureau of Energy under of the MOEA, and Directorate General of Highways under the Ministry of Transportation and Communications) [1,2,17]. In addition, information regarding the regulatory measures for the promotion of bioethanol use and the standards for E3 gasohol was accessed through official websites [18,19]. On the basis of these background data, accessed through official and open websites, the bottlenecks to domestic development of bioethanol were compiled and correlated with the official plans for bioethanol use since 2007. These plans included domestic production using the large local lignocellulosic resources (i.e., second-generation bioethanol feedstock such as rice straw and sorghum stalk) and research and development (R&D) for pretreatment and fermentation technologies by the universities and national institutes under the sponsorship of the National Energy Programs [7].

3. Results and Discussion

3.1. *Status of Motor Gasoline Supply/Consumption and Bioethanol Supply*

3.1.1. Trend Analysis of Motor Gasoline Supply/Consumption during 2000–2020

According to the statistical database by the Bureau of Energy (BOE) under the MOEA [1], the data on motor gasoline supply and consumption in Taiwan during the period of 2005–2020 are listed in Table 1 and also shown in Figure 1. The trend analysis was further addressed as follows:

1. In Taiwan, about 10 million kiloliters (KL) of motor gasoline per year were consumed over the past fifteen years. However, it indicated a V-type fluctuation during the period of 2005–2011, which should be attributed to the soaring oil prices (gasoline price thus increased) and economic recession due to the financial crisis of 2007–2008. By contrast, the data on motor gasoline supply indicated an up-and-down trend from 15.1 million KL in 2005 to 12.5 million KL in 2020. However, the motor gasoline supply showed a decreasing variation (i.e., 16.6 million KL in 2016, 15.5 million KL in 2017, 15.3 million KL in 2018, 14.7 million KL in 2019, and 12.5 million KL in 2020). There was no doubt that the impact of COVID-19 on motor gasoline supply was very obvious during the year 2020.
2. The ratio of motor gasoline consumption in the transportation sector to domestic gasoline consumption accounted for over 99%. Further, the trend of gasoline fuel consumption in the transportation sector was in accordance with the data on the amounts of registered gasoline motors during the period. For example, the amounts of newly registered passenger cars were 337,886 vehicles in 2015, as compared to 340,349 vehicles in 2020 [9]. In addition, the significant increase in fuel-efficient cars, diesel passenger cars, hybrid electric cars and electric vehicles in recent years also resulted in the suppression of gasoline motors growth [10], suggesting that the consumption of petroleum-based fuels will be on decreasing trend in the near future.

Table 1. Motor gasoline supply and consumption in Taiwan during the period of 2005–2020 [1].

Item	2005	2010	2015	2020
Total supply	15,109.4	14,869.3	15,790.8	12,502.8
Production	15,058.9	14,869.3	15,591.8	12,257.6
Export	4811.4	4947.0	5512.2	2324.9
Import	50.5	0.0	199.0	245.2
Domestic consumption	10,578.5	9784.5	10,155.5	10,170.5
Transportation	10,501.8	9713.3	10,097.1	10,105.9
Others [2]	76.7	71.2	58.4	64.6
Change in stocks	−280.5	137.8	123.1	7.4

[1] Source [3]; unit: 10^3 kiloliter (KL). [2] Including the energy (own use), industrial, agricultural, and service sectors.

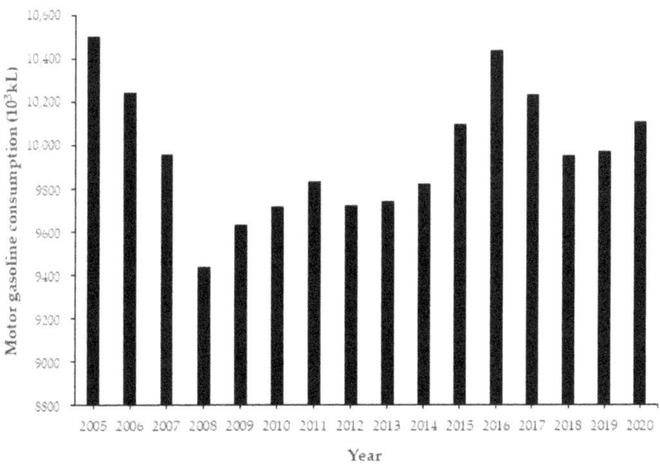

Figure 1. Variations of motor gasoline consumption during 2015–2020.

3.1.2. Trend Analysis of Bioethanol Supply during the Period of 2007–2020

After entering the World Trade Organization (WTO) on 1 January 2002, the production of bioethanol from sugar processing by-products (i.e., molasses) in Taiwan has been temporarily stopped by the Taiwan Sugar Corporation (TSC, one of the state-owned enterprises) since then [7]. However, the crude oil price has risen by the mid-2000s, becoming an unstable factor for economic development. The Taiwan government was facing pressure to reform the industrial structure and raise energy-saving technologies. Meanwhile, the Kyoto Protocol—put into force on 16 February 2005—pushed the government to mitigate greenhouse gas (GHG) emissions and also develop a renewable energy industry. Although Taiwan is not a signatory to the Protocol, this treaty, which is relevant to the environmental, economic, and energy issues, played a vital role in leading the national development because of its high dependence on imported energy (over 97%) [2]. A green policy was determined for promoting bioethanol in accordance with the direction concluded by the National Energy Conference in June 2005 [7]. The bioethanol in the E3 gasohol was supplied by imports in the early stage and mostly supplied by domestic producers in the later stage. In 2007, the MOEA launched the Bioethanol Execution Plan with several development stages, which will be described in Section 3.3. Furthermore, the Petroleum Management Act was subsequently revised in January 2008 to add new clauses concerning the blends of fossil fuels with biofuels at a fixed ratio (i.e., E3 gasohol, as addressed in Section 3.2.1). Figure 2 depicts the variations of bioethanol supply imported during the period of 2007–2020 [1]. Although the government provided a subsidy of NT$ 1.0–2.0 per liter based on the price of 95-unleaded gasoline, the bioethanol supply has indicated a decreasing demand since 2012.

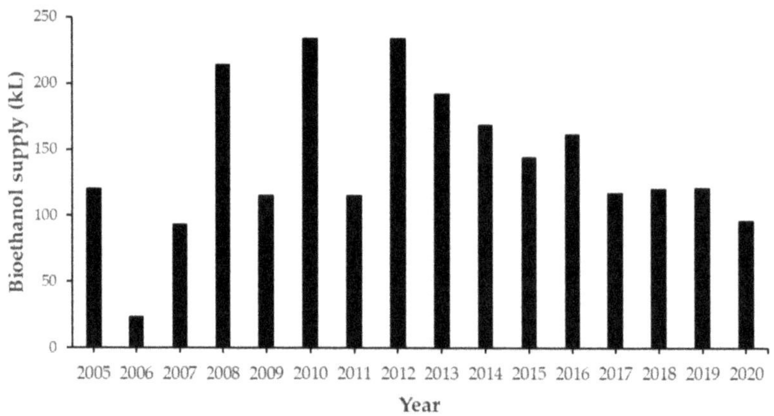

Figure 2. Variations of bioethanol supply during the period of 2007–2020 in Taiwan.

3.2. *Regulations Relevant to Bioethanol Production and Use*

3.2.1. Petroleum Management Act

For the main purposes of promoting the sound development (production, supply, sale, etc.) of the domestic oil industry and giving equal consideration to environmental protection, the Petroleum Management Act was initially promulgated on 11 October 2001 and recently revised on 4 June 2014. According to Article 29 of the Act, petroleum-based fuels can be imported or sold in the domestic market if they meet national standards. Therefore, the central competent authority (i.e., Bureau of Standards, Metrology, and Inspection, BSMI) of the MOEA established the national standard, which entitled "Unleaded Gasoline for Automobile" in the Chinese National Standards (CNS) code by CNS 12614. Table 2 lists the specifications of unleaded gasoline for automobiles based on the code of CNS 12,614 [19]. Table 3 shows the further specifications of bioethanol-E3 (95E3 gasohol), which was accessed on the website of the only gasohol supplier (i.e., Chinese Petroleum Corporation) [20]. In addition, the central competent agency (i.e., BOE) of the MOEA will stipulate the relevant measures for the businesses engaging in the production of the renewable energies of alcohol gasoline, biodiesel, or renewable oil products under Article 38 of the Act. The regulation ("Business Management Measures Governing the Production, Import, Blending and Sale of Alcohol Gasoline, Biodiesel, or Renewable Oil Products") was first announced on 12 December 2001 and recently revised on 3 March 2019. Based on the definitions by the regulation, alcohol gasoline (gasohol) is a fuel, which is prepared by blending fuel ethanol with fossil gasoline, or direct use of denatured fuel ethanol without blending fossil gasoline. Herein, the fuel ethanol means the pure ethanol (i.e., E100) with less than 0.5 vol% of water content, which is manufactured from the processing steps (i.e., fermentation, distillation, and dehydration) of biomass materials containing sugar, starch, or lignocellulosic residues.

Table 2. Specifications of unleaded gasoline for automobiles in Taiwan [1].

Property		Limit		Unit
		Minimal	Maximal	
Appearance and color		Bright and clear		- [2]
Density (at 15 °C)		0.720	0.775	g/mL
Copper strip corrosion			No. 1	Level
Oxidation stability (Induction period method, 100 °C)		240	-	min
Cleaning glue solvent content		-	4	mg/100 mL
Benzene content		-	0.9	vol%
Sulfur content		-	10.0	mg/kg
Oxygen content	Ethanol not contained	-	2.7	wt%
	Ethanol contained	-	3.24	wt%
Ethanol content		-	3.0	vol%
Lead content		-	5	mg/L
Aromatics content		-	35.0	vol%
Olefins content		-	18.0	vol%
Vapor pressure (at 37.8 °C)	Ethanol not contained	45	60	kPa
	Ethanol contained	45	66.9	
Distillation temperature	10 vol% vaporized (T_{10})	-	70	°C
	50 vol% vaporized (T_{50})	-	121	
	90 vol% vaporized (T_{90})	-	190	
	Final boiling point	-	225	
Distillation residue		-	2.0	vol%
Drivability index (DI)		-	597	°C

[1] Based on the code of CNS 12614. [2] Unit free.

Table 3. Specifications of denatured ethanol fuel for blending with gasoline for use in an automotive spark-ignition fuel engine in Taiwan [1].

Property	Performance Range		Unit
	Minimal	Maximal	
Appearance and color	Bright and clear		- [2]
Density (at 25 °C)	0.720	0.775	g/mL
Ethanol content	99.5	-	vol%
Methanol content	-	0.3	vol%
Water content	-	0.5	mg/kg
Copper content	-	0.07	mg/kg
Sulfur content	-	30.0	mg/kg
Acidity (as acetic acid)	-	30	mg/L
pH	6.5	9.0	-
Conductivity	-	500	μS/m
Denaturants content	2	5	vol%

[1] Based on the code of CNS 15109. [2] Unit free.

3.2.2. Renewable Energy Development Act

Regarding the policy and promotion for biomass energy, the Renewable Energy Development Act (REDA) should be the most important regulation [21], which was passed on 8 July 2009 and revised on 1 May 2019. According to the definition of biomass energy in the Act, it refers to the energy generated from direct use or treatment of vegetation, marsh gas (biogas), and domestic organic waste. Although the promotion incentives, such as feed-in-tariff (FIT), and support or subsidies played a determining role in the installation of renewable power system [18], these incentive measures in the Act are not beneficial for the industry development of biomass fuels, such as bioethanol and biodiesel. However, the REDA also provides the relevant measures concerning the biomass fuel, which were addressed as follows:

1. According to Article 6 of the Act, the central competent authority (i.e., MOEA) shall take into account the development potential of renewable energy and its impact on the domestic economy and stable power supply in order to set the promotion goals for renewable energy and the percentage of each category, and formulate and announce the development plans and initiatives for the next two years and by 2025. Based on the economic benefits, technological developments, relevant factors, and the promotion goals and schedules for heat utilization of renewable energy have been stipulated by the MOEA. However, biomass fuels are not included by the MOEA in the current promotion goals for renewable energy in Taiwan by 2025.
2. According to Article 6 of the Act, the MOEA shall consider reasonable costs and profits for the heat utilization of renewable energy (including biomass fuels and solar energy) and shall prescribe regulations on subsidies and rewards for heat utilization according to the effectiveness of their energy contribution. For the heat utilization in the biomass fuels, such subsidy expenses for the substituted portions of petroleum energy may be financed by the Petroleum Fund under the Petroleum Management Act. Furthermore, the reward expenses for the exploitation of fallow land or idle land for agriculture, forestry, animal husbandry to plant energy crops for producing biomass fuels will be financed by the Agricultural Development Fund. The regulations governing such reward eligibility, conditions and subsidy methods, and schedule shall be prescribed by the MOEA in conjunction with the Council of Agriculture (i.e., COA).

3.2.3. Air Pollution Control Act

In 1975, the Air Pollution Control Act (APCA) was initially enacted. With the progressive development of air pollution control and prevention measures, the Act was revised several times and recently amended on 1 August 2018. Under the authorization of the Act, there are two important regulations concerning the exhaust emissions from vehicles or motors and the composition/property standards for the use of bioethanol fuel.

1. "Exhaust Emission Standards of mobile Sources"

According to Article 36 of the Act, air pollutants emitted from mobile sources (e.g., vehicles or motors) should meet the exhaust emissions standards, which shall be stipulated by the central competent authority (i.e., EPA). Therefore, the regulation, named "Exhaust Emission Standards of mobile Sources", was first promulgated on 5 June 1980 and recently amended on 27 July 2020. In Provision 3 of the regulation, the exhaust emissions from motors or vehicles using gasoline and alternative clean fuel must comply with the Standards of carbon monoxide (CO), hydrocarbon (HC), nitrogen oxides (NOx), particulate matter (PM) and particulate number (PN) [18], as listed in Table 4. In view of the standards based on the driving cycle testing, the emission standards of new vehicles in Table 4 were implemented from 1 September 2019. In order to improve ambient air quality, the Standards will be further revised to reduce the emissions of HC and NOx from passenger cars in the future. In this regard, more energy-saving cars, hybrid motors, electric vehicles or electronic tolling devices entered the market in Taiwan [22,23].

2. "Standards for the Compositions of Fuels in Mobile Sources"

Under the authorization (i.e., Article 39) of the Act, the manufacture, import, sale, or use of fuel supplied to vehicles (or motors) shall comply with the composition standards for fuel types determined by the central competent authority (i.e., EPA). The standards (as listed in Table 5), named "Standards for the Compositions of Fuels in Mobile Sources" [18], were first promulgated on 15 December 1999 and recently amended on 20 March 2021. In Provision 3 of the regulation, the composition standards for vehicles (or motors) with gasoline (including E3 gasohol) cover the properties (or contents) of benzene, sulfur, oxygen, aromatics and olefins, and vapor pressure. It is noted that the benzene content in gasoline fuel has been reduced from 0.9 vol% in 2020 to 0.8 vol% in 2024. The stringent regulation will be expected to reduce the content of aromatics in motor engine fuels by adding clean fuels such as bioethanol or biobutanol, thus improving air quality due to the reduction in the air pollutants emitted from mobile sources [24].

Table 4. The standards of engine exhaust emissions from gasoline and clean alternative fuel in Taiwan [1].

<table>
<tr><th colspan="2" rowspan="2">Vehicle Classification</th><th colspan="6">Emission Standards (Based on Driving Cycle Testing) [2]</th></tr>
<tr><th>CO (mg/km)</th><th>THC (mg/km)</th><th>NMHC (mg/km)</th><th>NOx (mg/km)</th><th>PM (mg/km)</th><th>PN (p/km)</th></tr>
<tr><td colspan="2">Sedans and station Wagons</td><td>1000</td><td>100</td><td>68</td><td>60</td><td rowspan="4">4.5</td><td rowspan="4">6.0 × 10^11</td></tr>
<tr><td rowspan="3">Trucks, wagon passenger vehicles</td><td>Reference mass ≤1305 kg</td><td>1000</td><td>100</td><td>68</td><td>60</td></tr>
<tr><td>Reference mass >1305, ≤1760 kg</td><td>1000</td><td>130</td><td>90</td><td>75</td></tr>
<tr><td>Reference mass >1760 kg</td><td>2270</td><td>160</td><td>108</td><td>82</td></tr>
</table>

[1] The standards are applied to new vehicles since the promulgation date on 1 September 2019. [2] Notations: Carbon monoxide (CO), total hydrocarbons (THC), non-methane hydrocarbons (NMHC), nitrogen oxides (NOx), particulate matter (PM), and particulate number (PN).

Table 5. The standards for the composition of gasoline (including E3 gasohol) in Taiwan.

Property	Limit (Maximal)	
	Starting from 1 July 2020	Starting from 1 January 2024
Benzene content	0.9 vol%	0.8 vol%
Sulfur content	10 mg/kg	10 mg/kg
Vapor pressure (37.8 °C) [1]	60 kPa	60 kPa
Oxygen content [1]	2.7 wt%	2.7 wt%
Aromatics content	35 vol%	35 vol%
Olefins content	18 vol%	18 vol%

[1] Prior to the promulgation of E3 gasohol for all vehicles in Taiwan area, the maximal limits of vapor pressure and oxygen content are 66.9 kPa and 3.24 wt%, respectively.

3.3. *Official Plans for Bioethanol Use and Domestic Production Since 2007*

As described above, the policy for promoting the use of bioethanol and its domestic production was adopted in 2007 under the considerations of energy security, environmental sustainability and green economy. In February 2007, the MOEA launched the Bioethanol Execution Plan with three development stages (listed in Table 6), aiming at supplying E3 gasohol in all gas stations since 1 January 2011. Furthermore, the Petroleum Management Act was subsequently revised in January 2008. According to the Plan, the first stage was based on the "Green Public Vehicle Pilot Plan" that public vehicles in Taipei City must refuel E3 gasohol at eight designated gas stations. During the period, these gas stations also provided E3 gasohol for all private vehicles volunteering to refuel by subsidizing a discount rate at NT$ 1.0–2.0 per liter. The second stage was to promote the use of E3 gasohol for all public and private vehicles, which were refueled in Taipei City at eight designated gas stations and in Kaohsiung City at six designated gas stations. During the promotion period from July 2009 to September 2011, the total E3 gasohol supply amounted to about 11,400 KL, which was equivalent to the use of E100 bioethanol by 340 KL. Due to the insufficient supply of domestic production, the source of bioethanol in the E3 gasohol came from Asian countries such as Thailand during the first and second stage promotion period.

Table 6. Various stages of development plans for bioethanol in Taiwan.

Implementation Period	Promotion Plan	Promotion Measure
2007/9–2008/12	Pilot plan for green public vehicles	Public vehicles in Taipei City (capital city in Taiwan) must refuel E3 gasohol.
2009/1–2010/12	E3 gasohol plan in metropolitan area	E3 gasohol was supplied for all vehicles in Taipei City and Kaohsiung City [2]
2011/1–	E3 gasohol plan in all gas stations [1]	E3 gasohol was supplied for all vehicles in Taiwan area.
2015/1 (or 2020/1)–	E5 gasohol plan in all gas stations [1]	
2025/1–	E10 gasohol plan in all gas stations [1]	

[1] The bioethanol production plans have been halted by the Taiwan government, but the E3 gasohol plan still ran up to now. [2] E3 gasohol was only supplied by 14 gas stations in Taipei City and Kaohsiung City.

In order to be in accordance with the policy for promoting the use of bioethanol and its domestic production, the COA, in cooperation with the MOEA, subsidized a state-owned enterprise (i.e., Taiwan Tobacco and Liquor Corporation) to supply initial domestic demands by producing a small amount of bioethanol from the fermentation of sweet potatoes. It was planned that E3 gasohol would be gradually supplied by domestic production at the third and further promotion stages. The TSC (a state-owned enterprise that once produced bioethanol from molasses) and other private enterprises were also planning to establish the commercial bioethanol plants or bio-refineries using sugarcane juice, processing by-products (i.e., molasses, bagasse), or lignocellulosic residues (e.g., rice straw, sorghum stalk) in the fermentation process. According to the financial analysis by Liang and Jheng [25], the domestic oil price will rise 0.363–1.292% with the use of gasohol instead of gasoline at all gas stations, thus causing a fall by 0.009–0.03% in the gross domestic product (GDP) and a rise by 0.024–0.085% in consumer price index (CPI). Although the investment project for the mass production of bioethanol has been completed by the TSC and submitted to the central competent authority (i.e., MOEA) for approval in 2011 [7], the feasibility of establishing a new bioethanol plant seemed to be not profitable. This project withdrawn by the TSC can be attributed to the high domestic production cost (as compared to the imported price of bioethanol), insufficient support by the gasoline end users (E3 gasohol consumption not expected during the promotion period), and the imported crude oil price inclined significantly. Other private enterprises involving the domestic production of bioethanol also encountered similar bottlenecks. Finally, the MOEA decided to halt the mass production of bioethanol from lignocellulosic residues or agricultural by-products in Taiwan. Alternatively, the bioethanol supply must be dependent on the imports because the E3 gasohol plan still ran up to now.

3.4. *Bottlenecks to Domestic Development of Bioethanol*

3.4.1. Producer Side

With the international oil prices soaring and increasing threat of global warming during the early 2000s, the efforts to find clean energy sources to reduce GHG emissions became an urgent issue for government and industry. Under the policy support by the Taiwan government, a state-own company (i.e., Taiwan Sugar Corp.) and several private enterprises planned to invest local production of bioethanol from imported molasses, or sweet sorghum and other energy crops (e.g., napier grass) planted in fallow land. However, the domestic production of bioethanol from lignocellulosic residues (e.g., rice straw, sorghum stalk) or agricultural by-products (e.g., molasses, bagasse) in Taiwan still posed a higher cost as compared to the imported bioethanol. It was reported that the production cost of cellulosic ethanol will be as high as NT$35–38 (US$1.2–1.3) per liter [11]. This bottleneck was inherently caused by several economic factors, including interest rate, imported crude oil price, imported bioethanol price, labor cost, land cost (or land rent), harvesting and transporting cost for lignocellulosic materials, and capital cost for purchasing production machines and equipment. Among these factors, the international oil price may be the most significant one that indicated violet fluctuations decreasing from US$130 a barrel in July 2008 to US$80 a barrel in July 2009. Although the domestic bioethanol production would contribute to the reduction in GHG emissions and the reactivation in agricultural fallow lands by planting energy crops [7], the government also provided some subsidies under the Renewable Energy Development Act. Obviously, there are no enterprises willing to invest in the domestic bioethanol production from local mass non-food feedstocks such as rice straw and bagasse. During the past decade (2009–2018), the universities and research institutions were funded by the MOEA and the Ministry of Science and Technology (MOST) through the National Energy Program in the lab- and pilot-scale bioethanol production projects [7]. However, the research and development (R&D) and regulatory policies were not implemented in a complementary direction for commercial bioethanol production in Taiwan because it currently seemed not to be economically profitable. This situation was also reflected in the research publications and their corresponding project grants by the government based on the survey through the academic database such as the Web of Science.

3.4.2. Supply (Storage) Side

Although the use of E3 gasohol contributed to mitigate GHG emissions and improve urban air quality, its stability in the storage tank was relatively lower than gasoline due to the hygroscopic properties of ethanol [26]. This gasohol will be liable for absorbing moisture from the atmosphere or humid environments, thus causing phase separation. As Taiwan's climate features the tropical zone with high humidity, this situation will be more serious, forming distinct layers in the storage pool or tank, where a thicker layer of gasoline mixed with a little ethanol appears on top, and a thinner layer of water and more ethanol appears on the bottom. This process is unavoidable, and it can also be triggered by a drop in temperature. In addition, it is better to reduce the storage time in the E3 gasohol tank or pool at the gas stations because this biofuel could cause corrosion and rust inside the tanks and pipelines in the presence of water [27–29]. Because of this potential for phase separation at any ethanol level such as E3 gasohol, it is imperative that motor gasoline containing bioethanol is not exposed to water during its supply chain (distribution and storage) at the service stations for the purpose of preventing water contamination.

3.4.3. Consumer Side

Concerning the reasons for ceasing bioethanol promotion in Taiwan, the insufficient support by the motor fuel consumers should be the most important one. The author addressed the following viewpoints from the consumer side.

1. Although the government encouraged the use of E3 gasohol by subsidizing NT$ 1.0–2.0 per liter fueled, this price was still high when compared to that of 95-unleaded gasoline.
2. There are only 14 gas stations with the supply of E3 gasohol in metropolitan areas (one is Taipei City located in northern Taiwan, another is Kaohsiung City situated in southern Taiwan), indicating that it is very inconvenient to refuel the green motor gasoline.
3. As described above, the property of E3 gasohol is hygroscopic, easily causing corrosion and rust inside the pipelines and tending to clog fuel filters and lines. Sometimes, there are increased risks of detonation and engine durability due to the phase separation that occurred in the fuel tank [30,31].

4. Conclusions and Future Perspectives

Since the Kyoto Protocol came into force in 2005, the coupling of renewable energy development with GHG emission mitigation has become a sustainable development goal in government governance. In Taiwan, this challenge is extra important because it highly depends on imported energy. Since the early 2000s, the Taiwan government promulgated some regulatory measures and promotion plans (or programs) on bioethanol use and domestic production. Obviously, these efforts on E3 gasohol use were not successful during the period of 2007–2020. In addition, the plan for domestic bioethanol production from local lignocellulosic residues (i.e., rice straw and bagasse) or energy crops has decreased since 2011. It was concluded that the feasibility of establishing a new bioethanol plant seemed to be not currently profitable. It can be attributed to the high domestic production cost, insufficient support by the end users, and imported crude oil price inclined rapidly during the period of 2007–2009. On the other hand, the bottlenecks to gasohol use in Taiwan have been addressed from the sides of producer, seller, and consumer.

In Taiwan, domestic bioethanol production from lignocellulosic residues or energy crops currently seemed not to be highly profitable by international crude oil prices. However, with renewable energy sources for generating electricity and biofuels expanding rapidly and economically, reducing GHG emissions from our transportation system will be the next major hurdle we must overcome in order to meet the zero-emission target under the Paris Climate Agreement by 2050. In this regard, it is necessary to promote gasohol use as a transitional stage from clean fuel to the electric vehicle. In order to expand the gasohol consumption for reducing GHG emissions and criteria air pollutants in urban areas, the Taiwan government should subsidize more discounts for newer car owners (consumers) of fueling E3 gasohol. In response to the government's energy transformation policy, the CPC Corporation (one of the state-owned companies focusing on petroleum refinery and petroleum-based products) should take more corporate social responsibility (CSR) for providing this clean fuel at more gas stations.

Author Contributions: Conceptualization, W.-T.T. and Y.-R.L.; data collection, W.-T.T.; data analysis, W.-T.T.; writing—original draft preparation, W.-T.T.; writing—review and editing, W.-T.T. and Y.-R.L. Both authors have read and agreed to the published version of the manuscript.

Funding: This research received no external funding.

Institutional Review Board Statement: Not applicable.

Informed Consent Statement: Not applicable for studies not involving humans.

Data Availability Statement: Not applicable.

Conflicts of Interest: The authors declare no conflict of interest.

References

1. Energy Statistics Database (Ministry of Economic Affairs, Taiwan). Available online: http://www.esist.org.tw/database (accessed on 24 June 2021).
2. Ministry of Economic Affairs (MOEA). *Energy Statistics Handbook*; MOEA: Taipei, Taiwan, 2020.
3. Environmental Protection Administration (EPA). *Taiwan Greenhouse Gases Inventory*; EPA: Taipei, Taiwan, 2020.
4. Ministry of Economic Affairs (MOEA). *Guidelines on Energy Development*; MOEA: Taipei, Taiwan, 2017.
5. Tsai, W.T.; Lan, H.F.; Lin, D.T. An analysis of bioethanol utilized as renewable energy in the transportation sector in Taiwan. *Renew. Sustain. Energy Rev.* **2008**, *12*, 1364–1382. [CrossRef]
6. Liou, H.M. Policies and legislation driving Taiwan's development of renewable energy. *Renew. Sustain. Energy Rev.* **2010**, *14*, 1763–1781. [CrossRef]
7. Chung, C.C.; Yang, S.V. The emergence and challenging growth of the bio-ethanol innovation system in Taiwan (1949–2015). *Int. J. Environ. Res. Public Health* **2016**, *13*, 230. [CrossRef] [PubMed]
8. Liu, S.Y.; Lin, C.Y. Development and perspective of promising energy plants for bioethanol production in Taiwan. *Renew. Energy* **2009**, *34*, 1902–1907. [CrossRef]
9. Su, M.Y.; Tzeng, W.S.; Shyu, Y.T. An analysis of feasibility of bioethanol production from Taiwan sorghum liquor waste. *Bioresour. Technol.* **2010**, *101*, 6669–6675. [CrossRef]
10. Ko, C.H.; Wang, Y.N.; Chang, F.C.; Chen, J.J.; Chen, W.H.; Hwang, W.S. Potentials of lignocellulosic bioethanols produced from hardwood in Taiwan. *Energy* **2012**, *44*, 329–334. [CrossRef]
11. Wen, P.L.; Lin, J.X.; Lin, S.M.; Feng, C.C.; Ko, F.K. Optimal production of cellulosic ethanol from Taiwan's agricultural waste. *Energy* **2015**, *89*, 294–304. [CrossRef]
12. Su, M.H.; Tso, C.T. Life cycle assessment of environmental and economic benefits of bio-ethanol in Taiwan. *Int. J. Green Energy* **2011**, *3*, 339–354. [CrossRef]
13. Su, M.H.; Huang, C.H.; Lin, W.Y.; Tso, C.T.; Lur, H.S. A multi-years analysis on the energy balance, green gas emissions, and production costs of the first and second generation bioethanol. *Int. J. Green Energy* **2014**, *12*, 168–184.
14. Su, M.H.; Huang, C.H.; Li, W.Y.; Tso, C.T. Water footprint analysis of bioethanol energy crops in Taiwan. *J. Clean. Prod.* **2015**, *88*, 132–138. [CrossRef]
15. Chiu, C.C.; Shiang, W.J.; Lin, C.J.; Wang, C.H.; Chang, D.M. Water footprint analysis of second-generation bioethanol in Taiwan. *J. Clean. Prod.* **2015**, *101*, 271–277. [CrossRef]
16. Chang, F.C.; Lin, L.D.; Ko, C.H.; Hsieh, H.C.; Yang, B.Y.; Chen, W.H.; Hwang, W.S. Life cycle assessment of bioethanol production from three feedstocks and two fermentation waste reutilization schemes. *J. Clean. Prod.* **2017**, *143*, 973–979. [CrossRef]
17. Highway Statistical Data Query (Directorate General of Highways, Ministry of Transportation and Communications, Taiwan). Available online: https://stat.thb.gov.tw/hb01/webMain.aspx?sys=100&funid=e11100 (accessed on 27 June 2021).
18. Laws and Regulation Retrieving System (Ministry of Justice, Taiwan). Available online: https://law.moj.gov.tw/Eng/index.aspx (accessed on 24 June 2021).
19. CNS Online Service (Bureau of Standards, Metrology and Inspection, MOEA, Taiwan). Available online: https://www.cnsonline.com.tw/?locale=en_US (accessed on 21 June 2021).
20. Specifications for Motor Unleaded Gasoline (CPC Corporation, Taiwan). Available online: https://www.cpc.com.tw/cl.aspx?n=37 (accessed on 21 June 2021).
21. Gao, A.M.Z. Taiwan's recent efforts to promote renewable energy development: Policy measures, legal measures, challenges, and solutions in the post-Fukushima era. *Renew. Energy Law Policy Rev.* **2012**, *3*, 263–279.
22. Tsai, W.T. Trend analysis of Taiwan's greenhouse gas emissions from the energy sector and its mitigation strategies and promotion actions. *Atmosphere* **2021**, *12*. (in press) [CrossRef]

23. Lu, S.M. A low-carbon transport infrastructure in Taiwan based on the implementation of energy-saving measures. *Renew. Sustain. Energy Rev.* **2016**, *58*, 499–509. [CrossRef]
24. Tsai, J.H.; Ko, Y.L.; Huang, C.M.; Chiang, H.L. Effects of blending ethanol with gasoline on the performance of motorcycle catalysts and airborne pollutant emissions. *Aerosol Air Qual. Res.* **2019**, *19*, 2781–2792. [CrossRef]
25. Liang, C.Y.; Jheng, R.H. Long-term policy and strategies for bioethanol development in Taiwan. *Qual. Bank Taiwan* **2010**, *61*, 68–107. (In Chinese)
26. Spitzer, P.; Fisicaro, P.; Seitz, S.; Champion, R. pH and electrolytic conductivity as parameters to characterize bioethanol. *Accredit. Qual. Assur.* **2009**, *14*, 671–676. [CrossRef]
27. Thomson, J.K.; Pawel, S.J.; Wilson, D.F. Susceptibility of aluminum alloys to corrosion in simulated fuel blends containing ethanol. *Fuel* **2013**, *111*, 592–597. [CrossRef]
28. Rocabruno-Valdes, C.I.; Escobar-Jimenez, R.F.; Diaz-Blanco, Y.; Gomez-Aguilar, J.F.; Astorga-Zaragoza, C.M.; Uruchurtu-Chavarin, J. Corrosion evaluation of Aluminum 6061-T6 exposed to sugarcane bioethanol-gasoline blends using the Stockwell transform. *J. Electroanal. Chem.* **2020**, *878*, 114667. [CrossRef]
29. Thangavelu, S.K.; Ahmed, A.; Ani, F.N. Corrosive characteristics of bioethanol and gasoline blends for metals. *Int. J. Energy Res.* **2016**, *40*, 1704–1711. [CrossRef]
30. Yusoff, M.N.A.M.; Zulkifli, N.W.M.; Masum, B.M.; Masjuki, H.H. Feasibility of bioethanol and biobutanol as transportation fuel in spark-ignition engine: A review. *RSC Adv.* **2015**, *5*, 100184–100211. [CrossRef]
31. Thangavelu, S.K.; Ahmed, A.S.; Ani, F.N. Review on bioethanol as alternative fuel for spark ignition engines. *Renew. Sustain. Energy Rev.* **2016**, *56*, 820–835. [CrossRef]

Review

Valorisation of CO_2 into Value-Added Products via Microbial Electrosynthesis (MES) and Electro-Fermentation Technology

Marzuqa Quraishi [1,†], Kayinath Wani [2], Soumya Pandit [2,*], Piyush Kumar Gupta [2], Ashutosh Kumar Rai [3,†], Dibyajit Lahiri [4], Dipak A. Jadhav [5], Rina Rani Ray [6], Sokhee P. Jung [7], Vijay Kumar Thakur [8,9,10] and Ram Prasad [11,*]

1 Amity Institute of Biotechnology, Amity University, Mumbai 410206, India; marzuqa20@gmail.com
2 Department of Life Sciences, School of Basic Sciences and Research, Sharda University, Greater Noida 201310, India; kainaatwani38@gmail.com (K.W.); dr.piyushkgupta@gmail.com (P.K.G.)
3 Department of Biochemistry, College of Medicine, Imam Abdulrahman Bin Faisal University, Dammam 31441, Saudi Arabia; akraibiotech@gmail.com
4 Department of Biotechnology, University of Engineering & Management, Kolkata 700156, India; dibyajit.lahiri@uem.edu.in
5 Department of Agricultural Engineering, Maharashtra Institute of Technology, Aurangabad 431010, India; deepak.jadhav1795@gmail.com
6 Department of Biotechnology, Maulana Abul Kalam Azad University of Technology, Kolkata 700064, India; raypumicro@gmail.com
7 Department of Environment and Energy Engineering, Chonnam National University, Gwangju 61186, Korea; sokheejung@chonnam.ac.kr
8 Biorefining and Advanced Materials Research Centre, Scotland's Rural College, Edinburgh EH9 3JG, UK; Vijay.Thakur@sruc.ac.uk
9 School of Engineering, University of Petroleum & Energy Studies (UPES), Dehradun 248007, India
10 Department of Mechanical Engineering, School of Engineering, Shiv Nadar University, Noida 201314, India
11 Department of Botany, Mahatma Gandhi Central University, Motahari 845401, India
* Correspondence: sounip@gmail.com (S.P.); rpjnu2001@gmail.com (R.P.)
† These authors contributed equally to this work.

Citation: Quraishi, M.; Wani, K.; Pandit, S.; Gupta, P.K.; Rai, A.K.; Lahiri, D.; Jadhav, D.A.; Ray, R.R.; Jung, S.P.; Thakur, V.K.; et al. Valorisation of CO_2 into Value-Added Products via Microbial Electrosynthesis (MES) and Electro-Fermentation Technology. *Fermentation* **2021**, *7*, 291. https://doi.org/10.3390/fermentation7040291

Academic Editor: Konstantinos G. Kalogiannis

Received: 24 September 2021
Accepted: 26 November 2021
Published: 30 November 2021

Abstract: Microbial electrocatalysis reckons on microbes as catalysts for reactions occurring at electrodes. Microbial fuel cells and microbial electrolysis cells are well-known in this context; both prefer the oxidation of organic and inorganic matter for producing electricity. Notably, the synthesis of high energy-density chemicals (fuels) or their precursors by microorganisms using bio-cathode to yield electrical energy is called Microbial Electrosynthesis (MES), giving an exceptionally appealing novel way for producing beneficial products from electricity and wastewater. This review accentuates the concept, importance and opportunities of MES, as an emerging discipline at the nexus of microbiology and electrochemistry. Production of organic compounds from MES is considered as an effective technique for the generation of various beneficial reduced end-products (like acetate and butyrate) as well as in reducing the load of CO_2 from the atmosphere to mitigate the harmful effect of greenhouse gases in global warming. Although MES is still an emerging technology, this method is not thoroughly known. The authors have focused on MES, as it is the next transformative, viable alternative technology to decrease the repercussions of surplus carbon dioxide in the environment along with conserving energy.

Keywords: bioelectrochemical system (BES); carbon dioxide sequestration; extracellular electron transfer (EET); electroactive microorganisms; microbial biocatalyst; electro-fermentation; circular economy; downstream processing (DSP); gene manipulation

1. Introduction

Carbon dioxide is naturally abundant (about 0.03% to 0.04%) in the atmosphere and is eventually responsible for the ecological balance of the ecosystem [1,2]. However, the ever-increasing population and the energy demands have led to changes in the natural cycles of greenhouse gases (including CO_2). Industrial emissions and the misuse of fossil fuels have led to about a 40% upsurge in the total atmospheric CO_2 and about a 78% rise in the greenhouse gases concentration from 1990 to 2016. Hence, the accumulation of carbon dioxide has led to absorption and re-emission of heat, attributing to an additional warming of the planet [2–7]. The changes in the land-use practices predominantly, deforestation and more use of agricultural land, cement production, use of fossil fuels for energy generation and transportation are the major factors contributing to the carbon emissions [3–6]. CO_2 can be captured and converted into carbon-neutral value-added products via microbial electrosynthesis (MES) [8].

MES is a novel microbial electrochemical technology that supplies electrons to microorganisms via an electric current (biocathode-driven i.e., biofilm + cathode) inside an electrochemical cell. These microbes act as biocatalysts and use the electrons for reducing carbon dioxide to eventually yield industrially relevant products like transportation fuels [9,10]. It is a fascinating alternative for capturing and expanding the value of the electrical energy generated from recurrent renewable sources (like sun, geothermal, biomass, or wind) [8,11,12]. This interchange of energy to different usable carbon materials is the most riveting way for storing energy, its distribution and utilisation [10,13,14]. Production of organic compounds from MES is considered to be an effective technique for the inception of various beneficial multi-carbon reduced end-products like acetate and butyrate by the valorisation of low-value CO_2. Further, bio-production dependent on CO_2 is advantageous, as it uses less arable land and freshwater resources, has low CO_2 emissions, no major nutritional supplementation is needed, has excess substrate availability and lastly, chemical bonds can be employed for the storage of excess electrical energy [8,15,16]. Although MES technology is still in its infancy, it has been demonstrated as a promising green alternative for CO_2 sequestration and bioelectrosynthesis of high-valued multi-carbon organic compounds. The paucity of knowledge must be resolved before the commercialisation of MES technology [7,8,17–20].

The MES process imitates the natural photosynthesis process if the external power is supplied from a renewable solar source, depicting plenty of advantages as compared to the bioenergy procedure that depends on photosynthesis [9,10]. Some studies have shown that Gram-negative (most efficient being *Sporomusa ovata* DSM-2662) and Gram-positive (like *Moorella thermoacetica* and *Clostridium* spp.) acetogenic bacteria gain electrons from graphite electrodes and act as an electron (e^-) donor in the reduction of CO_2-producing multi-carbon compounds extracellularly. A strain named *Clostridium ljungdahlii* is capable of MES, which can be genetically controlled and can be used to generate high-valued commodities [21–25].

This review paper is a comprehensive analysis of numerous products obtained by the use of MES, including the downstream processing, its commercialisation potential and a few limitations. It further discusses the recent trends, emphasising MES and the role of electroactive microbes for their various applications including electricity production and wastewater treatment.

2. Bioelectrochemical System (BES)

Bioelectrochemical systems (BESs) are revolutionary novel bioengineering technology that has substantially diversified their scope over the past decade [26]. These are capable of converting electrical energy into chemical energy (like in microbial electrolytic cells (MECs)) and vice versa (like in microbial fuel cells (MFCs)) by degrading several organic compound substrates, especially lignocellulosic biomass derived from wastewater with the help of microbes or their enzymes to generate valuable products [2,27] such as methanol, ethanol, acetate, formate, or hydrocarbons; these commodities (being precursors) are later converted or directly used as a sustainable green alternative to fossil fuels (See Figure 1). The emerging MES process of producing high-value chemicals has greatly broadened the BES's scope. BES being an eco-friendly and energy-saving technology has gained much popularity, it revolves around e^- transfer and energy transformation. Researchers are now exploiting the design of electrochemical devices, electrodes, catalyst and separator material optimisation and screening of electroactive microorganisms [2,16,28–31].

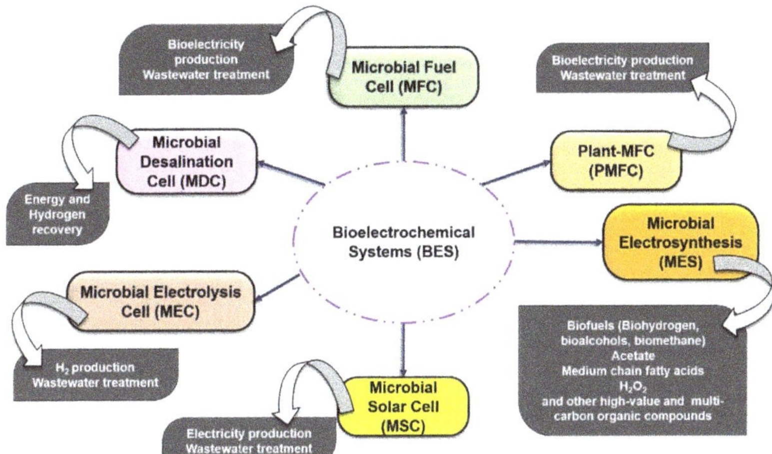

Figure 1. A schematic overview of multiple categories of BESs depending on the mode of their applications [28,29,31].

A typical BES consists of a cathode, anode and an optional membrane that separates the two of them. Figure 2 portrays a schematic representation of BES for e^- transfer from electrodes to microorganisms. Oxidation occurs in the anodic chamber (like the oxidation of acetate or water) and the reduction takes place in the cathodic chamber (like the O_2 reduction or H_2 evolution). At least one of these two half-reactions is biocatalysed, either by microbial cells, their enzymes or their organelles. The aqueous electrolyte solution surrounds the electrodes, where the reactants and products reside [16,27–30].

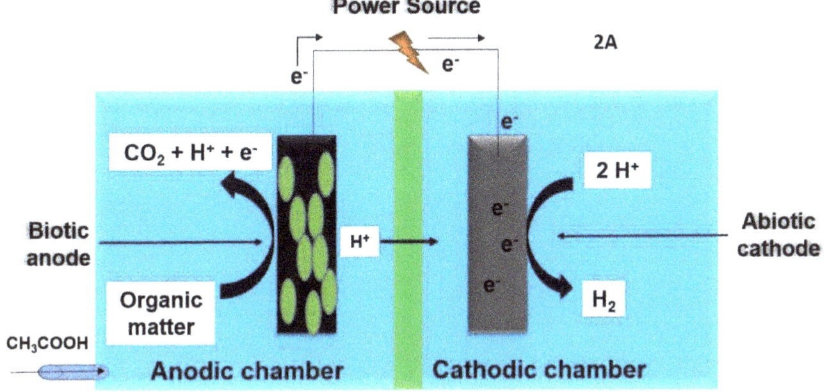

Figure 2. *Cont.*

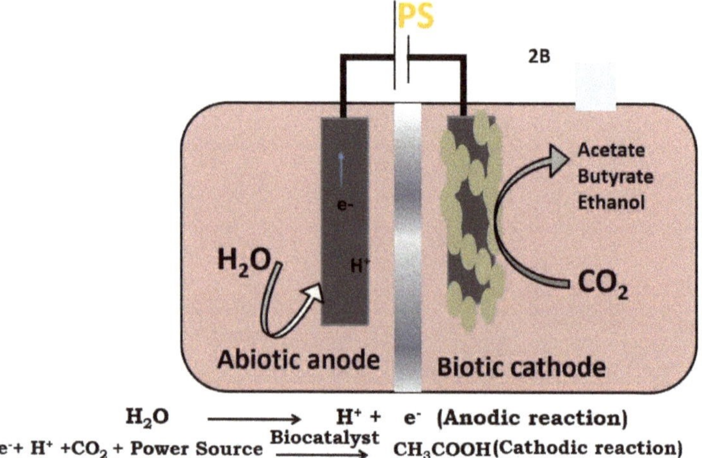

Figure 2. (**A**) A generalised schematic representation of a typical dual-chambered Bioelectrochemical System (BES) depicting its construction and the processes carried out in the system. On the anode, microbial biofilm comprising exoelectrogens oxidises the organic matter. Electrons and protons on being released travel via distinct paths to the cathode, where they are reduced to form H_2. The anode potential being higher than that of the cathode ensures a non-spontaneous reaction, making it essential to apply an external power supply to facilitate the bioelectrochemical reaction. (**B**) [32–36].

For didactical reasons, the overall process has been summed up using acetate as an example, being the most prominent carbon source for MFC's bioanode during laboratory studies [37].

$$\text{Anode: } CH_3COO^- + 4H_2O_{(l)} \rightarrow 2HCO_3^- + 9H^+ + 8e^- \tag{1}$$

$$\text{Cathode: } 8\,(H^+ + e^- \rightarrow \frac{1}{2}\,H_{2(g)}) \tag{2}$$

$$\text{Overall reaction: } CH_3COO^- + 4H_2O_{(l)} \rightarrow 2HCO_3^- + H^+ + 4H_{2(g)} \tag{3}$$

When electrical power is provided to the BES system, it is said to be in Microbial Electrolysis Cell mode. The extra power is supplied to intensify the reaction kinetics and to drive thermodynamically detrimental cathodic half-reactions. The bacteria employed in an MES are typically anaerobic homoacetogenic bacteria that employ reducing agents or electrons provided by the cathode to metabolically convert H_2 and CO_2 to acetate and other chemicals. A number of lithoautotrohphs are also utilised in the metabolic conversion of CO_2 to acetate and other organic molecules. The Wood–Ljungdahl or acetyl-CoA route follows the anaerobic conversion to acetate Figure 2B. Optionally, MFC can be applied to deliver the power to the electrochemical circuit [16,30,38,39].

MFCs trigger the chief growth and development of the microbial electrochemistry discipline and generate electricity utilising the microorganisms that are capable of handling and growing on the electrode (in this case, anode) surface, along with the ability to use electrodes as an e^- acceptor for the oxidation of organic compounds. In such systems, the electrical force is accumulated from the anodic response and the cathodic half-reactions take place simultaneously (like a decrease in O_2) [40].

BES system is available as planktonic microbial cells as well as a biofilm. The electroactive biofilm contains electrochemically active (EAM) and inactive microorganisms. This system having various functions like the breakdown of complex substrates proves to be beneficial. EAMs also empower the productive exchange of electrons from or towards solid-state electrodes to boost the current densities, improve the energy efficiencies and production in these systems. For the same purpose, extracellular electron transport (EET) is used to transport the e^- from or towards an insoluble e^- acceptor or donor (Figure 3) [41,42].

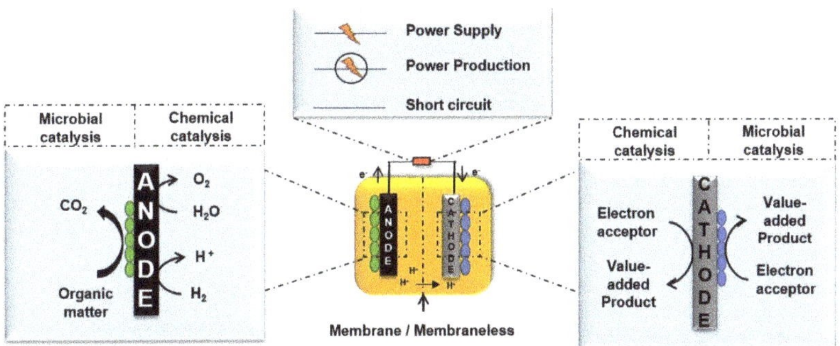

Figure 3. An in-depth overview of the concepts associated with the highly versatile microbe–electrode interaction-based microbial BES technology and a multitude of choices available to carry out a diverse range of processes at both anodic and cathodic chambers simultaneously [43].

2.1. Transmission of Electrons at the Anode

An ideal anode should have low resistance, large surface area, high electrical conductivity, anti-corrosiveness, strong mechanical strength, fouling resistance, chemical stability, good biocompatibility and scalability, preferably with ease of construction and mainly of low cost. The property of some EAM has been known over for a century to provide e^- to the electrodes. However, the mechanisms of anodic EET have most extensively been investigated in the last few years. The conversion of organic substrate to electric current does not only occur in fermentation reactions, but also in natural respiratory processes. This finding was a significant advancement that aided in discovering the mechanism behind electrons being drawn from microorganisms and the existence of two diverse EET pathways. The principal pathway of direct electron transfer (DET) includes the immediate contact between the electron transfer chain (ETC) of the microorganisms and the electrode surface. The other component, mediated electron transfer, shuttles e^- carriers and reversibly reduce and oxidise them in between electrodes and microbes. So far, only *Geobacter* spp., *Rhodoferax* spp. and *Shewanella* spp. have been widely researched to explore the EET mechanisms. The initial DET occurs through membrane-bound ETC proteins, like Cytochrome c (Cyt c), while the other carries e^- from the microbe to the surface of the electrode along conductive pili (also known as nanowires), which are attached to membrane-bound e^- transport compounds [44–46].

2.2. Transmission of Electrons at the Cathode

The cathode serves as a reservoir of e^- donors for microbes and thus influences the potency of the process [47]. The desired cathode should have the following properties for being used as a biocathode, it should have high productivity, excellent chemical stability, biocompatibility with high mechanical strength, surface area and low cost. The most extensively studied and the most frequently used end-product of CO_2 conversion for cathode efficiency assessment is acetate. Carbon-based compounds in the MES frameworks are extensively employed cathodes for CO_2 reduction [48].

The material of the cathode plays a pivotal role in electrohydrogenesis and electromethanogenesis. The latter need less energy input as compared to the former (−0.23 to −0.41 V vs. Standard Hydrogen Electrode (SHE)). The BES with microbial biocathode relies on microbes to receive the e^- from a solid electrode, acting as a donor for reducing the terminal e^- acceptor, these are known as electrotrophs. Researchers have mainly focused on understanding the mechanism of the anode while information on the reverse processes like the flow of e^- from electrodes to microorganisms was insubstantial. In 2004, cathodic DET flow was reported for the reduction of various forms of nitrogen (nitrate NO_3^- to nitrite NO_2^-) from the cathode and assured at about −0.34 V vs. SHE, it further enriched the substances along with microbial species [18,46,49]. Several *Desulfovibrio* sp. has been successfully employed for biohydrogen production with biocathode [50]. To date, the exact pathways of cathodic e^- transfer to an electroactive acetogen are still unknown [47]. Researchers have recommended numerous mechanisms that resemble the bioanode processes but possess different redox potentials.

Some general reactions carried out by anaerobic methanogens are [51]:

$$\text{Methanol: } 4CH_3OH \rightarrow 3CH_4 + CO_2 + 2H_2O \quad (4)$$

$$\text{Hydrogen: } 4H_2 + CO_2 \rightarrow CH_4 + 2H_2O \quad (5)$$

$$\text{Metals: } 4Me_0 + 8H^+ + CO_2 \rightarrow 4Me^{++} + CH_4 + 2H_2O \quad (6)$$

$$\text{Acetate: } CH_3COOH \rightarrow CH_4 + CO_2 \quad (7)$$

$$\text{Methylamine: } 4(CH_3)NH_2 + 2H_2O \rightarrow 3CH_4 + CO_2 + 4NH_3 \quad (8)$$

Cathodic e^- acceptors and their maximum power densities have been elaborated by Ucar et al. [35].

2.3. Electrosynthesis Assisted by Microbes

Using bioanode systems along with a chemical cathode in electrosynthesis is currently in demand [52]. H_2 can be generated with the help of platinum cathode MEC, the pH increases by consuming protons at the cathode. Similarly, hydrogen peroxide can be produced in BES via carbon cathode, which can later be used as a beneficial chemical. H_2O_2 thus produced can further be used for oxidation reactions, bioproduction and bioremediation processes [36,39,46,53–55], as well as in Fenton reaction [56]. Thus, microbial assisted electrosynthesis can efficiently be employed for the production of disinfectants or oxidants [39].

Some of the microorganisms used in MES for the production of various targets include *Sporomusa* sp., *Clostridium* sp., *Acetobacterium* sp., *Methanobacterium* sp. and many more [47]. The two main genera found to be dominant in this process include *Bacteroidetes* and *Proteobacteria*. Other than these, another prime genus of bacteria is *Firmicutes* [57]. For the production of methane, *Methanobacterium* sp. is considered to be a salient genus. Similar to the role of *Geobacter* sp. in bioanodes, these are considered vital for biocathode enrichment [58,59]. The microbes involved in the volatile fatty acid (VFA) and butyrate productions are *Megasphaera* sp. and *Clostridium* sp. In the case of hydrogen (H_2) production, acetogens are the superior community in the media, but H_2-producing bacteria control the biofilm production [60]. For the production of acetate, *Acetobacterium* sp. play a pivotal role. *Methanobacterium* sp. are also detected in acetate-producing biocathodes. Current experiments have demonstrated that microbes (predominantly *Firmicutes*) may be responsible for H_2 catalysis as they generate methane at low cathode potential, increasing the reducing power and production of organic compounds in MES [24,61].

Biofilm used in MES systems increases the efficiency and stability of the overall system in the long-run, by preventing the washing out of microorganisms. Nevertheless, this technology has drawbacks too, like microbe-electrode interactions and extracellular electron transfer related challenges that can be overcome by electrode engineering and optimisation [62,63]. Strategies such as modification of the electrode surface by the generation of 3D structures have been shown by Kerzenmacher [64]. Modifications such as enriching the surface of the electrode with positively charged molecules also upsurge the efficiency of the process. Alternatively, changes in the composition of the microbial community could also help. Intermixing of cultures or creating co-cultures can be used in biofilm-based MES to enhance the performance of the system as these cultures are robust to changes in the environment and are flexible with different types of substrates [65]. However, when present in abundant quantity, the species compete with one another for e^-, leading to a decrease in the product specificity [66]. A potential solution for the same is enriching specific species by using electrochemically-driven reactions in the long run, along with the addition of supplements. Biofilm-based MES highlighting the different process performances have been explained by Fruehauf [65].

2.4. Electroactive Microbes and Extracellular Electron Transfer

Electromicrobiology explores and exploits the microbe-electron (both donor and acceptor) interaction. Currently, the use of electroactive microorganisms has become fascinating in sustainable bioengineering practices. In these electroactive microbes, e^- transfer reactions encompass beyond the cell surface in a process called extracellular electron transfer (EET) [67,68], in MFCs, the electricigens (anodic catalysts) are employed. Electrogens are microbes that can release e^- onto an extracellular electrode (anode) surface, resulting in a positive electric current [69]. For instance, iron-reducing exoelectrogen bacteria *Geobacter sulfurreducens* produces high power density at moderate temperatures [28,68,70–75]. On the other hand, Electrotrophs retrieve e^- from an extracellular electrode (cathode) surface, resulting in an opposite (negative) electric current, like in MES [76]. For instance, Fe (II) and sulphur-oxidising bacteria *Acidithiobacillus ferrooxidans* switch their energy source from diffusible iron ions to direct e^- uptake from a polarised electrode [77].

In MES, the cathodic e^- autotrophic microbes convert CO_2 to fuels, chemicals, biodetergents, bioplastics and recover metals from metallurgy waste streams. This ability of theirs acts as an advantage for them in several environmental niches, one such distinct boon being selectively isolating rare strains, characterising them and utilising their characteristics in sustainable BES technologies [28,70,72–75]. EET characteristics and behaviour in BES of numerous organisms have been discussed by Kracke et al. [78].

Using advancing cross-disciplinary fields like material science, electrochemistry, biotechnology and MES, the ongoing energy issues can be ingeniously resolved. The main challenge in using this technology is shuttling electrons into microbes from the reductive cathode in sufficient quantities to produce products at a suitable level. This shuttling occurs in the form of redox-active compounds (e.g., the cofactor Flavin secreted by microbes), which transports e^-. These compounds are reduced by redox partaking enzymes such as Cyt c embedded on the surface of the microbial cell and then shuttled as electrons to the anode where they are oxidised. There are three mechanisms for the same; these are electrolysis of water, production of soluble e^- mediators and direct transfer of e^- (See, Table 1) [79,80].

The majority of the studies in Table 1 required a mediator to delegate microbial-anode interaction. When natural electroactive bacteria were employed as the producer strain or when the producer strain was co-cultured with an electroactive strain, an artificial e^- carrier was not needed. To date, only a handful of the bacterium has been evaluated as potential aspirants for anodic electro-fermentation. Another mutual feature shared by the microbes in the table is the use of carbon-based anodes with an average 0.4 V (vs. SHE) applied potential [81].

Two prime methods for increasing the EET of microbes are the introduction of EET mechanisms from other microbes capable of MES or the overexpression of the mechanism of the gene itself. The EET chain of *Shewanella oneidensis* comprising *MtrA*, *MtrB* and *MtrC* were inserted into *E. coli* as a representative example of the previous case to create an electrical conduit on the surface of its cell. The protein inserted through the heterologous pathway are then functionally expressed in the bacterium. Further, the interaction of protein *MtrA* and the bacteria aid in accelerating the process of reduction of soluble Fe(III). The modified strains reduce the metal and solid metal ions by about 8 and 4 folds (against the wild strain) respectively [62,63,102–104] The *G. sulfurreducens* strain can overexpress in both heterologous and homologous states. Gene expression of the *pilA* gene encoding for the structure of protein pilin can be spiked by interrupting the gene that encodes for the periplasmic Cyt c. Hence, this increases the rate of iron reduction. Two genes that code for Cyt c are *GSU1771* and *GSU3274*, the second one being a more prominent target for the movement of e^- [62,63,105–108].

Table 1. Synthesis of high-value chemicals through anodic electro-fermentation and microbial electrosynthesis (MES). Adapted from [79].

Microbe	Substrate	Product	Mechanisms of EET	Genetic Modification of Host	Yield (Y) and/or Titre (T)	Ref.
			Anodic Electro-Fermentation			
Shewanella oneidensis	Glucose	Acetate	Direct electron transfer (ET)	Introduction of *E. coli* galactose permease (*galP*) and glucose kinase (*glk*) genes.	No	[82]
	Glycerol	Ethanol; Acetate	Direct ET	Introduction of *Zymomonas mobilis* ethanol production module and *E. coli* glycerol utilisation module	Y = 52% ± 4% T = 1.28 ± 0.02 g L^{-1}; Y = 13% ± 6% T = 0.29 ± 0.08 g L^{-1}	[83]
	Lactate	Acetoin	Direct ET	Introduction of *Bacillus subtilis* acetolactate decarboxylase and acetolactate synthase; Deletion of genomic prophages; Knockout of the phosphotransacetylase and acetate kinase genes	Y = 52% T = 0.24 g L^{-1} Productivity = 0.91 mg h^{-1}	[84]
Actinobacillus succinogenes	Glycerol	Succinate; Acetate; Formate	Neutral red mediated ET	Transmembrane mediator transport was improved by atmospheric and room temperature plasma mutagenesis.	Y = 68% T = 23.92 ± 0.08 g L^{-1}; Y = 7% T = 1.15 ± 0.77 g L^{-1}; Y = 19% T = 2.57 ± 0.11 g L^{-1}	[85]
Klebsiella pneumoniae	Glycerol	Acetate; 3-Hydroxypropionic acid; 1,3-Propanediol	Direct ET	Not modified	T = 21.7 mM; T = 7.6 mM; T = 45.5 mM	[86]
Clostridium cellobioparum* + *Geobacter sulfurreducens	Glycerol	Ethanol	Direct ET	Adaptive evolution of *C. cellobioparum*	T = 10 g L^{-1}	[87]
Propionibacterium freudenreichii	Glycerol & propionate; Only Propionate; Lactate & propionate	Acetate	Ferricyanide mediated ET	Enhanced bacterial growth & substrate consumption	Y = 56% T = 0.38 g L^{-1}; Y = 68% T = 0.47 g L^{-1}; Y = 60% T = 0.42 g L^{-1}	[81]
***Enterobacter aerogens* NBRC 12010**	Glycerol	Ethanol; Hydrogen	Thionine mediated ET	Increased glycerol consumption	Y = 92% T = 3.93 g L^{-1}; Y = 74% T = 0.14 g L^{-1}	[81]
Cellulomonas uda* + *Geobacter sulfurreducens	Cellobiose	Ethanol	Direct ET	Adaptive evolution and deleted *G. sulfurreducens* hydrogenase gene	No	[88]

Table 1. *Cont.*

Microbe	Substrate	Product	Mechanisms of EET	Genetic Modification of Host	Yield (Y) and/or Titre (T)	Ref.
			Anodic Electro-Fermentation			
Ralstonia eutropha	Fructose	Poly hydroxy butyrate	poly (2-methacryloyloxyethyl phosphorylcholine-co-vinyl-ferrocene)-mediated ET	Not modified	No	[89]
Escherichia coli	Lactate	Acetate; Ethanol	Direct ET	Introduction of *S. oneidensis* MR-1 *Mtr* pathway	Productivity = 0.038 mM day^{-1}; T = 40 ± 3 μM	[90]
Escherichia coli* + *Methano bacterium formicicum	Glycerol	Ethanol; Acetate	Methylene blue-mediated ET	Cyt c introduction—*CymA, MtrA* and *STC* from *S. oneidensis*	Y = 35% ± 5% T = 55.25 ± 7.76 g L^{-1} Productivity = 12.12 ± 1.70 mg h^{-1}; Y = 20% ± 1% T = 40.75 ± 2.37 g L^{-1} Productivity = 8.94 ± 0.52 mg h^{-1}	[91]
Pseudomonas putida F1	Glucose	2-Keto-gluconate	7 different mediators-based mediated ET	Not modified	Y = 90% ± 2% T = 1.47 ± 0.27 g L^{-1} Productivity = 1.75 ± 0.33 mg h^{-1}	[92]
Pseudomonas putida	Glucose	2-ketoglu conic acid	Direct ET	Overexpression of periplasmic glucose dehydrogenase	Productivity = 0.25 ± 0.02 mmol g_{CDW}^{-1} h^{-1}	[93]
Corynebacterium glutamicum* + *Zymomonas mobilis	Glucose Glucose	L-lysine; Ethanol	Ferricyanide-mediated ET Methyl naphthoquinone, humic acid, methylene blue, neutral red, 1,4-riboflavin, butane-disulfonate and tempol-mediated ET	Feedback deregulated mutant and overexpressed redox-related genes—*ZMO0899, ZMO1116* and *ZMO1885*	T = 2.9 Mm Productivity = 0.2 mmol L^{-1} h^{-1} Bioelectricity generation = 2.0 m Wm^{-2}; T = ~ 42.5 g L^{-1}	[94]
			Microbial electrosynthesis (MES)			
Clostridium pasteurianumDSM 525	Glucose; Glycerol	Butanol; 1,3-propanediol	Direct ET	Not modified	T = 1.00 ± 0.20 g L^{-1}; T = 4.74 g L^{-1}	[47]
Geobacter sulfurreducens	CO_2; Succinate	Glycerol	Direct ET	Not modified	T = 8.7 ± 0.3 mM	[95]
Sporomusa ovate* + *Methanococcus maripaludis	CO_2	Acetate; CH_4	H_2-mediated ET	Not modified	T = 0.2 to 0.3 mM; T = 0.2 to 0.3 mM;	[34]
Sporomusa ovate	CO_2	Acetate	Direct ET	Not modified	No	[96]

Table 1. *Cont.*

Microbe	Substrate	Product	Mechanisms of EET	Genetic Modification of Host	Yield (Y) and/or Titre (T)	Ref.
Shewanella onedensis MR-1	Acetoin	2,3-butanediol	Direct ET	Heterologous expression of butanediol dehydrogenase (*Bdh*) gene along with a light-driven proton pump and hydrogenase gene Δ*hyaB*Δ*hydA* knockout	T = 0.03 mM	[97]
			Anodic Electro-Fermentation			
Clostridium pasteurianum	Glycerol	1,3-propanediol; n-butanol	Neutral red and brilliant blue-mediated ET	Not modified	Y = 0.41 mol mol^{-1} glycerol in brilliant blue-mediated ET; Y = 0.35 mol mol^{-1} glycerol in Neutral red-mediated ET	[98]
Saccharomyces cerevisiae	Dhea	7α–OH–DHEA	Neutral red and 7α-hydroxylase-mediated ET	Heterogenous expression of 7α-hydroxylase	T = 288.6 ± 7.8 mg L^{-1}	[99]
Ralstonia eutropha	CO_2	Iso-propanol	H_2-mediated ET	Not modified	T = 216 mg L^{-1}	[100]
		3-methyl-1-butanol; Isobutanol	Formate mediated ET	Introduction of genes *alsS*, *ilvC*, *ilvD*, *kivd* and *yqhD*; Knockout of polyhydroxy butyrate synthesis gene cluster (*phaC1*, *phaA* and *phaB1*)	Both depicted a titre of 140 mg L^{-1}	[79]
Xanthobacter autotrophicus	N_2 and H_2O	NH_3	H_2-mediated ET	Not modified	T = ~ 0.8 mM	[101]

2.5. Increasing Electrode Interaction

The relationship between microbes and electrodes relies mainly on the cohesive nature of the biofilm, the electrode and how these species interact with the electrode [109]. To implement a variety of microbes, the electrodes have been improvised, yet, further research on strain modification for pure cultures is still necessary [110]. Two frequently used strains for understanding electrode interactions are *S. oneidensis* and *G. sulfurreducens*. In a related review, the latter bacterium was altered by deleting the genes that encode for a protein controlling the *Pilz* domain, forming a more coherent and conductive biofilm. Further, this mutant produced 6-fold more conductive biofilm when compared to its wild type. More production of pili led to a smaller potential loss. In another experiment, the former bacterium was altered to allow the production of biofilm with the help of heterologous overexpression of the cyclic di-GMP pathway gene that originated from *E. coli*. Hence, after a few hours, the collection of cells and electrolytes from the electrode depicted a significant change against the wild type of enhanced electrode [68,111–113].

3. Techniques for Improving MES Performance

It is imperative to enhance the MES's performance and optimise it while maintaining a low budget [114]. During the anodic and cathodic processes, the electron transfer system of bacteria is likely to follow their different path. So, the electrode materials don't need to yield good results in both the microbial cathode and microbial anode equally [115]. Various factors (physical, chemical and biological) affect MES differently, these have been depicted in Figure 4.

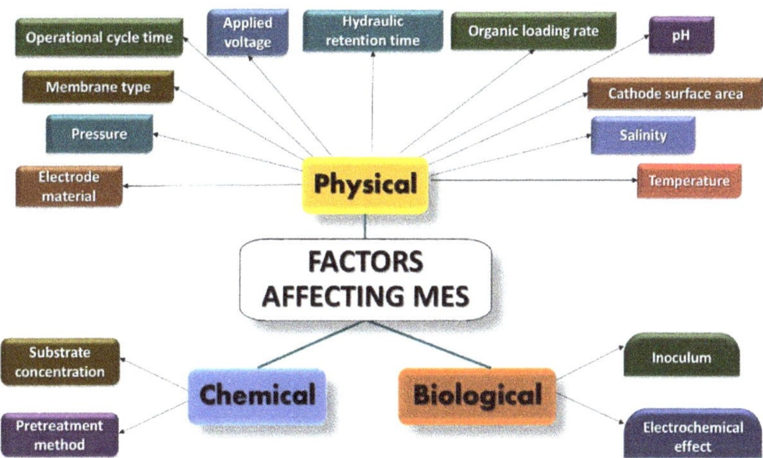

Figure 4. Several physical, chemical and biological factors affecting MES.

3.1. Cathode Fabrication

Since 2010, numerous commercially accessible carbon electrodes of various shapes and forms (like rods, fabric, block, AC (activated carbon) plates, gas diffused AC, reticulated vitreous carbon (RVC) and granules of fibre felt) have been developed for direct CO_2 reduction. Important features like chemical tolerance to degradation, cost-effectiveness, biocompatibility with long term and proven application of bioanodes make the cathodes more efficient and drive a better and broader usage in the long run in MES processes. For instance, industrially commercialised carbon material like graphite, carbon plate and carbon fabric and 3D structures such as carbon felt and carbon fibre rod electrodes are used in MES processes [2,48,61,116].

The key reason behind introducing surface modified materials is to promote cathodic biofilm for efficient CO_2 reduction in MES. The key to enhanced bioproduction lies in the interaction between the bacteria and the cathode, therefore it is a requisite to choose the material and build a suitable cathode [117]. Carbon felt, mesh and cloth are commonly used to create surface-modified cathodes due to their various advantages, including high porosity, resilience, larger surface area and many more. Other features and characteristics of cathode and design modification to produce charge are complicated because bacterial attachment, electrode microbe rate of electron transfer, selective development of biofilm and maximum production rate cannot be sufficiently increased by treating the electrode with melamine and ammonia, against unmodified carbon cloth [118,119]. Compared to untreated graphite, the microwave treatment of nickel nanowire coated graphite increases the surface roughness by about 50 folds. Other modified electrodes include carbon cloth modified by utilising nanoparticles like gold, palladium, nickel or cotton and polyester-modified carbon nanotubes-coated cathodes (considered the most promising materials) or NanoWeb-RVC-cathodes that improve acetate output by 33.3 folds relative to unmodified carbon plates [48,118–121].

Biocathodes are electrodes enriched with microorganisms, with whose help they perform reductive reactions for various substrates. These electrodes use the metabolic activity of the microbes for the same purpose. Cai et al. [122], in a bioelectrochemical method, implemented a biocathode to increase the oxygen reduction and to generate electricity in air cathode MFC. They used it with water for the removal of contaminants present in the catholyte solution and used it for the production of target commodities such as H_2 production using protons as e^- acceptors. Kondaveeti and their team [123] utilised biocathode in BES to reduce CO_2 to form products (methane, VFAs and alcohols). Tahir and his team [124] used MXene–coated biochar as an MES biocathode for selective VFA production. Improvement of the biofilm development, attachment of bacterial cell, rate of e^- transfer at cathode surface, as well as the rate of chemical production, requires key elements like best cathode materials, selective microbial groups and well-organised reactor design. There are various reports and studies on new materials of electrode discovery and modification systems of surface for anodic process development (See Table 2) [115].

Table 2. It depicts the carbon cloth cathode treatment and acetate production rate for each day including consumption density [16].

Carbon Cloth Cathode Treatment	Average Current Consumption Density (mA m^{-2})	Acetate (mM m^{-2} day^{-1})	Coulombic Efficiency
Carbon cloth	-71 ± 11	30 ± 7	76 ± 14
3-Aminopropyltriethoxysilane	-206 ± 11	95 ± 20	82 ± 11
Ni	-302 ± 48	136 ± 33	80 ± 15
Melamine	-69 ± 9	31 ± 08	80 ± 15
Carbon Nanotube-cotton	-220 ± 1	102 ± 25	83 ± 10
Cyanuric chloride	-451 ± 79	205 ± 50	81 ± 16
Ammonia	-60 ± 21	28 ± 14	82 ± 8
Pd	-320 ± 64	141 ± 35	79 ± 16
Chitosan	-475 ± 18	229 ± 56	86 ± 12
Polyaniline	-189 ± 18	90 ± 22	85 ± 7
Au	-388 ± 43	181 ± 44	83 ± 14
Carbon Nanotube-polyester	-210 ± 13	96 ± 24	82 ± 8

3.2. Anode Fabrication

According to various studies, carbonaceous materials like carbon cloth, glassy carbon, graphite felt, carbon, granules of graphite and rods are mainly used as the anode material and are available commercially. These materials have advantages in chemical stability and good conductivity of electricity. There are various methods available to enhance the formation of biofilm and MFC performance. The primary is to increase the attachment of bacteria with a bacterial electrode via EET. In this electron transfer, the potential is kept at enzyme redox potentials rather than the addition of mediator potential and exogenous mediators. For example, the node is pre-treated with ammonia gas, aqua fortis and ethylenediamine at 700 °C, with HNO_3 and quinone or quinoids, showing a spike in anode density of the microbial cell at a successful rate. HNO_3 and hydrazine can also be used as alternatives [125,126].

Second is the electrode pretreatment before displaying them to MFC operation and biofilm development. Other pretreatment methods are also available that aim at changing the surface of the electrode with different redox molecules, this helps in the transfer of an e^- from the microorganism to the electrode. Biofilm formation is also a difficult process and affects different factors like charge, hydrophobicity, topography along with bacterial properties and environmental factors [127,128]. Several studies suggest that the positively charged surface of anode allows for higher attachment of bacteria and biofilm activity. When charged bacterial surfaces are suspended in the aqueous suspension, then the charged surface attracts more bacterial cells. The positive surface charge on the carbon clothes was extended from 0.38 to 3.99 meq m^{-2} when treated with ammonia which successfully reduced the acclimation time of microbial by 50% and maximal power density increased. Bacteria are more attracted to groove or braided surface rather than a smooth surface to colonise porously and adhere easily to the surface as rough surface is the more favourable site for colonisation [16,129–132]. MFC performance can be enhanced by expanding the available expanse for biofilm growth using rough or porous materials [128].

3.3. MES and Gene Manipulation

A gene manipulation technique via systems biology approach must be carefully designed and tested to determine which gene of the microbe should be altered and for what purpose to enhance MES. The strategy which can be adopted is the microbe's EET efficiency improvement to increase the cathodic chamber activity. Alternatively, a different pathway can be used for yielding valuable products by choosing the heterologous or homologous expression. The appropriate method can be dependent on whether or not the target strain is simple to adjust (based on the tools available). One of the most representative MES capable acetogen is *Sporomusa ovata*, but in a recent study, *C. ljungdahlii* were applied [24,133,134].

3.3.1. Modification of Pathways for Generating Value-Added Products

When the microbe capable of binding to the electrode is incompetent to create enough of the desired output but can use another pathway to produce the desired result, modification of the pathway for the processing of value-added commodities becomes essential. The alternate method could be exogenous, producing a heterologous effect or competitiveness, suppressing its gene system. Microbe such as *Clostridium ljungdahlii* has often been used due to the presence of a well-established tool that allows deletion of the gene [135].

Genetically modified (GM) microorganisms mentioned in this subsection have not been introduced in MES systems yet. Nevertheless, these still have enough potential of being MES-capable microbes because they contain most of the e^- transfer systems proposed in *C. ljungdahlii*. Through heterologous expression, this strain was altered for generating butanol from the initial *Clostridium acetobutylicum*. The resulting strain increased the production of butanol significantly. Nevertheless, *C. ljungdahlii* transformed butanol to butyrate and so no butanol was found at the end of the process. In the experiment, the presence of butanol was confirmed due to the active expression of microbe genes [136–138]. In another research, a lactose inducible mechanism designed for *Clostridium perfringens* was incorporated into *C. ljungdahlii*. This arrangement improved the production of ethanol by 30-fold [139]. Apart from advantages like easy optimisation for highest yields, high selectivity, providing resistance against system fluctuations and O_2 intrusion and facilitating a wider product spectrum of high-value molecules; GM microbes also have a few limitations. The most prominent limitation being questionable societal acceptance and grant of approval by the government [8].

3.3.2. Host Cell Selection

Concerning adaptation to other areas or even new trials, GM MES is only in its early stages, holding an infinite scope of research. Genetic modification is a method used in biotechnology to recreate host cell DNA and perform insertion and deletion of genes or producing point mutation through homologous or heterologous expression. Therefore, when analysing the possibility of GM in MES, the preference of the host cell is a primary concern. The selected host organism that is to be subjected to genetic modification should have a simple and fully sequenced genome along with the required genetic tool. Additionally, microorganisms that have more advantages and all the other necessary features are favoured for MES. Three principal aspirants from our perspective for MES are *E. coli*, *C. ljungdahlii* and *Cyanobacteria* [18,62,63,140]

Escherichia coli

One of the most commonly used laboratory microbes in the field of biotechnology is *E. coli*. GM *E. coli* is a crucial microorganism that plays a central role in the generation of heterologous proteins, like the development of vaccines, bioremediation and much more [141,142]. Genetic tools present in this organism can be exploited in the field of MES. Neutral red is used as an electron shuttle in *E. coli*, aiding in generating more products such as pyruvate, lactic acid and succinic acid when compared to usual conditions [62,63,91,143]. Further, this bacterium is commonly utilised for heterologous expression to find out the features of redox protein that take part in the process of e^- transfer [144]. Hence, suggesting that *E. coli* has ample capabilities to function as a host cell in the MES field.

Clostridium ljungdahlii

C. ljungdahlii holds high worth as a host cell not only because it is a potential MES-capable microbe, but also due to its ability to reduce CO_2, treat waste gas and produce various products, like ethanol and acetic acid via fermentation [135,145]. Hence, *C. ljungdahlii*-driven MES can be used as a cell factory with continuous treatment of industrial waste gas. Modifying its gene proves that it is capable of generating diverse multi-carbon products as described earlier [139].

Cyanobacteria (Cyano)

The only photosynthetic prokaryote capable of extracting oxygen by splitting water is cyanobacteria, which shares several benefits with microalgae [146]. Additionally, this blue-green algae are a crucial third-generation biomass producer because of its capability to photosynthesise (oxygenic), non-food-based feedstock, high per-acre productivity and land-independent growth [147,148]. Research based on intracellular and extracellular e^- transfer in Cyano has revealed that they vastly differ from other microbes used in MES [149]. Hence, as a host cell for MES, these next-generation biomass producers hold drastic value in the field of research. Appropriate and enough data is needed to develop an efficient strategy. Nevertheless, it is almost impossible to obtain nearly all the in vitro and in vivo data due to a lack of room and adequate time. A recent development in the bioinformatics field offers the perfect method known as "in-silico" to solve issues related to the collection of data. Numerous aspects and features of the microbe could be analysed using the in-silico method. These research-based in-silico methods present the forecast results on ATP yield, biomass, CO_2 fixing pathway, reduction degree of the product and substrate and fluctuations in the flux via specific pathways [150]. Therefore, evaluation based on in-silico methods can help in providing vital data to develop an efficient strategy of metabolism.

4. MES Allows Biocatalysts to Utilise CO_2 and Generate Electricity

These days, the most important global problem is the elevated CO_2 emissions that cause a spike in the average global temperature. Greenhouse gases play a central role in emergent global warming. There are many greenhouse gases responsible, but among those, the sole contribution of CO_2 is 63%, which is quite high. This has led to the emergence of the use of different CO_2 capturing and storing techniques. Various CO_2 utilisation procedures have been discovered that have been used for hoarding and transforming CO_2 into high-valuable products via MES by applying bio-electrochemical techniques using electricity as the source of energy [8–10,151–154].

MES focuses on using renewable sources rather than using non-renewable sources such as crude oil, decreasing the usage of naphtha-based chemicals. MES is not harmful to the environment, making it useful for future production of CO_2 and the protection of the environment by CO_2 sequestration, averting various environmental issues. We are aware of the thermodynamic stability of CO_2, it requires a supply of external energy for the activation and various conversion reactions. MES set-up usually comprises a couple of chambers called abiotic anodic chamber and biotic cathodic chamber that contains proton exchange membrane (pEM) aiding in the movement of protons across the two chambers (See Figures 2 and 3). In the anodic chamber, water molecules split into protons (p^+), e^- and gaseous O_2 is released. The p^+ are transferred to the cathodic chamber via the pEM and the e^- are drawn to the cathode through an external circuit. In the cathodic chamber, the e^- and p^+ or energy carriers such as H_2 and CO_2 are integrated by biocatalysts to produce volatile fatty acids (VFAs), alcohols, butyrate, formate, acetate, etc. via H_2 mediated e^- transfer or DET. Generally, the acetogens employ the Wood–Ljungdahl (WL) pathway for CO_2 fixation and are used as biocatalysts at the cathode, yielding acetate as the chief MES product, nevertheless, other organic chemicals with more carbon content, like butyrate, caproate, ethanol, caprylate, propionate and isopropanol can also be produced [25,114,155].

For breaking the double bonds of CO_2 a massive quantity of energy is needed as CO_2 is thermodynamically very stable. Therefore, metal catalysts can be implemented for reducing the energy to be employed for cleaving the bonds. This bottleneck can be overcome by using different and better catalysts that are easy to manipulate or biocatalysts in the form of enzymes or microbes, reducing the economic feasibility of the process. Microorganisms responsible for reducing CO_2 into organic compounds by consuming e^- are termed electroautotrophs. These types of microorganisms can survive as biofilms and also as planktonic cells in the bulk phase. Using biocatalysts for CO_2 reduction can be beneficial in operating the process and capital costs in the process. MES also has a few drawbacks, it currently has low product yield, the product once obtained needs to undergo downstream processing (purification and separation); its high capital cost makes scaling up challenging; and lastly, longer carbon chain chemicals have a low production rate [2,8,19]. A comparative synopsis of several microbial catalysts used, cathode materials employed and yield of product obtained via MES has been depicted by Jourdin and his team [19].

5. Diverse Products Obtained from CO_2

Technologies like the MES are the need of the hour, as they not only focus on the high efficiency of the system but also aid in the production of a variety of products by reducing and converting CO_2 in MES [2,8,10,15,17,156]. One of the most common by-products is acetate [157]. Alternatively, butyrate, oxobutyrate, ethanol and isopropanol can also be produced using MES [20,158]. It was also observed that on further reduction of acetate that has been generated and accumulated in the system, more commodities can be produced [157]. In the case of ethanol, it was observed that organisms such as *S. ovata*, when kept under highly reductive conditions in the presence of excess O_2, generated ethanol [22,159]. Often a variety of alcohols and VFAs have been observed during MES production [160,161].

Acetyl-CoA is the primary precursor in CO_2 reduction that takes place via the Wood-Ljungdahl (WL) pathway [162]. This precursor molecule then undergoes several steps and metabolic changes to produce diverse chemicals [163]. Acetogenic microbes are used for this process to reduce CO_2, using H_2 as the e^- donor. By modifying the conditions under which the system is operated, mechanisms such as solventogenesis metabolism can also be introduced in MES [164]. An important microorganism used for this process is *Clostridium ljungdahlii*. Although pure cultures can be used for this process, even mixed cultures are an amazing alternative, as portrayed in Table 3 [25,145,165,166]. Apart from ethanol, butyrate and acetate, lactate and succinate can also be produced from intermediate products generated in the system via the Krebs cycle. To summarise, MES produced precursor compounds such as acetate, which can further be upgraded to longer chain fatty acid, biofuels, bioplastics and so on, via multi-step conversions. For example, acetate on being upgraded can produce butyrate and caproate via the chain elongation method in MES. Some other examples include the production of methanol, formaldehyde and ethylene from methane and the production of single-cell proteins and polyhydroxyalkanoates from short-chain fatty acids [20,114,158,163,167].

Table 3. Microorganisms were tested by Nevin and his team [96] to check the solid-state electrode's ability to receive electrons and to use CO_2 as the terminal electron acceptor, following the Wood-Ljungdahl pathway.

Species	Electron Consumption? (EC)	EC Rate vs. *S. ovata*	Electron Recovery in Products	Products Formed
Moorella thermoacetica	Yes	-	85% ± 7% (n = 3)	Acetate
Clostridium lijungdahli	Yes	-	82% ± 10% (n = 3)	Acetate + Minor formate and 2-oxobutyrate over time
Sporomusa ovata	Yes	100%	86% ± 21%	Acetate + Trace of 2-oxobutyrate
Acetobacterium woodii	No	-	-	-
Sporomusa silvacetica	Yes	10%	48% ± 6%	Acetate + Trace of 2-oxobutyrate + non-identified products
Clostridium aceticum	Yes	-	53% ± 4% (n = 2)	2-oxobutyrate and acetate as prime products and other non-identified
Sporomusa sphaeroides	Yes	5%	84% ± 26% (n = 3)	Acetate

Nevin and his team [96] were the first to prove that biocathode systems can be used for the reduction of CO_2 and to yield acetate. Some microorganisms are known to obtain the carbon for metabolic processes from the atmospheric CO_2 and some from inorganic sources. Both pure and mixed cultures can be used in this process, but mixed cultures are more preferred as they can be obtained in large quantities simultaneously and can easily tolerate environmental conditions as compared to pure cultures which need a specific growth medium and are vulnerable to system fluctuations and O_2 intrusion [2,8]. Till now, both mixed and pure cultures have shown similar recovery of e^-. CO_2 can be reduced to acetate with the aid of acetogenic bacteria that uses hydrogen as an electron donor. The pure culture of this bacteria is poured into the cathodic compartment that has been already filled with a mixture of gas and various e^- donors for increased and sufficient growth of the culture on the electrodes. H_2 production was controlled by applying −400 mV potential (vs. SHE) to the cathode, meanwhile, after switching the gas feed to N_2-CO_2, the acetate was produced along with the small volume of 2-oxobutyrate that had e^- recovery up to 85%. The cathode biofilms used were long-lasting as they were capable of accepting the e^- after 3 months and could also produce acetate but, the acetogenic microorganisms lack this property and gained very little energy by the reduction of CO_2, which implies the use of substantial energy inputs or expensive catalysts which seems impractical. The researchers further tried experimenting with other microbial species to check whether they were capable of MES or not, several other acetogenic bacteria were able to gain electrons at the electrode. As seen in Table 3. the acetogen *Sporomosa ovata*, a close relative to *M. thermoacetica* was able to directly accept e^- from the cathode and transform CO_2 to acetate and 2-oxobutyrate, whereas *A. woodii*, was unable to do so as it lacked Cyt c and relies upon the sodium gradient which is coupled to the WL pathway, thereby, reflecting a different behaviour as compared to other acetogens in Nevins experiment [24,25,78].

5.1. H_2 Production via MES

Hydrogen is a valuable fuel that can be produced efficiently by MEC. Materials used as cathode catalysts include platinum for the production of H_2 from MES, but because of its high cost, it is not viable economically, so other alternative materials like stainless steel and nickel are used as these have low cost, stability and low over-potentials. For H_2 production, nickel and stainless steel have more efficiency against platinum owing to their low voltage and cheap cost [32,34,168–170].

But enzymatic biocathodes are unstable and not self-generating, hence they lose their activity of catalysis over time. The study suggested that H_2 production is successfully catalysed by immobilising the enzymes responsible for catalysing the reversible reaction at carbon electrodes. *Desulfovibrio* species (hydrogenases processing microbe) are used for hydrogen production by immobilising methyl viologen that acts as a redox mediator. Mixed cultures can be employed to enhance microbial H_2 production as they show more desirable characteristics like steadiness and relevancy in BES. Acetate and hydrogen are used as an e^- donor that changes the electrode polarity and with anode attached biochemically-active biofilm reverses the biocathode's mode for H_2 production [171–173]. Examples of studies on H_2 production using several substrates and VFA mixtures in MECs have been reviewed by Rivera et al. [37] and Cardeña and team [36].

5.2. Acetate Production via MES

For acetic acid, the production of the NanoWeb-RVC (carbon nanotubes on reticulated vitreous carbon) showed very high efficiency as a biocathode component. This type of electrode is advantageous for macro structured RVC and nanostructured surface modification. Effective mass transfer is ensured to and from the biocatalyst due to the high surface area to volume ratio of the macroporous RVC. The carbon nanostructure increases microbial EET, improves the interaction between microbe and electrode, enhances the development of microbes and helps in bacterial attachment. So NanoWeb-RVC displays a high intrinsic performance as a biocathode component for MES and is considered an effective material from an engineering perspective.

Electrophoretic deposition (EPD) is a method in which colloidal solution is utilised to make thin films. It has also been used on a large scale to make highly porous electrodes for electrochemical applications from the deposition of carbon nanotubes (CNT). For processing the CNTs, EPD is the easiest process to operate that employs simple equipment. However, it can also produce narrow films from colloidal suspensions on substrates irregular in shape. An increase in production can be achieved just by expanding the dimensions of the existing substrate to be coated. So EPD demands bulky and industrial-scale manufacturing of porous electrodes. The MES has been recorded to achieve a high acetic acid production rate of up to 685 g m^{-2} day^{-1} from CO_2, using enriched microbial culture and a newly synthesised material for the electrode [16,61,174,175].

The study by Tian et al. [168] demonstrated that the MES performance can improve by hydrogen evolution reaction catalyst (HER). This involved the construction of a molybdenum carbide (Mo_2C) modified electrode, an active HER electrocatalyst, the final acetate yield rate of MES is much higher. Electrochemical studies and analysis also suggested that Mo_2C can be induced for the production of H_2 and help in the biofilm formation and monitor the mixed culture of microbes. It shows an electronic structure similar to the metal group like platinum, considering the high performing HER electrocatalyst. The presence of molybdenum carbide in carbon felt (Mo_2C-CF) results in increased evolution of H_2 in the MES, which averagely shows 12.7 times higher than CF without Mo_2C. The presence of

Mo_2C also helps to regulate the mixed culture of microbes in biofilms and the planktonic cells in microbial electrosynthesis. Some of the microbes involved in the MES system are namely *Acetobacterium*, *Citrobacter*, *Arcobacter*, etc. H_2 acts as the e^- carrier and helps in e^- transport through a hydrogen-related metabolic system due to the presence of HER electrocatalyst cathode in the MES. This also helps in the CO_2 reduction step in MES due to the coupling of an active HER cathode. Refer to Table 4 to compare the yield of acetate when a mixed microbial flora is employed for MES.

The coupling of molybdenum carbide in CF cathode is one of the most vital, rapid and simple studies that efficiently improve the MES system. To develop a highly efficient H_2 catalyst, a neutral condition but the HER electrocatalyst of Mo_2C reported the advantages of hydrogen evolution even in the acidic condition. Therefore, the presence of active HER catalysts like Mo_2C increases the release of hydrogen, which helps the growth of the biofilm of mixed microbial culture, and thus resulted in a higher reduction rate of CO_2 and generation of acetate in the MES system [168,176].

Table 4. Review of literature on the yield of acetate via mixed microbial flora in MES.

Cathode Material	$E_{cathode}$ (V vs. SHE)	Current Density ($A\ m^{-2}$)	Volumetric Production Rate ($g\ L^{-1}\ day^{-1}$)	Maximum Acetate Titre ($g\ L^{-1}$)	Coulombic Efficiency (%)	Ref.
12 mg cm^{-2} Mo_2C	−0.85	−5.2	0.19	5.72	64	[168]
NanoWeb-RVC	−0.85	−37	0.03	1.65	70	[61]
Graphene-nickel foam	−0.85	−10.2	0.19	5.46	70	[177]
VITO-CoRE™ electrode fabricated with activated carbon	−0.6	−0.069	0.14	4.97	45.5	[178]
Carbon felt (CF)	−1.26	−5.0	0.06	1.29	58	[179]
	−0.903	−2.96	−0.14	4.7	89.5	[180]
CF and stainless steel	−0.78	−15	0.14	2	22.5	[7]
	−0.9	−10	1.3	0.6	40	[181]
RVC-EPD	−0.85	−102	-	11	100	[182]
rGO-CF	−0.85	−4.9	0.17	7.1	77	[177]
CF with fluidised GAC (16 g L^{-1})	−0.85	−4.08	0.14	3.9	65	[183]
Graphite stick-graphite felt	−0.8	−20	0.14	8.28	-	[184]
Graphite granules	−0.6	-	1.0	10.5	69	[185]

5.3. Formic Acid Production via MES

CO_2 can be converted into liquid formic acid with the help of sustainable electricity, which later can serve as a chemical for preserving food, alternative future fuel and an energy storage molecule. Formic acid has been derived primarily from fossil reserves, which are estimated to get depleted, to solve this problem green alternative ways have been discovered through microbial transformations. Formic acid production was earlier carried at laboratory scale using CO_2 and electricity, a direct electrochemical conversion where H_2O is dissociated into H_2 and O_2, and the former is then used to reduce CO_2 into formic acid. Production of formic acid requires less energy as compared to methane and methanol production against CO_2 [154,186,187].

At a commercial scale, formic acid can be produced from methanol via multiple pathways, initially, it is transformed to methyl formate followed by hydrolysis to produce formate. Later on, formate production via direct conversion is analysed through hydrogenation; nevertheless, the end product is formate. Therefore, a single-step chemical reaction has been recently discovered, where formate is produced via electrochemical reduction of CO_2 by H_2 generated from H_2O. The electrochemical set-up consists of an anode and cathode, where the hydrolysis of water and formation of formate takes place. Different electrode systems have been used for direct electrochemical reduction of CO_2, including metals, nonmetals and bioelectrodes. The electroreduction potential of −1.85 V vs. SHE is required for yielding formic acid. However, different types of compounds can be formed in MES, including various hydrocarbons like alcohols and carboxylic acids such as butyric acid. Various studies have revealed the selective production of formic acid from CO_2 using different electrode materials, but the products generated are mostly non-specific. The properties of formate of high solubility, easy conversion into other compounds and its decomposition at the anode are responsible for the lower yield and increase in separation cost. These limitations have been resolved by utilising enzymatic CO_2 electroreduction for the generation of formic acid. The key enzyme involved is formate dehydrogenase, it catalyses the oxidation of formate to CO_2, and reduces CO_2 to formate, which is chaperoned by the NAD^+ to NADH redox cycle [188–190].

5.4. *Syngas Production via MES*

To satisfy the potential demand for biomethane and energy, the present supply of organic waste is not sufficient, hence it is necessary to increase the output to fulfil this demand [191,192]. The organic sources of methane are limited, although enough CO_2 is present from the industrial exhaust, electrons can also be acquired from water, sulphides and ammonium. Further, methane can directly be converted to syngas through hydrogenotrophic methanogenesis or it can also be produced indirectly by using acetate as an intermediate [192,193]. Syngas or synthetic gas is primarily a combination of gases such as H_2, CO and sometimes CO_2 and is used for electricity generation [191,194]. The anodic and cathodic chambers were constructed keeping in mind the reactor volume. This would provide an optimum surface area for the system, hence making the process more efficient [192].

Initially, when syngas fermentation was combined with a single cell anaerobic digestor (AD-MES system) [192], an experiment of three phases was conducted where the anodic and cathodic chambers were set up considering the volume of the reactor. In phase I, the experiment was conducted inside a glass reactor which was a lab-scale fermentation reactor. The II phase triggered the open circuit in which electrodes were established in the glass reactor through phase 1 and fresh inoculum was added. In phase III voltage of -0.8V vs. SHE was applied to the syngas from phase II for the production of gases like methane, carbon dioxide and hydrogen [195–197]. The conclusion of this experiment built the starting point of combining two processes i.e., syngas fermentation with a single cell AD-MES system [195]. Examples of production of syngas (such as biomethane and biohydrogen) and value-added biochemicals (such as H_2O_2, bioalcohols, acetate and VFAs) using BESs have been briefly summarised by Kumar et al. [198].

Coupling Anaerobic digestion (AD) with MES is one of the most novel technologies through which CO_2 can be generated and can further be reduced to methane with the help of microbes [195]. Similarly, using the same system, the CO_2 present in the biogas can be converted to acetic acid or other chemicals with the help of chemolithoautotrophic microorganisms. The main benefit of using this combination (AD-MES) is that the system is cost-effective and requires low capital, simultaneously the system also continuously produces and upgrades biogas while using only small energy input. The biogas generated by anaerobic digestion contains approximately 40–60% of CO_2, which can be utilised as a feedstock in the MES to generate diverse chemicals by reducing CO_2. Integrating both these processes has been proven to enhance the production of methane, hence enriching the composition of methane in biogas, as well as producing other value-added chemical commodities. Thus, offering additional economic benefits [33,48,51,199–201].

6. MES Enhancement

The CO_2 utilisation and the production of unsaturated VFAs on electrophoretic deposition and 3D-reactors, were continuously produced till the termination end of the trial, but the sole chemical compound generated was acetate, however, there was no accumulation of any other compounds like alcohols or VFAs. In the beginning, when the culture was transferred to the reactor, a steady production of acetate was observed concerning CO_2 consumption. The max average CO_2 utilisation rate observed was about 24.8 mol m^{-2} day^{-1}

and the rate of acetic acid production was around 11.6 mol m^{-2} day^{-1} was reached, from 7 weeks onwards on EPD-3D [182]. Given a carbon balance, 94 ± 2% of CO_2 was discovered to be changed over to acetic acid derivation (with the rest of the carbon probably being utilised for biomass creation), while an e^- balance uncovered a much all the more hitting result with 100 ± 4% of the electrons expended being recouped as acetic acid derivation. The changed regulations and item virtue accomplished in these investigations are outstandingly high, particularly for blended societies, which makes it intriguing for possible huge scope creation applications and downstream handling. Moreover, the accomplished acetic acid derivation production rate was around 685 ± 30 g m^{-2} day^{-1} is about 3.6 fold higher than the most noteworthy production rate [169] (Refer Table 4). Besides, a genuinely high acetic acid production titre of up to 11 g L^{-1} was obtained, without any indications of item restraint of the dynamic microbes by then. It is along these lines very possible that the titre would have extended much higher qualities had the experiment not been halted. A high titre is a basic trademark for forthcoming enormous scope usage as it delivers the downstream processing a lot simpler than when the item fixation was low [169,182].

MES performance can be evaluated utilising a few key boundaries; these are recorded in Table 5, for most MES to acetic acid derivation is reported to date. The results summed up in the table are gathered from mixed as well as pure cultures of microorganisms in fed-batch or continuous mode. There are various types of materials of cathode and cathode applied. The potentials make it difficult to differentiate between the studies. For modern industries, the bioproduction process, for instance, in the fermentation production rate of 2 to 4 g L^{-1} h^{-1} having 99% yield is necessary for process feasibility. In the past decade, researchers have tried to access the scale-up viability of BESs specifically. However, considering the 3-dimensional nature of the electrode and its total surface area to volume unit of 2620 m^2 m^{-3}, attained 72% [19,169,177,182].

Table 5. Major performance factors of most MES to acetate researches reported until now. Adapted from [16,19,177,202].

Microbial Inoculum	Cathode Material	$E_{cathode}$ (V vs. SHE)	Current Density (A m^{-2})	Acetate Production (g^{-2} day^{-1})	Max Acetate (g L^{-2})	Electron Recovery into Acetate %
S. ovata (continuous)	Graphite rods	−0.4	−0.208	1.3	0.063	86 ± 21
C. ljungdahlii (continuous)	Graphite rods	−0.4	−0.029	0.14	-	72
Brewery WW sludge (fed-batch)	Graphite granules	−0.590	-	-	1.71	67
Enriched Brewery WW sludge (fed-batch)	Graphite granules	−0.590	-	-	10.5	69
Enriched WWTP sludge (fed-batch)	Carbon felt	−0.9	-	34.5	-	89.5
S. ovata (continuous)	Carbon cloth chitosan	−0.4	−0.475	2.7	0.118	86 ± 12
S. ovata (continuous)	CNT cotton CNT polyester	−0.4	−0.215	~1.2	0.059	83 ± 10
S. ovate (continuous)	Network coated graphite	−0.4	~−0.625	3.3	-	82 ± 14
Enriched Brewery WW sludge (fed-batch)	Graphite rods	−0.6	−0.92 ± 0.12	8.56 ± 3.22	-	40
Mesophilic Brewery WW anaerobic sludge (fed-batch)	Graphite felt	−1.1	~−2.8	10.1	1.4	65

Table 5. *Cont.*

Microbial Inoculum	Cathode Material	$E_{cathode}$ (V vs. SHE)	Current Density (A m^{-2})	Acetate Production (g^{-2} day^{-1})	Max Acetate (g L^{-2})	Electron Recovery into Acetate %
Anaerobic digester	Graphite granules	−0.6	-	-	-	28.9 ± 6.1
Mixed natural & engineered sludge (fed-batch)	NanoWeb-RVC	−0.85	−37	192	1.65	70 ± 11
Enriched Mixed natural & engineered sludge (fed-batch)	EPD-3D	−0.85	−102	685	11	100 ± 4

7. Downstream Processes Involved in MES

Downstream processing of aimed complexes includes microbe, media and other by-products from the catholyte (electrolyte in the cathodic chamber). Hence, scholars have come up with new technology for distinguishing products, usually acetate extraction from the whole catholyte. Anion exchange resin (AER) is employed in extracting acetate in MES from the catholyte, the ratio of acetate in the solution was 16:4. AER can absorb 10 to 20 mg g^{-1} acetate in just a single day from a catholyte broth comprising several kinds of compounds. Another technique is employed for the separation of butyrate from catholyte by using a membrane with a hollow fibre made up of propylene. Acetate can be separated by an alternative extraction method using an extraction chamber placed between an anodic and a cathodic chamber. Anion exchange membrane and PEM distinguishes the extraction chamber from the prior mentioned ones, only allowing passage of carboxylic acid to get deposited inside the extraction chamber. The concomitant production and separation of carboxylic acid have many privileges over the high capital and operational costs of using more than one set-up, providing economic sustainability to MES [19,31,161,203–206].

7.1. Process for Conventional Separation

The most commonly used process for separating organic acids is adsorption, here the ions are exchanged between carboxylate groups and functionalised solid sorbents. The activity of sorbents relies on the pH of the solution; when the pH is in intermediate capacity (~6.5), the adsorption increases, however at increasing pH the concentration of ionised acid will rise along with the decline in protonated amine concentration. Another excellent alternative technique employed in the conventional separation of organic acids is liquid-liquid separation. Several extracts like aliphatic amines and tri-n-octyl-phosphine are preferred for this process [207,208].

7.2. Pressure and Concentration-Driven Separation Process

In another extraction process (protraction) involving the immobilisation of organic solvents with the help of capillary action into the small pores of the hydrophobic micro-filtration membrane, the feed gets separated from the permeate. Diffusion of organic compounds occurs rapidly via organic solvent onto the membrane which can be extracted on the permeate side. This membrane only provides mechanical support, the extracts perform the chief function. The process of protraction is generally employed for extracting VFAs. It is the favoured process over liquid-liquid extraction as it is a simultaneous process, solvent stripping occurs, it is not expensive and the amount of solvent required is also little [209].

Higher mass transfer rates are achieved by changing the configuration of the membrane set-up oppositely while using hollow fibre membranes, where the shell side is for feeding of organic phase while the tube side is for the aqueous phase. Nowadays, silicon membranes are widely used for the extraction processes in which the water is utilised as an extract and has portrayed excellent selectivity towards VFAs based on hydrophobicity. The process is overcome at a low pH as only un-dissociated acids are extracted. The selective extraction of butyric acid was achieved over acetic acid and propionic acid. However, this process is also reliable for alcohol extraction as the nutrients are preserved, making them feasible for extracting products from MES catholyte [155,210,211].

In the process of pervaporation, the process of partial evaporation is used to separate the compounds where the permeate side is kept under the influence of a vacuum. The extracts most commonly used in this process are the high-molecular-weight alkyl amines like tri-n-octyl phosphine oxide, trioctylamine and trilaurylamine [8]. The process of nanofiltration (NF) has been studied extensively for the extraction of VFAs. In NF, the effect of pH has been observed on the membrane charges as well as the degree of ionisation of acid. An increase in rejection of acetic acid has been seen (from 0–65%) when pH increases (3 to 7). The NF membrane is negatively charged, thereby, when the pH is increased the membrane restricts the carboxylate ions because of ongoing electrostatic effects. This makes a low pH for NF more beneficial for separating acetic acid [210,211].

7.3. Process of Reactive Extraction

To extract the organic acids, ionic liquids (ILs) have been studied. ILs were used to coincidently concentrate and esterify the acetic acid that was previously extracted via membrane electrolysis by Anderson and his fellow researchers [212]. He used bis (trifluoromethylsulfonyl)-imide IL, for concentrating acetic acid up to 80 Mm, when ethanol was added (max conversion of 90%), it was esterified to ethyl acetate. There is one other technique for the separation of VFAs, it is the usage of organic solvents in the presence of supercritical CO_2. For the extraction of propionic acid, trioctylamine was used in supercritical CO_2, high extraction productivity was achieved (97%-propionic acid), only a small portion vanished acid-amine complex formation. Pressure and temperature conditions need to be maintained as they have a high effect on productivity [213–215].

8. Advancing towards Sustainable Development of MES

8.1. Uses of Renewable Sources of Energy and Integrated Hybrid Systems

The best solution to the present-day challenges including resource scarcity, waste generation and sustaining economic benefits is an economy that is environmentally and economically regenerative like that of a circular bio-based economy depicted by MES [8,25,114]. Using alternative energy resources for the generation of biofuel are becoming an increasing trend. The idea of biorefinery has been suggested to encourage a bio-based economy to promote the use of renewable sources like biomass; to produce fuels, generate electricity, heat and other beneficial chemicals in a circular economy model stimulating material reusing and recycling [25,216]. This feedstock primarily comprises energy-generating crops and waste biomass, which is a readily available alternative that can partially substitute the current reliance on fossils providing a green source of feedstock for chemicals and fuel [217]. Renewable biomass can provide reliable, stable and sustainable energy, being easily accessible and continually replenished [218].

The decreased expense of facilities and installations for clean energy generation is a crucial factor in the ongoing development of the plant. Remarkable global investments in trending sustainable renewable technologies such as photovoltaics (PV), turbines that run on wind, hydro and biomass have technically made it easier and cost-effective to generate 1.9 to 6.3 fold more energy than the global energy demand from the renewable sources of biomass [219–222]. In addition, the finance division has provided low-interest rates on investment in clean energy [218]. As a result, energy needs to be retained and stored for a future supply-demand [223]. In addition, energy cannot be specifically cohered into chemical-based devices or fuel. Biomass is the only alternative green option, yet their production is limited because of lower efficiency [224]. Therefore, novel techniques that could directly turn electricity into fuel and chemicals are required. Hence, microbial electrosynthesis (MES) technology proves to be a promising solution [225].

Reducing carbon dioxide into fuels and multi-carbon organic chemicals has been described as an appealing method for the transformation of solar energy. However, non-biological electrochemical CO_2 removal is challenging [96]. The findings indicate that microbial catalysts could be a feasible solution and the current-driven microbial carbon dioxide reduction reveals an entirely different mode of photosynthesis when combined with PV [17]. Compared to traditional biomass-based methods, this turns solar energy more effectively into organic compounds. In this review paper, the authors have analysed and discussed the fundamentals of MES.

The bioelectrochemical processes require electrical energy. For instance, production of 1 kg of acetic acid consumed around an operational voltage of 3 V, and caproic acid requires double such power consumption. Without the cost of maintenance and DSP, the cost of electricity in producing 1 kg of acetic acids would be higher compared to its commercial value. The integration of energy from fossil fuels has various disadvantages in net generation and reduction of carbon emission. Therefore, the development of low-cost renewable energy resources is considered vital in the sustainable biorefineries development of MES. This development plan is less expensive and decreases the cost of electricity price by around 20 to 30% and is increasingly innovative compared to other sources, including energy based on fossil fuels. Renewable energy provides clean electric energy from natural recurrent sources like solar, wind, hydro, geothermal. Figure 5 depicts how these energy sources can be assimilated either indirectly to power MES or directly for product transformation using photoactive electrodes [8].

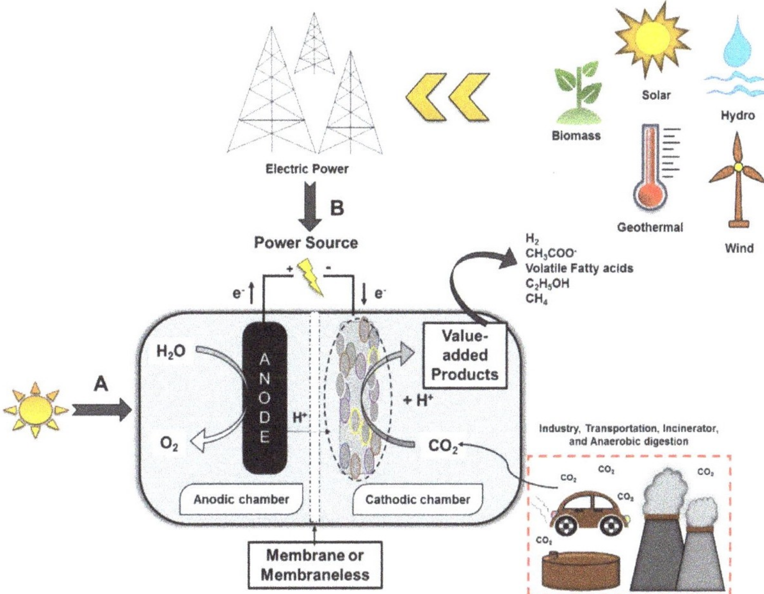

Figure 5. Microbial Electrosynthesis power supply with (**A**) Direct and (**B**) Indirect source. Production of high value-added chemicals and corresponding extracellular electron transfer (EET) processes interpreting the flow of electrons from the cathodic electrode to CO_2, the terminal electron acceptor [8].

Indirect renewable energy supply is considered more convenient as the excess power produced at a low cost by naturally fluctuating renewable sources can be stored as multi-carbon chemicals. Certain disadvantages of this strategy include the temporary decrease of MES production rate due to the fluctuating supply of electricity, and sometimes the metabolic pathway goes astray from carboxylic production to methane production. Exclusive electronic circuits or batteries can be used to provide a constant current to the reactor which helps in avoiding these related issues in MES production.

The advantage of direct energy supply is that it is a self-supporting electricity source. This can be achieved by using light-emitting diodes like LEDs and a power storage unit for constant delivery of electric current to the cathode. Sufficient currents can be delivered using an advanced method like photo-electrochemical cells for water splitting to achieve CO_2 reduction electrochemically and also the production of methane, acetic acid and isopropanol from CO_2 using photo-electrochemical anodes coupled with biological cathodes as portrayed in Table 6 below. Other than this, enzymatic biocathodes are also used in CO_2 reduction, but they cost higher and require periodic regeneration, making them less convenient for MES production [5,8].

Table 6. Hybrid system devices used in MES for wastewater treatment and CO_2 recycling. Adapted from [8].

Inoculum	Cell Design	Cathode	Anode	Current (mA/cm^2)	Main Product (Yield/Final Concentration)	Coulombic Efficiency (%)	Solar Conversion Efficiency (%)
Engineered *Ralstonia eutropha*	Single chamber	NiMoZn or stainless steel	CoP_i	0.5–1.1	Isopropanol (216 mg/L)	3.9	0.7
Enriched methanogenic community	Dual chambers	Carbon cloth	TiO_2 nanowire array	0.07–0.09	Methane (1.92 L/(m^2 d))	95	0.1
Sporomusa ovata	Dual chamber	Si and TiO_2 nanowires arrays	TiO_2 nanowires	0.3	Acetic acid (6 g/L)	86	0.38
Effluent from methanogenic MES	Dual chamber	Chitosan modified carbon cloth	TiO_2/CdS on Fluorine-doped tin oxide (with copper zinc tin sulphide sensitiser)	0.6	Methane (15 L/(m^2 d), 20.8 L/(m^2 d) with copper zinc tin sulphide)	93	0.62 (0.86 with copper zinc tin sulphide)

Reducing the emission of CO_2 and wastewater treatment, both are challenges faced by industries. A combined process of oxidation and reduction at simultaneous electrodes, of pollutants and CO_2 respectively, are reported in solving these problems with the addition of decreasing the energy demand of MES. The reaction of oxygen evolution obtained in the process of wastewater oxidation at the anode and reduction of CO_2 at cathode require expensive and high potential catalysts, the oxygen produced at the anode inhibits the strictly anaerobic microorganisms by O_2 diffusing towards the cathode. The wastewater treatment at the anode produces carbon dioxides that reprocess at the cathode chamber and act as a precursor in the production of chemicals. Selective oxidation target compounds like alkene can be done by photocatalytic oxidation. The usefulness of this method is required to study further with the practical use of wastewater, but enhancement of both the anodic-cathodic reaction in combined systems can be challenging [195].

The sustainability analysis shown in Figure 6 was conducted for MES by which numerous crucial parameters such as production rates and energy usage can reduce the techno-economic and sustainable viability of biochemical production [17].

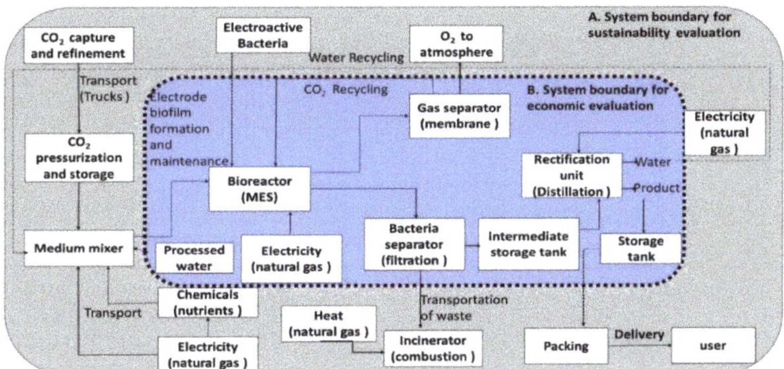

Figure 6. Schematic representation of sustainability and techno-economic assessment of MES.

8.2. Electronic Design and Energy Storage for MES

As mentioned above MES requires a supply of electricity constantly for 24 by 7 functioning, but the reactor shows fluctuations of current flow. When there is a need for fluctuating energy sources to be used as a power source to MES, additional expenditure is required for the storage system for recovering the excessive energy for gaining and delivering that energy to MES reactors constantly. In the past 5 years, the cost price of energy storage systems like batteries have decreased by 60% and is expected to decrease further in the next 4 years duration. So installation of batteries for solar power harvesting is low at the cost, but other alternatives are available like, the energy storage devices that can be charged from the electric energy produced from wastewater treatment by MFC and this will eventually discharge to the power reactor of the MES system.

A change in a metabolic pathway or CO_2 recycling occurs when microbial cultures shift while mixed culture is being used, when there are non-uniform potentials between the stack of the cells, this can be created by an uneven distribution of charge on the MES electrode caused by the use of inhomogeneous microbial catalysts. The potential control in MES parallel stacks is difficult, this balancing can be achieved by a power management (PM) system present in the cell stacks. This PM system is used to switch off the connection by detecting the overloading and over-voltage occurring in the system, this reduces the stress on battery units and increases the lifespan, allowing stable chemical output in Despite the technological advances of the field, production rates far beyond the current record are required for MES commercialisation. Other similar switching systems are implemented to improve the output energy of MFC grids. Some of them include potentiostatic cell control which is expensive but effectively control electrode potentials and galvanostatic cell control, which is cheaper but shows fluctuation and division [226,227].

8.3. Commercialisation of MES

Despite the technological advancements in this area, the productivity (both yield as well as cost) required for the massive scale preparation of MES is still beyond the current record, hampering its commercialisation. MES surely depicts a promising green future, but further studies, trials and researches are some of its prerequisites to accomplish the present need. Moreover, a reproducible, durable plan is needed for the carboxylate molecules' production, so far only acetic acid has been developed yet. Therefore, the paucity of knowledge must be resolved as a priority [7,17].

Industries such as dry cleaning, welding, preparation and processing of foaming agents and soft drinks utilise carbon dioxide [228]. Although, the amount of CO_2 used by these industries is negligible and does not significantly affect the levels of carbon emission in nature. International companies such as PRAXAIR are working on turning ambient carbon dioxide into a readily usable form by collecting and turning it into chemicals. Other companies working on the same subject include Novomer, Newlight and Algenol, which turn atmospheric carbon dioxide into polypropylene carbonate, AirCarbon plastics and ethanol. Firms like Phycal are testing out unique approaches by opting for biological pathways in place of chemical pathways by culturing algae in ponds for CO_2 sequestration [152,229–231]. Further, the cultured algae are processed for producing oils and biofuels. Biofuels are advantageous since they do not require large masses of agricultural land or freshwater for cultivation, contrary to biomass-based fuel production. Table 7 describes various patents involved in microbial electrosynthesis technology.

Table 7. Several existing patents concerning microbial electrosynthesis (MES).

Patent No.	Description	Ref.
US9856449B2	The innovation offers mechanisms and approaches for the generation of various organic compounds by utilising CO_2 as an origin of carbon and electricity as an energy's point of origin. A reaction cell is supplied with an anode and a cathode (containing microbial biofilm) distinguished by the utilisation of a selectively porous membrane and conjugated to an electrical power source. The microbial biofilm contains an electrogen that can accept electrons and, in a cathode half-reaction, it can convert CO_2 to an organic compound and water, which is then decomposed into free molecular O_2 and p^+ in an anode half-reaction. The half-reactions are powered by electricity from an external source. Butanol, ethanol, formate, acetate and 2-oxobutyrate are the compounds that can be produced using this technology.	[232]
KR101892982B1	As per the current innovation, a traditional carbon electrode surface has been altered with a positive amine compound to surge the amount of adsorption of an EAM biofilm on the surface while simultaneously, improving the efficiency of transfer of electrons by adding metal nanoparticles, thereby maximising the generation of several biofuels such as biomethanol and hexanol.	[233]
US10494596B2	In specific, the system refers to MES, via which a microbial strain capable of collecting electrons from an electrode is used to generate CO_2, formate or H_2 in co-cultivation with a strain of microbial development such as methanogen, acetogen or other microbes capable of producing those products.	[234]
US20190301029A1	A system of bioelectric processing of organic molecules like acetate is studied in this current disclosure. In addition, it also proposes strategies for generating a hydrocarbon-based product using CO_2 as the source of carbon.	[235]
WO2020053529A1	The innovation discloses the process for the regeneration of the reactor's bioanode operation and the application of the reactor for the electrosynthesis of organic acids and organic waste alcohols.	[236]
CN111961691A	This innovation discloses the utilisation of a biocathode in MES for catalytic CO_2 reduction and synthesis of organic compounds. Preparation of the biocathode with *Ruminococcus*, *Clostridium* and *Lachnospiraceae* culture and injection of CO_2 into the cathode chamber, circulating aeration and setting the theoretical range for polarisation to be −0.8 V to 1.2 V (vs. Ag/AgCl). The invention reduces the CO_2 content drastically with a simultaneous high synthesising rate of organic compounds.	[237]
US10711318B2	In comparison to the wild type of strain, a GM *Geobacter sulfurreducens* strain demonstrates enhanced functionality as a cathode biofilm. This strain is effective in utilising CO_2 as an origin of carbon and electricity as an origin of energy, employing a reverse tricarboxylic acid mechanism to produce a carbonaceous chemical.	[238]
CN110528017B	This paper demonstrates the bubbling tower of an electrolytic H_2 MES reactor. An electrolytic bath configured below the reactor supplies the bubble tower with micro-nano H_2 bubbles. H_2 and CO_2 are supplied by the microbes suspended in the bubble tower and then processed into organic compounds. This innovation is ideal for the method of H_2 Induced microbial CO_2 fixation, which is also relevant to the process of H_2-driven microbial sewage denitrification. This has perks of high coulombic performance, fast reactor start time, high current density, high output intensity, high system stability, compared to the conventional MES system dependent on the electroactive surface biofilm.	[239]

The energy required in these systems is supplied from PV cells, which makes the system imitate the process of photosynthesis. The utilisation of carbon dioxide as a feedstock poses various challenges and drawbacks due to its inertness. Due to a low Gibbs energy value and relatively low or no reactivity these systems tend to demand more energy for

CO_2 sequestration into valuable products [186]. The biocatalysts used in MES not only make the system more efficient and cost-effective but also lower the energy needed to carry out the process. These microorganisms are functional even in mild environmental conditions, making the process more sustainable. Nevertheless, these biocatalysts can be a challenge when considering sensitivity and nutrient requirements at the field scale. Excess power and energy generated using renewable sources can later be stored using MES technology [240].

9. Prospects

The transition from traditional technologies to MES is a vital element of sustaining and protecting the ecosystem for the future and aiding the production of value-added commodities. Nevertheless, one cannot neglect that the parameters influencing the efficacy of such systems are yet to be optimised. When compared to other traditional technologies MES is facing serious issues with production yield. Another issue is in its electron transfer mechanism, which is a specifically crucial step for CO_2 reduction. Further study and involvement in topics such as the design, low-cost manufacturing and GM microbes and their metabolic pathways used in MES are therefore required to make MES technology suitable for large-scale implementation. To understand the potential improvements to make the device more successful, factors such as the pH, partial pressure, cultivation substrate, electrode potential, the architecture and feeding conditions of the reactor when optimised increase the efficiency of MES systems. Microbes present in the MES are also critical for the efficient functioning of the system; it should be noted that they are directly dependent on the mentioned parameters. The prime issue faced in MES and electro-fermentation is the variance in the e^- transport mechanisms in the microbes, making it problematic to identify a universal model organism.

The scope of EET has increased in the past decade and has led to the utilisation of electric current with microbes impelling swift advancement in this field. The chief bottlenecks of electro-fermentation are the scarcity of available gene-editing tools to engineer metabolic pathways of target commodities, and electrode materials and operation of the BES reactors, which all together limits the implementation of electro-fermentation at scaled-up levels.

MES can also be employed for developing bio-refineries to store power in the form of organic compounds that can be used later. Hence, whatever energy is lost can be recovered and reused from the energy stored. Another bonus of MES technology is that it does not require agricultural land for the generation of biofuels. Future developments include the improvement of MES by using hybrid systems wherein MES systems are integrated with already established technologies such as an AD or a PV cell, increasing the bio-production process by about 9 folds. Therefore, such hybrid systems should be promoted and worked upon. Hence, using MES and optimising its parameters, more sustainable and environmentally friendly bio-refinery industries can be set up.

The trend of integrating several technologies is rapidly being accepted globally due to its resulting doubled advantages. MES can be integrated with other advanced technologies such as anaerobic fermentation, membrane electrolysis, CO_2 membrane separation, membrane contactors, microalgal photobioreactors and enzyme-assisted MES, to complement and improve the performance of CO_2 sequestration and make its conversion realistic and practicable.

10. Conclusions

- MES and Electro-fermentation are innovations that not only aim at minimising the emissions of greenhouse gases but also contributes to low manufacturing prices boosting the circular bioeconomy, offering a practical solution to lighten the ever-expanding global issues.
- Both these processes provide a plethora of premium products like biofuels, bioenergy and can also perform concurrent valorisation of CO_2 and wastewater.

- Recently, there have been multiple strategies in optimising the MES process and improving its efficiency, including treating the cultivation substrate to include adequate nutrients, enhancing the architecture and feeding conditions of the reactor, enriching the inoculum mix culture, running reactors in optimised conditions and also boosting the microbial interactions by spatially organising the cathode.
- However, significant challenges need to be tackled before commercialisation. Both these technologies, in general, are still far from practical application and further research into basic operational variables, long-term stability, continuous production, modelling, repeatability and scalability is still necessary.
- Overall, this review paper promotes further studies on promising microbial aspirants to aid advancement in this emergent field, with the subsequent aim of bringing this sustainable technology one step closer to real-world applications.

Author Contributions: Conceptualisation, S.P., D.L. and D.A.J.; software, K.W. and M.Q.; validation, S.P., R.R.R. and R.P.; writing—original draft preparation, M.Q., K.W. and S.P.; writing—review and editing, S.P., M.Q., D.A.J., A.K.R. and P.K.G.; supervision, S.P.J., V.K.T., R.R.R. and R.P. All authors have read and agreed to the published version of the manuscript.

Funding: Authors duly acknowledge the grant received from Sharda University seed grant project (SUSF2001/01).

Institutional Review Board Statement: Not applicable.

Informed Consent Statement: Not applicable.

Data Availability Statement: Not applicable.

Conflicts of Interest: The authors declare no conflict of interest.

References

1. Inglezakis, V.J. Extraterrestrial Environment. *Environ. Dev. Basic Princ. Hum. Act. Environ. Implic.* **2016**, 453–498. [CrossRef]
2. Zhang, S.; Jiang, J.; Wang, H.; Li, F.; Hua, T.; Wang, W. A review of microbial electrosynthesis applied to carbon dioxide capture and conversion: The basic principles, electrode materials, and bioproducts. *J. CO_2 Util.* **2021**, *51*, 101640. [CrossRef]
3. Fawzy, S.; Osman, A.I.; Doran, J.; Rooney, D.W. Strategies for mitigation of climate change: A review. *Environ. Chem. Lett.* **2020**, *18*, 2069–2094. [CrossRef]
4. Brack, D. *Background Analytical Study Forests and Climate Change;* United Nations Forum on Forest: New York, NY, USA, 2019.
5. Gielen, D.; Boshell, F.; Saygin, D.; Bazilian, M.D.; Wagner, N.; Gorini, R. The role of renewable energy in the global energy transformation. *Energy Strateg. Rev.* **2019**, *24*, 38–50. [CrossRef]
6. Perera, F. Pollution from fossil-fuel combustion is the leading environmental threat to global pediatric health and equity: Solutions exist. *Int. J. Environ. Res. Public Health* **2018**, *15*, 16. [CrossRef]
7. Bajracharya, S.; Vanbroekhoven, K.; Buisman, C.J.N.; Strik, D.P.B.T.B.; Pant, D. Bioelectrochemical conversion of CO_2 to chemicals: CO_2 as a next-generation feedstock for electricity-driven bioproduction in batch and continuous modes. *Faraday Discuss.* **2017**, *202*, 433–449. [CrossRef] [PubMed]
8. Dessì, P.; Rovira-Alsina, L.; Sánchez, C.; Dinesh, G.K.; Tong, W.; Chatterjee, P.; Tedesco, M.; Farràs, P.; Hamelers, H.M.V.; Puig, S. Microbial electrosynthesis: Towards sustainable biorefineries for the production of green chemicals from CO_2 emissions. *Biotechnol. Adv.* **2021**, *46*, 107675. [CrossRef] [PubMed]
9. Ben-Iwo, J.; Manovic, V.; Longhurst, P. Biomass resources and biofuels potential for the production of transportation fuels in Nigeria. *Renew. Sustain. Energy Rev.* **2016**, *63*, 172–192. [CrossRef]
10. Kumaravel, V.; Bartlett, J.; Pillai, S.C. Photoelectrochemical Conversion of Carbon Dioxide (CO_2) into Fuels and Value-Added Products. *ACS Energy Lett.* **2020**, 486–519. [CrossRef]
11. Kumar, M. Social, Economic, and Environmental Impacts of Renewable Energy Resources. In *Wind Solar Hybrid Renewable Energy System;* IntechOpen: London, UK, 2020. [CrossRef]
12. Cabau-Peinado, O.; Straathof, A.J.J.; Jourdin, L. A General Model for Biofilm-Driven Microbial Electrosynthesis of Carboxylates From CO_2. *Front. Microbiol.* **2021**, *12*, 669218. [CrossRef]
13. Badwal, S.P.S.; Giddey, S.S.; Munnings, C.; Bhatt, A.I.; Hollenkamp, A.F. Emerging electrochemical energy conversion and storage technologies. *Front. Chem.* **2014**, *2*, 79. [CrossRef] [PubMed]
14. Enescu, D.; Chicco, G.; Porumb, R.; Seritan, G. Thermal energy storage for grid applications: Current status and emerging trends. *Energies* **2020**, *13*, 340. [CrossRef]
15. Kondaveeti, S.; Abu-Reesh, I.M.; Mohanakrishna, G.; Bulut, M.; Pant, D. Advanced Routes of Biological and Bio-electrocatalytic Carbon Dioxide (CO_2) Mitigation Toward Carbon Neutrality. *Front. Energy Res.* **2020**, *8*, 94. [CrossRef]

16. Jourdin, L. *Microbial Electrosynthesis from Carbon Dioxide: Performance Enhancement and Elucidation of Mechanisms*; The University of Queensland: Brisbane, Australia, 2016. [CrossRef]
17. Christodoulou, X.; Okoroafor, T.; Parry, S.; Velasquez-Orta, S.B. The use of carbon dioxide in microbial electrosynthesis: Advancements, sustainability and economic feasibility. *J. CO_2 Util.* **2017**, *18*, 390–399. [CrossRef]
18. Jourdin, L.; Burdyny, T. Microbial Electrosynthesis: Where Do We Go from Here? *Trends Biotechnol.* **2021**, *39*, 359–369. [CrossRef]
19. Das, S.; Diels, L.; Pant, D.; Patil, S.A.; Ghangrekar, M.M. Review—Microbial Electrosynthesis: A Way Towards The Production of Electro-Commodities Through Carbon Sequestration with Microbes as Biocatalysts. *J. Electrochem. Soc.* **2020**, *167*, 155510. [CrossRef]
20. Vassilev, I.; Hernandez, P.A.; Batlle-Vilanova, P.; Freguia, S.; Krömer, J.O.; Keller, J.; Ledezma, P.; Virdis, B. Microbial Electrosynthesis of Isobutyric, Butyric, Caproic Acids, and Corresponding Alcohols from Carbon Dioxide. *ACS Sustain. Chem. Eng.* **2018**, *6*, 8485–8493. [CrossRef]
21. Shin, H.J.; Jung, K.A.; Nam, C.W.; Park, J.M. A genetic approach for microbial electrosynthesis system as biocommodities production platform. *Bioresour. Technol.* **2017**, *245*, 1421–1429. [CrossRef]
22. Ammam, F.; Tremblay, P.L.; Lizak, D.M.; Zhang, T. Effect of tungstate on acetate and ethanol production by the electrosynthetic bacterium Sporomusa ovata. *Biotechnol. Biofuels* **2016**, *9*, 163. [CrossRef]
23. Jiang, Y.; Jianxiong Zeng, R. Expanding the product spectrum of value-added chemicals in microbial electrosynthesis through integrated process design—A review. *Bioresour. Technol.* **2018**, *269*, 503–512. [CrossRef]
24. Tremblay, P.L.; Zhang, T. Electrifying microbes for the production of chemicals. *Front. Microbiol.* **2015**, *6*, 201. [CrossRef]
25. Wood, J.C.; Grové, J.; Marcellin, E.; Heffernan, J.K.; Hu, S.; Yuan, Z.; Virdis, B. Strategies to improve the viability of a circular carbon bioeconomy-A techno-economic review of microbial electrosynthesis and gas fermentation. *Water Res.* **2021**, *201*, 117306. [CrossRef]
26. Syed, Z.; Sogani, M.; Dongre, A.; Kumar, A.; Sonu, K.; Sharma, G.; Gupta, A.B. Bioelectrochemical systems for environmental remediation of estrogens: A review and way forward. *Sci. Total Environ.* **2021**, *780*, 146544. [CrossRef]
27. Pant, D.; Singh, A.; van Bogaert, G.; Olsen, S.I.; Nigam, P.S.; Diels, L.; Vanbroekhoven, K. Bioelectrochemical systems (BES) for sustainable energy production and product recovery from organic wastes and industrial wastewaters. *RSC Adv.* **2012**, *2*, 1248–1263. [CrossRef]
28. Bajracharya, S.; Sharma, M.; Mohanakrishna, G.; Dominguez Benneton, X.; Strik, D.P.B.T.B.; Sarma, P.M.; Pant, D. An overview on emerging bioelectrochemical systems (BESs): Technology for sustainable electricity, waste remediation, resource recovery, chemical production and beyond. *Renew. Energy* **2016**, *98*, 153–170. [CrossRef]
29. Shaikh, R.; Rizvi, A.; Quraishi, M.; Pandit, S.; Mathuriya, A.S.; Gupta, P.K.; Singh, J.; Prasad, R. Bioelectricity production using plant-microbial fuel cell: Present state of art. *S. African J. Bot.* **2020**. [CrossRef]
30. Zheng, T.; Li, J.; Ji, Y.; Zhang, W.; Fang, Y.; Xin, F.; Dong, W.; Wei, P.; Ma, J.; Jiang, M. Progress and Prospects of Bioelectrochemical Systems: Electron Transfer and Its Applications in the Microbial Metabolism. *Front. Bioeng. Biotechnol.* **2020**, *8*, 10. [CrossRef] [PubMed]
31. Li, S.; Chen, G. Factors affecting the effectiveness of bioelectrochemical system applications: Data synthesis and meta-analysis. *Batteries* **2018**, *4*, 34. [CrossRef]
32. Kadier, A.; Simayi, Y.; Abdeshahian, P.; Azman, N.F.; Chandrasekhar, K.; Kalil, M.S. A comprehensive review of microbial electrolysis cells (MEC) reactor designs and configurations for sustainable hydrogen gas production. *Alexandria Eng. J.* **2016**, *55*, 427–443. [CrossRef]
33. Anukam, A.; Mohammadi, A.; Naqvi, M.; Granström, K. A review of the chemistry of anaerobic digestion: Methods of accelerating and optimizing process efficiency. *Processes* **2019**, *7*, 504. [CrossRef]
34. Kracke, F.; Wong, A.B.; Maegaard, K.; Deutzmann, J.S.; Hubert, M.K.A.; Hahn, C.; Jaramillo, T.F.; Spormann, A.M. Robust and biocompatible catalysts for efficient hydrogen-driven microbial electrosynthesis. *Commun. Chem.* **2019**, *2*, 1–9. [CrossRef]
35. Ucar, D.; Zhang, Y.; Angelidaki, I. An overview of electron acceptors in microbial fuel cells. *Front. Microbiol.* **2017**, *8*, 643. [CrossRef]
36. Cardeña, R.; Cercado, B.; Buitrón, G. Microbial Electrolysis Cell for Biohydrogen Production. In *Biohydrogen*; Elsevier: Paris, France, 2019; pp. 159–185. [CrossRef]
37. Rivera, I.; Schröder, U.; Patil, S.A. Microbial electrolysis for biohydrogen production: Technical aspects and scale-up experiences. In *Biomass, Biofuels, Biochemicals: Microbial Electrochemical Technology: Sustainable Platform for Fuels, Chemicals and Remediation*; Elsevier: Gurgaon, India, 2018; pp. 871–898. [CrossRef]
38. Caizán-Juanarena, L.; Borsje, C.; Sleutels, T.; Yntema, D.; Santoro, C.; Ieropoulos, I.; Soavi, F.; Ter Heijne, A. Combination of bioelectrochemical systems and electrochemical capacitors: Principles, analysis and opportunities. *Biotechnol. Adv.* **2020**, *39*, 107456. [CrossRef] [PubMed]
39. Rabaey, K.; Rozendal, R.A. Microbial electrosynthesis-Revisiting the electrical route for microbial production. *Nat. Rev. Microbiol.* **2010**, *8*, 706–716. [CrossRef] [PubMed]
40. Drendel, G.; Mathews, E.R.; Semenec, L.; Franks, A.E. Microbial Fuel Cells, Related Technologies, and Their Applications. *Appl. Sci.* **2018**, *8*, 2384. [CrossRef]
41. Borole, A.P.; Reguera, G.; Ringeisen, B.; Wang, Z.W.; Feng, Y.; Kim, B.H. Electroactive biofilms: Current status and future research needs. *Energy Environ. Sci.* **2011**, *4*, 4813–4834. [CrossRef]

42. ter Heijne, A.; Pereira, M.A.; Pereira, J.; Sleutels, T. Electron Storage in Electroactive Biofilms. *Trends Biotechnol.* **2021**, *39*, 34–42. [CrossRef]
43. Patil, S.A.; Hägerhäll, C.; Gorton, L. Electron transfer mechanisms between microorganisms and electrodes in bioelectrochemical systems. *Bioanal. Rev.* **2014**, *1*, 71–129. [CrossRef]
44. Mao, L.; Verwoerd, W.S. Selection of organisms for systems biology study of microbial electricity generation: A review. *Int. J. Energy Environ. Eng.* **2013**, *4*, 1–18. [CrossRef]
45. Kumar, R.; Singh, L.; Zularisam, A.W.; Hai, F.I. Microbial fuel cell is emerging as a versatile technology: A review on its possible applications, challenges and strategies to improve the performances. *Int. J. Energy Res.* **2018**, *42*, 369–394. [CrossRef]
46. Pawar, A.A.; Karthic, A.; Lee, S.; Pandit, S.; Jung, S.P. Microbial electrolysis cells for electromethanogenesis: Materials, configurations and operations. *Environ. Eng. Res.* **2022**, *27*, 53–57. [CrossRef]
47. Choi, O.; Sang, B.I. Extracellular electron transfer from cathode to microbes: Application for biofuel production. *Biotechnol. Biofuels* **2016**, *9*, 11. [CrossRef]
48. Aryal, N.; Ammam, F.; Patil, S.A.; Pant, D. An overview of cathode materials for microbial electrosynthesis of chemicals from carbon dioxide. *Green Chem.* **2017**, *19*, 5748–5760. [CrossRef]
49. Werth, C.J.; Yan, C.; Troutman, J.P. Factors Impeding Replacement of Ion Exchange with (Electro)Catalytic Treatment for Nitrate Removal from Drinking Water. *ACS EST Eng.* **2021**, *1*, 6–20. [CrossRef]
50. Lim, S.S.; Kim, B.H.; Li, D.; Feng, Y.; Daud, W.R.W.; Scott, K.; Yu, E.H. Effects of Applied Potential and Reactants to Hydrogen-Producing Biocathode in a Microbial Electrolysis Cell. *Front. Chem.* **2018**, *6*, 318. [CrossRef] [PubMed]
51. Ali Shah, F.; Mahmood, Q.; Maroof Shah, M.; Pervez, A.; Ahmad Asad, S. Microbial ecology of anaerobic digesters: The key players of anaerobiosis. *Sci. World J.* **2014**, *2014*. [CrossRef] [PubMed]
52. Spiess, S.; Kucera, J.; Seelajaroen, H.; Sasiain, A.; Thallner, S.; Kremser, K.; Novak, D.; Guebitz, G.M.; Haberbauer, M. Impact of Carbon Felt Electrode Pretreatment on Anodic Biofilm Composition in Microbial Electrolysis Cells. *Biosensors* **2021**, *11*, 170. [CrossRef]
53. Dincer, I.; Zamfirescu, C. *Sustainable Hydrogen Production*; Elsevier Inc.: North York, ON, Canada, 2016; ISBN 9780128017487.
54. Krishnan, S.; Kadier, A.; Fadhil Bin MD Din, M.; Nasrullah, M.; Najiha, N.N.; Taib, S.M.; Wahid, Z.A.; Li, Y.Y.; Qin, Y.; Pant, K.K.; et al. Application of bioelectrochemical systems in wastewater treatment and hydrogen production. In *Delivering Low-Carbon Biofuels with Bioproduct Recovery*; Elsevier: Gurgaon, India, 2021; pp. 31–44. [CrossRef]
55. Hussain, S.; Aneggi, E.; Goi, D. Catalytic activity of metals in heterogeneous Fenton-like oxidation of wastewater contaminants: A review. *Environ. Chem. Lett.* **2021**, *19*, 2405–2424. [CrossRef]
56. Hassan, M.; Olvera-Vargas, H.; Zhu, X.; Zhang, B.; He, Y. Microbial electro-Fenton: An emerging and energy-efficient platform for environmental remediation. *J. Power Source* **2019**, *424*, 220–244. [CrossRef]
57. Song, T.; Wang, G.; Wang, H.; Huang, Q.; Xie, J. Experimental evaluation of the influential factors of acetate production driven by a DC power system via CO_2 reduction through microbial electrosynthesis. *Bioresour. Bioprocess.* **2019**, *6*, 1–10. [CrossRef]
58. Saheb-Alam, S.; Singh, A.; Hermansson, M.; Persson, F.; Schnürer, A.; Wilén, B.M.; Modin, O. Effect of start-up strategies and electrode materials on carbon dioxide reduction on biocathodes. *Appl. Environ. Microbiol.* **2018**, *84*. [CrossRef]
59. Thatikayala, D.; Min, B. Copper ferrite supported reduced graphene oxide as cathode materials to enhance microbial electrosynthesis of volatile fatty acids from CO_2. *Sci. Total Environ.* **2021**, *768*, 144477. [CrossRef] [PubMed]
60. Kumar, G.; Cho, S.K.; Sivagurunathan, P.; Anburajan, P.; Mahapatra, D.M.; Park, J.H.; Pugazhendhi, A. Insights into evolutionary trends in molecular biology tools in microbial screening for biohydrogen production through dark fermentation. *Int. J. Hydrogen Energy* **2018**, *43*, 19885–19901. [CrossRef]
61. Jourdin, L.; Freguia, S.; Donose, B.C.; Chen, J.; Wallace, G.G.; Keller, J.; Flexer, V. A novel carbon nanotube modified scaffold as an efficient biocathode material for improved microbial electrosynthesis. *J. Mater. Chem. A* **2014**, *2*, 13093–13102. [CrossRef]
62. Zhang, T.; Ghosh, D.; Tremblay, P.-L. Synthetic Biology Strategies to Improve Electron Transfer Rate at the Microbe–Anode Interface in Microbial Fuel Cells. *Bioelectrochem. Interface Eng.* **2019**, 187–208. [CrossRef]
63. Alfonta, L. Genetically Engineered Microbial Fuel Cells. *Electroanalysis* **2010**, *22*, 822–831. [CrossRef]
64. Kerzenmacher, S. Engineering of microbial electrodes. In *Advances in Biochemical Engineering/Biotechnology*; Harnisch, F., Holtmann, D., Eds.; Springer: Cham, Switzerland, 2019; Volume 167, pp. 135–180. [CrossRef]
65. Fruehauf, H.M.; Enzmann, F.; Harnisch, F.; Ulber, R.; Holtmann, D. Microbial Electrosynthesis—An Inventory on Technology Readiness Level and Performance of Different Process Variants. *Biotechnol. J.* **2020**, *15*, 2000066. [CrossRef]
66. Simões, M.; Simões, L.C.; Vieira, M.J. A review of current and emergent biofilm control strategies. *LWT Food Sci. Technol.* **2010**, *43*, 573–583. [CrossRef]
67. Nealson, K.H.; Rowe, A.R. Electromicrobiology: Realities, grand challenges, goals and predictions. *Microb. Biotechnol.* **2016**, *9*, 595. [CrossRef] [PubMed]
68. Ali, J.; Sohail, A.; Wang, L.; Haider, M.R.; Mulk, S.; Pan, G. Electro-microbiology as a promising approach towards renewable energy and environmental sustainability. *Energies* **2018**, *11*, 1822. [CrossRef]
69. Kumar, P.; Chandrasekhar, K.; Kumari, A.; Sathiyamoorthi, E.; Kim, B.S. Electro-Fermentation in Aid of Bioenergy and Biopolymers. *Energies* **2018**, *11*, 343. [CrossRef]

70. Ishii, T.; Kawaichi, S.; Nakagawa, H.; Hashimoto, K.; Nakamura, R. From chemolithoautotrophs to electrolithoautotrophs: CO_2 fixation by Fe(II)-oxidizing bacteria coupled with direct uptake of electrons from solid electron sources. *Front. Microbiol.* **2015**, *6*, 994. [CrossRef] [PubMed]
71. Kumar, R.; Singh, L.; Zularisam, A.W. Exoelectrogens: Recent advances in molecular drivers involved in extracellular electron transfer and strategies used to improve it for microbial fuel cell applications. *Renew. Sustain. Energy Rev.* **2016**, *56*, 1322–1336. [CrossRef]
72. Pillot, G.; Davidson, S.; Shintu, L.; Amin Ali, O.X.; Godfroy, A.; Combet-Blanc, Y.; Liebgott, P.-P. Electrotrophy as potential primary metabolism for colonization of conductive surfaces in deep-sea hydrothermal chimneys. 2 Key-words. *bioRxiv* **2020**. [CrossRef]
73. Yee, M.O.; Deutzmann, J.; Spormann, A.; Rotaru, A.E. Cultivating electroactive microbes-from field to bench. *Nanotechnology* **2020**, *31*, 174003. [CrossRef]
74. Ajunwa, O.M.; Audu, J.O.; Kumar, P.; Marsili, E.; Onilude, A.A. Electrotrophs and Electricigens; Key Players in Microbial Electrophysiology. In *Bioelectrochemical Systems*; Springer: Singapore, 2020; pp. 299–326. [CrossRef]
75. Cao, Y.; Mu, H.; Liu, W.; Zhang, R.; Guo, J.; Xian, M.; Liu, H. Electricigens in the anode of microbial fuel cells: Pure cultures versus mixed communities. *Microb. Cell Fact.* **2019**, *18*, 1–14. [CrossRef]
76. Hernandez, C.A.; Osma, J.F. Microbial Electrochemical Systems: Deriving Future Trends From Historical Perspectives and Characterization Strategies. *Front. Environ. Sci.* **2020**, *8*, 44. [CrossRef]
77. Saavedra, A.; Aguirre, P.; Gentina, J.C. Biooxidation of Iron by *Acidithiobacillus ferrooxidans* in the Presence of D-Galactose: Understanding Its Influence on the Production of EPS and Cell Tolerance to High Concentrations of Iron. *Front. Microbiol.* **2020**, *11*, 759. [CrossRef]
78. Kracke, F.; Vassilev, I.; Krömer, J.O. Microbial electron transport and energy conservation-The foundation for optimizing bioelectrochemical systems. *Front. Microbiol.* **2015**, *6*. [CrossRef] [PubMed]
79. Gong, Z.; Yu, H.; Zhang, J.; Li, F.; Song, H. Microbial electro-fermentation for the synthesis of chemicals and biofuels driven by bi-directional extracellular electron transfer. *Synth. Syst. Biotechnol.* **2020**, *5*, 304–313. [CrossRef]
80. Zhao, J.; Li, F.; Cao, Y.; Zhang, X.; Chen, T.; Song, H.; Wang, Z. Microbial extracellular electron transfer and strategies for engineering electroactive microorganisms. *Biotechnol. Adv.* **2020**, 107682. [CrossRef]
81. Vassilev, I.; Averesch, N.J.H.; Ledezma, P.; Kokko, M. Anodic electro-fermentation: Empowering anaerobic production processes via anodic respiration. *Biotechnol. Adv.* **2021**, *48*, 107728. [CrossRef] [PubMed]
82. Nakagawa, G.; Kouzuma, A.; Hirose, A.; Kasai, T.; Yoshida, G.; Watanabe, K. Metabolic Characteristics of a Glucose-Utilizing *Shewanella oneidensis* Strain Grown under Electrode-Respiring Conditions. *PLoS ONE* **2015**, *10*, e0138813. [CrossRef]
83. Flynn, J.M.; Ross, D.E.; Hunt, K.A.; Bond, D.R.; Gralnick, J.A. Enabling unbalanced fermentations by using engineered electrode-interfaced bacteria. *MBio* **2010**, *1*. [CrossRef] [PubMed]
84. Bursac, T.; Gralnick, J.A.; Gescher, J. Acetoin production via unbalanced fermentation in *Shewanella oneidensis*. *Biotechnol. Bioeng.* **2017**, *114*, 1283–1289. [CrossRef]
85. Zheng, T.; Xu, B.; Ji, Y.; Zhang, W.; Xin, F.; Dong, W.; Wei, P.; Ma, J.; Jiang, M. Microbial fuel cell-assisted utilization of glycerol for succinate production by mutant of *Actinobacillus succinogenes*. *Biotechnol. Biofuels* **2021**, *14*, 1–10. [CrossRef]
86. Kim, M.Y.; Kim, C.; Ainala, S.K.; Bae, H.; Jeon, B.H.; Park, S.; Kim, J.R. Metabolic shift of *Klebsiella pneumoniae* L17 by electrode-based electron transfer using glycerol in a microbial fuel cell. *Bioelectrochemistry* **2019**, *125*, 1–7. [CrossRef]
87. Speers, A.M.; Young, J.M.; Reguera, G. Fermentation of Glycerol into Ethanol in a Microbial Electrolysis Cell Driven by a Customized Consortium. *Environ. Sci. Technol.* **2014**, *48*, 6350–6358. [CrossRef]
88. Awate, B.; Steidl, R.J.; Hamlischer, T.; Reguera, G. Stimulation of electro-fermentation in single-chamber microbial electrolysis cells driven by genetically engineered anode biofilms. *J. Power Source* **2017**, *356*, 510–518. [CrossRef]
89. Nishio, K.; Kimoto, Y.; Song, J.; Konno, T.; Ishihara, K.; Kato, S.; Hashimoto, K.; Nakanishi, S. Extracellular Electron Transfer Enhances Polyhydroxybutyrate Productivity in *Ralstonia eutropha*. *Environ. Sci. Technol. Lett.* **2013**, *1*, 40–43. [CrossRef]
90. TerAvest, M.A.; Zajdel, T.J.; Ajo-Franklin, C.M. The Mtr Pathway of *Shewanella oneidensis* MR-1 Couples Substrate Utilization to Current Production in *Escherichia coli*. *ChemElectroChem* **2014**, *1*, 1874–1879. [CrossRef]
91. Sturm-Richter, K.; Golitsch, F.; Sturm, G.; Kipf, E.; Dittrich, A.; Beblawy, S.; Kerzenmacher, S.; Gescher, J. Unbalanced fermentation of glycerol in Escherichia coli via heterologous production of an electron transport chain and electrode interaction in microbial electrochemical cells. *Bioresour. Technol.* **2015**, *186*, 89–96. [CrossRef]
92. Lai, B.; Yu, S.; Bernhardt, P.V.; Rabaey, K.; Virdis, B.; Krömer, J.O. Anoxic metabolism and biochemical production in Pseudomonas putida F1 driven by a bioelectrochemical system. *Biotechnol. Biofuels* **2016**, *9*, 1–13. [CrossRef]
93. Yu, S.; Lai, B.; Plan, M.R.; Hodson, M.P.; Lestari, E.A.; Song, H.; Krömer, J.O. Improved performance of Pseudomonas putida in a bioelectrochemical system through overexpression of periplasmic glucose dehydrogenase. *Biotechnol. Bioeng.* **2018**, *115*, 145–155. [CrossRef]
94. Vassilev, I.; Gießelmann, G.; Schwechheimer, S.K.; Wittmann, C.; Virdis, B.; Krömer, J.O. Anodic electro-fermentation: Anaerobic production of L-Lysine by recombinant Corynebacterium glutamicum. *Biotechnol. Bioeng.* **2018**, *115*, 1499–1508. [CrossRef]
95. Soussan, L.; Riess, J.; Erable, B.; Delia, M.L.; Bergel, A. Electrochemical reduction of CO_2 catalysed by Geobacter sulfurreducens grown on polarized stainless steel cathodes. *Electrochem. Commun.* **2013**, *28*, 27–30. [CrossRef]

96. Nevin, K.P.; Woodard, T.L.; Franks, A.E.; Summers, Z.M.; Lovley, D.R. Microbial electrosynthesis: Feeding microbes electricity to convert carbon dioxide and water to multicarbon extracellular organic compounds. *MBio* **2010**, *1*. [CrossRef]
97. Tefft, N.M.; TerAvest, M.A. Reversing an Extracellular Electron Transfer Pathway for Electrode-Driven Acetoin Reduction. *ACS Synth. Biol.* **2019**, *8*, 1590–1600. [CrossRef] [PubMed]
98. Utesch, T.; Sabra, W.; Prescher, C.; Baur, J.; Arbter, P.; Zeng, A.-P. Enhanced electron transfer of different mediators for strictly opposite shifting of metabolism in *Clostridium pasteurianum* grown on glycerol in a new electrochemical bioreactor. *Biotechnol. Bioeng.* **2019**, *116*, 1627–1643. [CrossRef]
99. Zhang, Z.; Li, F.; Cao, Y.; Tian, Y.; Li, J.; Zong, Y.; Song, H. Electricity-driven 7α-hydroxylation of a steroid catalyzed by a cytochrome P450 monooxygenase in engineered yeast. *Catal. Sci. Technol.* **2019**, *9*, 4877–4887. [CrossRef]
100. Torella, J.P.; Gagliardi, C.J.; Chen, J.S.; Bediako, D.K.; Colón, B.; Way, J.C.; Silver, P.A.; Nocera, D.G. Efficient solar-to-fuels production from a hybrid microbial–water-splitting catalyst system. *Proc. Natl. Acad. Sci. USA* **2015**, *112*, 2337–2342. [CrossRef] [PubMed]
101. Liu, C.; Sakimoto, K.K.; Colón, B.C.; Silver, P.A.; Nocera, D.G. Ambient nitrogen reduction cycle using a hybrid inorganic–biological system. *Proc. Natl. Acad. Sci. USA* **2017**, *114*, 6450–6455. [CrossRef]
102. Kouzuma, A.; Kasai, T.; Hirose, A.; Watanabe, K. Catabolic and regulatory systems in *Shewanella oneidensis* MR-1 involved in electricity generation in microbial fuel cells. *Front. Microbiol.* **2015**, *6*, 609. [CrossRef]
103. Li, F.; Li, Y.X.; Cao, Y.X.; Wang, L.; Liu, C.G.; Shi, L.; Song, H. Modular engineering to increase intracellular NAD(H/+) promotes rate of extracellular electron transfer of *Shewanella oneidensis*. *Nat. Commun.* **2018**, *9*, 1–13. [CrossRef] [PubMed]
104. Goldbeck, C.P.; Jensen, H.M.; TerAvest, M.A.; Beedle, N.; Appling, Y.; Hepler, M.; Cambray, G.; Mutalik, V.; Angenent, L.T.; Ajo-Franklin, C.M. Tuning Promoter Strengths for Improved Synthesis and Function of Electron Conduits in *Escherichia coli*. *ACS Synth. Biol.* **2013**, *2*, 150–159. [CrossRef] [PubMed]
105. Andrade, A.; Hernández-Eligio, A.; Tirado, A.L.; Vega-Alvarado, L.; Olvera, M.; Morett, E.; Juárez, K. Specialization of the Reiterated Copies of the Heterodimeric Integration Host Factor Genes in Geobacter sulfurreducens. *Front. Microbiol.* **2021**, *12*, 626443. [CrossRef]
106. Hernández-Eligio, A.; Pat-Espadas, A.; Vega-Alvarado, L.; Huerta-Amparán, M.; Cervantes, F.; Juárez, K. Global transcriptional analysis of Geobacter sulfurreducens under palladium reducing conditions reveals new key cytochromes involved. *bioRxiv* **2018**, 319194. [CrossRef] [PubMed]
107. Mevers, E.; Su, L.; Pishchany, G.; Baruch, M.; Cornejo, J.; Hobert, E.; Dimise, E.; Ajo-Franklin, C.M.; Clardy, J. An elusive electron shuttle from a facultative anaerobe. *Elife* **2019**, *8*. [CrossRef]
108. Thirumurthy, M.A.; Jones, A.K. Geobacter cytochrome OmcZs binds riboflavin: Implications for extracellular electron transfer. *Nanotechnology* **2020**, *31*, 124001. [CrossRef]
109. Hassan, R.Y.A.; Febbraio, F.; Andreescu, S. Microbial electrochemical systems: Principles, construction and biosensing applications. *Sensors* **2021**, *21*, 1279. [CrossRef]
110. Logan, B.E.; Rossi, R.; Ragab, A.; Saikaly, P.E. Electroactive microorganisms in bioelectrochemical systems. *Nat. Rev. Microbiol.* **2019**, *17*, 307–319. [CrossRef]
111. Hirose, A.; Kasai, T.; Koga, R.; Suzuki, Y.; Kouzuma, A.; Watanabe, K. Understanding and engineering electrochemically active bacteria for sustainable biotechnology. *Bioresour. Bioprocess.* **2019**, *6*, 1–15. [CrossRef]
112. Leang, C.; Malvankar, N.S.; Franks, A.E.; Nevin, K.P.; Lovley, D.R. Engineering Geobacter sulfurreducens to produce a highly cohesive conductive matrix with enhanced capacity for current production. *Energy Environ. Sci.* **2013**, *6*, 1901–1908. [CrossRef]
113. Kouzuma, A.; Oba, H.; Tajima, N.; Hashimoto, K.; Watanabe, K. Electrochemical selection and characterization of a high current-generating Shewanella oneidensis mutant with altered cell-surface morphology and biofilm-related gene expression. *BMC Microbiol.* **2014**, *14*, 190. [CrossRef] [PubMed]
114. Jourdin, L.; Sousa, J.; van Stralen, N.; Strik, D.P.B.T.B. Techno-economic assessment of microbial electrosynthesis from CO_2 and/or organics: An interdisciplinary roadmap towards future research and application. *Appl. Energy* **2020**, *279*, 115775. [CrossRef]
115. Mahadevan, A.; Gunawardena, D.A.; Fernando, S. Biochemical and Electrochemical Perspectives of the Anode of a Microbial Fuel Cell. In *Technology and Application of Microbial Fuel Cells*; Wang, C.-T., Ed.; InTech: Hsinchu City, Taiwan, 2014. [CrossRef]
116. Aryal, N.; Kvist, T.; Ammam, F.; Pant, D.; Ottosen, L.D.M. An overview of microbial biogas enrichment. *Bioresour. Technol.* **2018**, *264*, 359–369. [CrossRef]
117. Selembo, P.A.; Merrill, M.D.; Logan, B.E. Hydrogen production with nickel powder cathode catalysts in microbial electrolysis cells. *Int. J. Hydrogen Energy* **2010**, *35*, 428–437. [CrossRef]
118. Yaqoob, A.A.; Ibrahim, M.N.M.; Rodríguez-Couto, S. Development and modification of materials to build cost-effective anodes for microbial fuel cells (MFCs): An overview. *Biochem. Eng. J.* **2020**, *164*, 107779. [CrossRef]
119. Nie, H.; Zhang, T.; Cui, M.; Lu, H.; Lovley, D.R.; Russell, T.P. Improved cathode for high efficient microbial-catalyzed reduction in microbial electrosynthesis cells. *Phys. Chem. Chem. Phys.* **2013**, *15*, 14290–14294. [CrossRef]
120. Kellon, J.E.; Young, S.L.; Hutchison, J.E. Engineering the Nanoparticle–Electrode Interface. *Chem. Mater.* **2019**, *31*, 2685–2701. [CrossRef]
121. Bian, B.; Alqahtani, M.F.; Katuri, K.P.; Liu, D.; Bajracharya, S.; Lai, Z.; Rabaey, K.; Saikaly, P.E. Porous nickel hollow fiber cathodes coated with CNTs for efficient microbial electrosynthesis of acetate from CO_2 using Sporomusa ovata. *J. Mater. Chem. A* **2018**, *6*, 17201–17211. [CrossRef]

122. Cai, P.J.; Xiao, X.; He, Y.R.; Li, W.W.; Zang, G.L.; Sheng, G.P.; Hon-Wah Lam, M.; Yu, L.; Yu, H.Q. Reactive oxygen species (ROS) generated by cyanobacteria act as an electron acceptor in the biocathode of a bio-electrochemical system. *Biosens. Bioelectron.* **2013**, *39*, 306–310. [CrossRef]
123. Kondaveeti, S.; Min, B. Bioelectrochemical reduction of volatile fatty acids in anaerobic digestion effluent for the production of biofuels. *Water Res.* **2015**, *87*, 137–144. [CrossRef]
124. Tahir, K.; Miran, W.; Jang, J.; Maile, N.; Shahzad, A.; Moztahida, M.; Ghani, A.A.; Kim, B.; Jeon, H.; Lee, D.S. MXene-coated biochar as potential biocathode for improved microbial electrosynthesis system. *Sci. Total Environ.* **2021**, *773*, 145677. [CrossRef]
125. Yaqoob, A.A.; Ibrahim, M.N.M.; Rafatullah, M.; Chua, Y.S.; Ahmad, A.; Umar, K. Recent advances in anodes for microbial fuel cells: An overview. *Materials* **2020**, *13*, 2078. [CrossRef]
126. Saito, T.; Mehanna, M.; Wang, X.; Cusick, R.D.; Feng, Y.; Hickner, M.A.; Logan, B.E. Effect of nitrogen addition on the performance of microbial fuel cell anodes. *Bioresour. Technol.* **2011**, *102*, 395–398. [CrossRef]
127. Nakanishi, E.Y.; Palacios, J.H.; Godbout, S.; Fournel, S. Interaction between Biofilm Formation, Surface Material and Cleanability Considering Different Materials Used in Pig Facilities—An Overview. *Sustainability* **2021**, *13*, 5836. [CrossRef]
128. Angelaalincy, M.J.; Navanietha Krishnaraj, R.; Shakambari, G.; Ashokkumar, B.; Kathiresan, S.; Varalakshmi, P. Biofilm Engineering Approaches for Improving the Performance of Microbial Fuel Cells and Bioelectrochemical Systems. *Front. Energy Res.* **2018**, *6*, 63. [CrossRef]
129. Uria, N.; Ferrera, I.; Mas, J. Electrochemical performance and microbial community profiles in microbial fuel cells in relation to electron transfer mechanisms. *BMC Microbiol.* **2017**, *17*, 208. [CrossRef]
130. Tuson, H.H.; Weibel, D.B. Bacteria-surface interactions. *Soft Matter* **2013**, *9*, 4368–4380. [CrossRef]
131. Schechter, M.; Schechter, A.; Rozenfeld, S.; Efrat, E.; Cahan, R. Anode Biofilm. In *Technology and Application of Microbial Fuel Cells*; Wang, C.-T., Ed.; InTech: Hsinchu City, Taiwan, 2014. [CrossRef]
132. Zhang, T.; Nie, H.; Bain, T.S.; Lu, H.; Cui, M.; Snoeyenbos-West, O.L.; Franks, A.E.; Nevin, K.P.; Russell, T.P.; Lovley, D.R. Improved cathode materials for microbial electrosynthesis. *Energy Environ. Sci.* **2012**, *6*, 217–224. [CrossRef]
133. Nevin, K.P.; Hensley, S.A.; Franks, A.E.; Summers, Z.M.; Ou, J.; Woodard, T.L.; Snoeyenbos-West, O.L.; Lovley, D.R. Electrosynthesis of organic compounds from carbon dioxide is catalyzed by a diversity of acetogenic microorganisms. *Appl. Environ. Microbiol.* **2011**, *77*, 2882–2886. [CrossRef]
134. Bhagchandanii, D.D.; Babu, R.P.; Sonawane, J.M.; Khanna, N.; Pandit, S.; Jadhav, D.A.; Khilari, S.; Prasad, R. A comprehensive understanding of electro-fermentation. *Fermentation* **2020**, *6*, 92. [CrossRef]
135. Leang, C.; Ueki, T.; Nevin, K.P.; Lovley, D.R. A Genetic system for *Clostridium ljungdahlii*: A chassis for autotrophic production of biocommodities and a model homoacetogen. *Appl. Environ. Microbiol.* **2013**, *79*, 1102–1109. [CrossRef]
136. Köpke, M.; Held, C.; Hujer, S.; Liesegang, H.; Wiezer, A.; Wollherr, A.; Ehrenreich, A.; Liebl, W.; Gottschalk, G.; Dürre, P. *Clostridium ljungdahlii* represents a microbial production platform based on syngas. *Proc. Natl. Acad. Sci. USA* **2010**, *107*, 13087–13092. [CrossRef]
137. Kolesinska, B.; Fraczyk, J.; Binczarski, M.; Modelska, M.; Berlowska, J.; Dziugan, P.; Antolak, H.; Kaminski, Z.J.; Witonska, I.A.; Kregiel, D. Butanol synthesis routes for biofuel production: Trends and perspectives. *Materials* **2019**, *12*, 350. [CrossRef] [PubMed]
138. Li, S.; Huang, L.; Ke, C.; Pang, Z.; Liu, L. Pathway dissection, regulation, engineering and application: Lessons learned from biobutanol production by solventogenic clostridia. *Biotechnol. Biofuels* **2020**, *13*, 1–25. [CrossRef] [PubMed]
139. Banerjee, A.; Leang, C.; Ueki, T.; Nevin, K.P.; Lovley, D.R. Lactose-inducible system for metabolic engineering of *Clostridium ljungdahlii*. *Appl. Environ. Microbiol.* **2014**, *80*, 2410–2416. [CrossRef]
140. Kracke, F.; Krömer, J.O. Identifying target processes for microbial electrosynthesis by elementary mode analysis. *BMC Bioinformatics* **2014**, *15*, 410. [CrossRef]
141. Khan, S.; Ullah, M.W.; Siddique, R.; Nabi, G.; Manan, S.; Yousaf, M.; Hou, H. Role of recombinant DNA technology to improve life. *Int. J. Genom.* **2016**, *2016*. [CrossRef]
142. Adrio, J.L.; Demain, A.L. Microbial enzymes: Tools for biotechnological processes. *Biomolecules* **2014**, *4*, 117–139. [CrossRef] [PubMed]
143. Wu, Z.; Wang, J.; Liu, J.; Wang, Y.; Bi, C.; Zhang, X. Engineering an electroactive *Escherichia coli* for the microbial electrosynthesis of succinate from glucose and CO_2. *Microb. Cell Fact.* **2019**, *18*, 1–14. [CrossRef]
144. Costa, N.L.; Carlson, H.K.; Coates, J.D.; Louro, R.O.; Paquete, C.M. Heterologous expression and purification of a multiheme cytochrome from a Gram-positive bacterium capable of performing extracellular respiration. *Protein Expr. Purif.* **2015**, *111*, 48–52. [CrossRef] [PubMed]
145. Phillips, J.R.; Huhnke, R.L.; Atiyeh, H.K. Syngas fermentation: A microbial conversion process of gaseous substrates to various products. *Fermentation* **2017**, *3*, 28. [CrossRef]
146. Hamilton, T.L.; Bryant, D.A.; Macalady, J.L. The role of biology in planetary evolution: Cyanobacterial primary production in low-oxygen Proterozoic oceans. *Environ. Microbiol.* **2016**, *18*, 325–340. [CrossRef]
147. Parmar, A.; Singh, N.K.; Pandey, A.; Gnansounou, E.; Madamwar, D. Cyanobacteria and microalgae: A positive prospect for biofuels. *Bioresour. Technol.* **2011**, *102*, 10163–10172. [CrossRef]
148. Alqahtani, M.F.; Katuri, K.P.; Bajracharya, S.; Yu, Y.; Lai, Z.; Saikaly, P.E. Porous Hollow Fiber Nickel Electrodes for Effective Supply and Reduction of Carbon Dioxide to Methane through Microbial Electrosynthesis. *Adv. Funct. Mater.* **2018**, *28*, 1804860. [CrossRef]

149. Bradley, R.W.; Bombelli, P.; Rowden, S.J.L.; Howe, C.J. Biological photovoltaics: Intra-and extra-cellular electron transport by cyanobacteria. *Biochem. Soc. Trans.* **2012**, *40*, 1302–1307. [CrossRef]
150. Pandit, A.V.; Mahadevan, R. In silico characterization of microbial electrosynthesis for metabolic engineering of biochemicals. *Microb. Cell Fact.* **2011**, *10*, 1–14. [CrossRef]
151. Cassia, R.; Nocioni, M.; Correa-Aragunde, N.; Lamattina, L. Climate change and the impact of greenhouse gasses: CO_2 and NO, friends and foes of plant oxidative stress. *Front. Plant Sci.* **2018**, *9*, 273. [CrossRef]
152. Das, S.; Das, S.; Das, I.; Ghangrekar, M.M. Application of bioelectrochemical systems for carbon dioxide sequestration and concomitant valuable recovery: A review. *Mater. Sci. Energy Technol.* **2019**, *2*, 687–696. [CrossRef]
153. Butti, S.K.; Mohan, S.V. Autotrophic biorefinery: Dawn of the gaseous carbon feedstock. *FEMS Microbiol. Lett.* **2017**, *364*, 166. [CrossRef]
154. Nitopi, S.; Bertheussen, E.; Scott, S.B.; Liu, X.; Engstfeld, A.K.; Horch, S.; Seger, B.; Stephens, I.E.L.; Chan, K.; Hahn, C.; et al. Progress and Perspectives of Electrochemical CO_2 Reduction on Copper in Aqueous Electrolyte. *Chem. Rev.* **2019**, *119*, 7610–7672. [CrossRef] [PubMed]
155. Tahir, K.; Miran, W.; Jang, J.; Woo, S.H.; Lee, D.S. Enhanced product selectivity in the microbial electrosynthesis of butyrate using a nickel ferrite-coated biocathode. *Environ. Res.* **2021**, *196*, 110907. [CrossRef]
156. del Pilar Anzola Rojas, M.; Mateos, R.; Sotres, A.; Zaiat, M.; Gonzalez, E.R.; Escapa, A.; De Wever, H.; Pant, D. Microbial electrosynthesis (MES) from CO_2 is resilient to fluctuations in renewable energy supply. *Energy Convers. Manag.* **2018**, *177*, 272–279. [CrossRef]
157. Mateos, R.; Sotres, A.; Alonso, R.M.; Morán, A.; Escapa, A. Enhanced CO_2 Conversion to Acetate through Microbial Electrosynthesis (MES) by Continuous Headspace Gas Recirculation. *Energies* **2019**, *12*, 3297. [CrossRef]
158. Batlle-Vilanova, P.; Ganigué, R.; Ramió-Pujol, S.; Bañeras, L.; Jiménez, G.; Hidalgo, M.; Balaguer, M.D.; Colprim, J.; Puig, S. Microbial electrosynthesis of butyrate from carbon dioxide: Production and extraction. *Bioelectrochemistry* **2017**, *117*, 57–64. [CrossRef]
159. Igarashi, K.; Kato, S. Reductive Transformation of Fe(III) (oxyhydr) Oxides by Mesophilic Homoacetogens in the Genus Sporomusa. *Front. Microbiol.* **2021**, *12*, 28. [CrossRef]
160. Gavilanes, J.; Noori, M.T.; Min, B. Enhancing bio-alcohol production from volatile fatty acids by suppressing methanogenic activity in single chamber microbial electrosynthesis cells (SCMECs). *Bioresour. Technol. Rep.* **2019**, *7*, 100292. [CrossRef]
161. Gavilanes, J.; Reddy, C.N.; Min, B. Microbial Electrosynthesis of Bioalcohols through Reduction of High Concentrations of Volatile Fatty Acids. *Energy Fuels* **2019**, *33*, 4264–4271. [CrossRef]
162. Zhuang, W.-Q.; Yi, S.; Bill, M.; Brisson, V.L.; Feng, X.; Men, Y.; Conrad, M.E.; Tang, Y.J.; Alvarez-Cohen, L. Incomplete Wood–Ljungdahl pathway facilitates one-carbon metabolism in organohalide-respiring Dehalococcoides mccartyi. *Proc. Natl. Acad. Sci. USA* **2014**, *111*, 6419–6424. [CrossRef]
163. Philips, J. Extracellular Electron Uptake by Acetogenic Bacteria: Does H_2 Consumption Favor the H_2 Evolution Reaction on a Cathode or Metallic Iron? *Front. Microbiol.* **2020**, *10*, 2997. [CrossRef] [PubMed]
164. Vassilev, I.; Kracke, F.; Freguia, S.; Keller, J.; Krömer, J.O.; Ledezma, P.; Virdis, B. Microbial electrosynthesis system with dual biocathode arrangement for simultaneous acetogenesis, solventogenesis and carbon chain elongation. *Chem. Commun.* **2019**, *55*, 4351–4354. [CrossRef]
165. Müller, V.; Wiechmann, A. Synthesis of Acetyl-CoA from Carbon Dioxide in Acetogenic Bacteria. In *Biogenesis of Fatty Acids, Lipids and Membranes*; Geiger, O., Ed.; Springer: Cham, Switzerland, 2017; pp. 1–18. [CrossRef]
166. Liu, K.; Atiyeh, H.K.; Stevenson, B.S.; Tanner, R.S.; Wilkins, M.R.; Huhnke, R.L. Mixed culture syngas fermentation and conversion of carboxylic acids into alcohols. *Bioresour. Technol.* **2014**, *152*, 337–346. [CrossRef]
167. Jin, S.; Jeon, Y.; Jeon, M.S.; Shin, J.; Song, Y.; Kang, S.; Bae, J.; Cho, S.; Lee, J.-K.; Kim, D.R.; et al. Acetogenic bacteria utilize light-driven electrons as an energy source for autotrophic growth. *Proc. Natl. Acad. Sci. USA* **2021**, *118*, 2020552118. [CrossRef]
168. Tian, S.; Wang, H.; Dong, Z.; Yang, Y.; Yuan, H.; Huang, Q.; Song, T.S.; Xie, J. Mo_2C-induced hydrogen production enhances microbial electrosynthesis of acetate from CO_2 reduction. *Biotechnol. Biofuels* **2019**, *12*, 71. [CrossRef]
169. Jourdin, L.; Lu, Y.; Flexer, V.; Keller, J.; Freguia, S. Biologically Induced Hydrogen Production Drives High Rate/High Efficiency Microbial Electrosynthesis of Acetate from Carbon Dioxide. *ChemElectroChem* **2016**, *3*, 581–591. [CrossRef]
170. Perona-Vico, E.; Feliu-Paradeda, L.; Puig, S.; Bañeras, L. Bacteria coated cathodes as an in-situ hydrogen evolving platform for microbial electrosynthesis. *Sci. Rep.* **2020**, *10*, 1–11. [CrossRef]
171. Chenevier, P.; Mugherli, L.; Darbe, S.; Darchy, L.; Dimanno, S.; Tran, P.D.; Valentino, F.; Iannello, M.; Volbeda, A.; Cavazza, C.; et al. Hydrogenase enzymes: Application in biofuel cells and inspiration for the design of noble-metal free catalysts for H_2 oxidation. *C. R. Chim.* **2013**, *16*, 491–505. [CrossRef]
172. Chen, H.; Dong, F.; Minteer, S.D. The progress and outlook of bioelectrocatalysis for the production of chemicals, fuels and materials. *Nat. Catal.* **2020**, *3*, 225–244. [CrossRef]
173. ter Heijne, A.; Geppert, F.; Sleutels, T.H.J.A.; Batlle-Vilanova, P.; Liu, D.; Puig, S. Mixed culture biocathodes for production of hydrogen, methane, and carboxylates. In *Advances in Biochemical Engineering/Biotechnology*; Harnisch, F., Holtmann, D., Eds.; Springer: Cham, Switzerland, 2019; Volume 167, pp. 203–229. [CrossRef]
174. Lalau, C.C.; Low, C.T.J. Electrophoretic Deposition for Lithium-Ion Battery Electrode Manufacture. *Batter. Supercaps* **2019**, *2*, 551–559. [CrossRef]

175. Atiq Ur Rehman, M.; Chen, Q.; Braem, A.; Shaffer, M.S.P.; Boccaccini, A.R. Electrophoretic deposition of carbon nanotubes: Recent progress and remaining challenges. *Int. Mater. Rev.* **2020**. [CrossRef]
176. Huang, J.; Wang, J.; Xie, R.; Tian, Z.; Chai, G.; Zhang, Y.; Lai, F.; He, G.; Liu, C.; Liu, T.; et al. A universal pH range and a highly efficient Mo2C-based electrocatalyst for the hydrogen evolution reaction. *J. Mater. Chem. A* **2020**, *8*, 19879–19886. [CrossRef]
177. Song, T.; Fei, K.; Zhang, H.; Yuan, H.; Yang, Y.; Ouyang, P.; Xie, J. High efficiency microbial electrosynthesis of acetate from carbon dioxide using a novel graphene–nickel foam as cathode. *J. Chem. Technol. Biotechnol.* **2018**, *93*, 457–466. [CrossRef]
178. Bajracharya, S.; van den Burg, B.; Vanbroekhoven, K.; De Wever, H.; Buisman, C.J.N.; Pant, D.; Strik, D.P.B.T.B. In situ acetate separation in microbial electrosynthesis from CO_2 using ion-exchange resin. *Electrochim. Acta* **2017**, *237*, 267–275. [CrossRef]
179. Patil, S.A.; Arends, J.B.A.; Vanwonterghem, I.; van Meerbergen, J.; Guo, K.; Tyson, G.W.; Rabaey, K. Selective Enrichment Establishes a Stable Performing Community for Microbial Electrosynthesis of Acetate from CO_2. *Environ. Sci. Technol.* **2015**, *49*, 8833–8843. [CrossRef]
180. Mohanakrishna, G.; Vanbroekhoven, K.; Pant, D. Impact of dissolved carbon dioxide concentration on the process parameters during its conversion to acetate through microbial electrosynthesis. *React. Chem. Eng.* **2018**, *3*, 371–378. [CrossRef]
181. Bajracharya, S.; Ter Heijne, A.; Dominguez Benetton, X.; Vanbroekhoven, K.; Buisman, C.J.N.; Strik, D.P.B.T.B.; Pant, D. Carbon dioxide reduction by mixed and pure cultures in microbial electrosynthesis using an assembly of graphite felt and stainless steel as a cathode. *Bioresour. Technol.* **2015**, *195*, 14–24. [CrossRef]
182. Jourdin, L.; Grieger, T.; Monetti, J.; Flexer, V.; Freguia, S.; Lu, Y.; Chen, J.; Romano, M.; Wallace, G.G.; Keller, J. High Acetic Acid Production Rate Obtained by Microbial Electrosynthesis from Carbon Dioxide. *Environ. Sci. Technol.* **2015**, *49*, 13566–13574. [CrossRef]
183. Dong, Z.; Wang, H.; Tian, S.; Yang, Y.; Yuan, H.; Huang, Q.; Song, T.S.; Xie, J. Fluidized granular activated carbon electrode for efficient microbial electrosynthesis of acetate from carbon dioxide. *Bioresour. Technol.* **2018**, *269*, 203–209. [CrossRef]
184. Xie, S.; Liang, P.; Chen, Y.; Xia, X.; Huang, X. Simultaneous carbon and nitrogen removal using an oxic/anoxic-biocathode microbial fuel cells coupled system. *Bioresour. Technol.* **2011**, *102*, 348–354. [CrossRef]
185. Marshall, C.W.; Ross, D.E.; Fichot, E.B.; Norman, R.S.; May, H.D. Long-term operation of microbial electrosynthesis systems improves acetate production by autotrophic microbiomes. *Environ. Sci. Technol.* **2013**, *47*, 6023–6029. [CrossRef]
186. Ganesh, I. Conversion of carbon dioxide into methanol-A potential liquid fuel: Fundamental challenges and opportunities (a review). *Renew. Sustain. Energy Rev.* **2014**, *31*, 221–257. [CrossRef]
187. Centi, G.; Quadrelli, E.A.; Perathoner, S. Catalysis for CO_2 conversion: A key technology for rapid introduction of renewable energy in the value chain of chemical industries. *Energy Environ. Sci.* **2013**, *6*, 1711–1731. [CrossRef]
188. Reymond, H.; Corral-Pérez, J.J.; Urakawa, A.; Rudolf Von Rohr, P. Towards a continuous formic acid synthesis: A two-step carbon dioxide hydrogenation in flow. *React. Chem. Eng.* **2018**, *3*, 912–919. [CrossRef]
189. Rumayor, M.; Dominguez-Ramos, A.; Irabien, A. Formic Acid manufacture: Carbon dioxide utilization alternatives. *Appl. Sci.* **2018**, *8*, 914. [CrossRef]
190. Álvarez, A.; Bansode, A.; Urakawa, A.; Bavykina, A.V.; Wezendonk, T.A.; Makkee, M.; Gascon, J.; Kapteijn, F. Challenges in the Greener Production of Formates/Formic Acid, Methanol, and DME by Heterogeneously Catalyzed CO_2 Hydrogenation Processes. *Chem. Rev.* **2017**, *117*, 9804–9838. [CrossRef]
191. Bahmanpour, A.M.; Signorile, M.; Kröcher, O. Recent progress in syngas production via catalytic CO_2 hydrogenation reaction. *Appl. Catal. B Environ.* **2021**, *295*, 120319. [CrossRef]
192. Sivalingam, V.; Ahmadi, V.; Babafemi, O.; Dinamarca, C. Integrating syngas fermentation into a single-cell microbial electrosynthesis (MES) reactor. *Catalysts* **2021**, *11*, 40. [CrossRef]
193. Bustan, M.D.; Haryati, S.; Hadiah, F.; Selpiana, S.; Huda, A. Syngas Production Improvement of Sugarcane Bagasse Conversion Using an Electromagnetic Modified Vacuum Pyrolysis Reactor. *Processes* **2020**, *8*, 252. [CrossRef]
194. El-Nagar, R.A.; Ghanem, A.A. Syngas Production, Properties, and Its Importance. *Sustain. Altern. Syngas Fuel* **2019**. [CrossRef]
195. Nelabhotla, A.B.T.; Dinamarca, C. Bioelectrochemical CO2 Reduction to Methane: MES Integration in Biogas Production Processes. *Appl. Sci.* **2019**, *9*, 1056. [CrossRef]
196. Enzmann, F.; Mayer, F.; Rother, M.; Holtmann, D. Methanogens: Biochemical background and biotechnological applications. *AMB Express* **2018**, *8*, 1. [CrossRef]
197. Munasinghe, P.C.; Khanal, S.K. Syngas fermentation to biofuel: Evaluation of carbon monoxide mass transfer coefficient (kLa) in different reactor configurations. *Biotechnol. Prog.* **2010**, *26*, 1616–1621. [CrossRef]
198. Kumar, G.; Saratale, R.G.; Kadier, A.; Sivagurunathan, P.; Zhen, G.; Kim, S.H.; Saratale, G.D. A review on bio-electrochemical systems (BESs) for the syngas and value-added biochemicals production. *Chemosphere* **2017**, *177*, 84–92. [CrossRef]
199. Annie Modestra, J.; Katakojwala, R.; Venkata Mohan, S. CO_2 fermentation to short-chain fatty acids using selectively enriched chemolithoautotrophic acetogenic bacteria. *Chem. Eng. J.* **2020**, *394*, 124759. [CrossRef]
200. Aryal, N.; Ghimire, N.; Bajracharya, S. Chapter 3—Coupling of microbial electrosynthesis with anaerobic digestion for waste valorization. In *Advances in Bioenergy*; Li, Y., Khanal, S.K., Eds.; Elsevier: Amsterdam, The Netherlands, 2020; pp. 101–127. [CrossRef]
201. Vu, H.T.; Min, B. Enhanced methane fermentation of municipal sewage sludge by microbial electrochemical systems integrated with anaerobic digestion. *Int. J. Hydrogen Energy* **2019**, *44*, 30357–30366. [CrossRef]

202. Yang, H.Y.; Hou, N.N.; Wang, Y.X.; Liu, J.; He, C.S.; Wang, Y.R.; Li, W.H.; Mu, Y. Mixed-culture biocathodes for acetate production from CO_2 reduction in the microbial electrosynthesis: Impact of temperature. *Sci. Total Environ.* **2021**, *790*, 148128. [CrossRef]
203. Names, A.; Whitman, J.D.; Hiatt, J.; Mowery, C.T.; Shy, B.R.; Yu, R.; Yamamoto, T.N.; Rathore, U.; Goldgof, G.M.; Whitty, C.; et al. Title: Test performance evaluation of SARS-CoV-2 serological assays. *medRxiv* **2020**, *29*, 30. [CrossRef]
204. Satinover, S.J.; Rodriguez, M.; Campa, M.F.; Hazen, T.C.; Borole, A.P. Performance and community structure dynamics of microbial electrolysis cells operated on multiple complex feedstocks. *Biotechnol. Biofuels* **2020**, *13*, 1–21. [CrossRef]
205. Saboe, P.O.; Manker, L.P.; Monroe, H.R.; Michener, W.E.; Haugen, S.; Tan, E.C.D.; Prestangen, R.L.; Beckham, G.T.; Karp, E.M. Energy and techno-economic analysis of bio-based carboxylic acid recovery by adsorption. *Green Chem.* **2021**, *23*, 4386–4402. [CrossRef]
206. Clarke, C.J.; Tu, W.C.; Levers, O.; Bröhl, A.; Hallett, J.P. Green and Sustainable Solvents in Chemical Processes. *Chem. Rev.* **2018**, *118*, 747–800. [CrossRef] [PubMed]
207. Keshav, A.; Wasewar, K.L.; Chand, S. Extraction of propionic acid using different extractants (tri-n-butylphosphate, tri-n-octylamine, and Aliquat 336). *Ind. Eng. Chem. Res.* **2008**, *47*, 6192–6196. [CrossRef]
208. Li, Q.-Z.; Jiang, X.-L.; Feng, X.-J.; Wang, J.-M.; Sun, C.; Zhang, H.-B.; Xian, M.; Liu, H.-Z. Recovery Processes of Organic Acids from Fermentation Broths in the Biomass-Based Industry. *J. Microbiol. Biotechnol* **2016**, *26*, 1–8. [CrossRef]
209. Woo, H.C.; Kim, Y.H. Eco-efficient recovery of bio-based volatile C2-6 fatty acids. *Biotechnol. Biofuels* **2019**, *12*, 1–11. [CrossRef]
210. Zhu, X.; Leininger, A.; Jassby, D.; Tsesmetzis, N.; Ren, Z.J. Will Membranes Break Barriers on Volatile Fatty Acid Recovery from Anaerobic Digestion? *ACS EST Eng.* **2021**, *1*, 141–153. [CrossRef]
211. Zhu, Y. Evaluation of Nanofiltration for the Extraction of Volatile Fatty Acids from Fermentation Broth. 2020. Available online: https://tel.archives-ouvertes.fr/tel-03008506 (accessed on 8 August 2021).
212. Andersen, S.J.; Berton, J.K.E.T.; Naert, P.; Gildemyn, S.; Rabaey, K.; Stevens, C.V. Extraction and esterification of low-titer short-chain volatile fatty acids from anaerobic fermentation with ionic liquids. *ChemSusChem* **2016**, *9*, 2059–2063. [CrossRef]
213. Montesantos, N.; Maschietti, M. Supercritical carbon dioxide extraction of lignocellulosic bio-oils: The potential of fuel upgrading and chemical recovery. *Energies* **2020**, *13*, 1600. [CrossRef]
214. Murali, N.; Srinivas, K.; Ahring, B.K. Increasing the production of volatile fatty acids from corn stover using bioaugmentation of a mixed rumen culture with homoacetogenic bacteria. *Microorganisms* **2021**, *9*, 337. [CrossRef]
215. Sprakel, L.M.J.; Schuur, B. Solvent developments for liquid-liquid extraction of carboxylic acids in perspective. *Sep. Purif. Technol.* **2019**, *211*, 935–957. [CrossRef]
216. Naik, S.N.; Goud, V.V.; Rout, P.K.; Dalai, A.K. Production of first and second-generation biofuels: A comprehensive review. *Renew. Sustain. Energy Rev.* **2010**, *14*, 578–597. [CrossRef]
217. Ho, D.P.; Ngo, H.H.; Guo, W. A mini-review on renewable sources for biofuel. *Bioresour. Technol.* **2014**, *169*, 742–749. [CrossRef]
218. Savla, N.; Shinde, A.; Sonawane, K.; Mekuto, L.; Chowdhary, P.; Pandit, S. Microbial hydrogen production: Fundamentals to application. In *Microorganisms for Sustainable Environment and Health*; Elsevier: Gurgaon, India, 2020; pp. 343–365. [CrossRef]
219. Committee of America's Energy Future. *America's Energy Future: Technology and Transformation*; National Academies Press: Washington, DC, USA, 2010; ISBN 0309116023.
220. IRENA. *Renewable Power Generation Costs in 2019*; International Renewable Energy Agency: Abu Dhabi, United Arab Emirates, 2020; ISBN 978-92-9260-244-4.
221. Owusu, P.A.; Asumadu-Sarkodie, S. A review of renewable energy sources, sustainability issues and climate change mitigation. *Cogent Eng.* **2016**, *3*. [CrossRef]
222. Zsiborács, H.; Baranyai, N.H.; Vincze, A.; Zentkó, L.; Birkner, Z.; Máté, K.; Pintér, G. Intermittent renewable energy sources: The role of energy storage in the european power system of 2040. *Electronics* **2019**, *8*, 729. [CrossRef]
223. Lund, H.; Kempton, W. Integration of renewable energy into the transport and electricity sectors through V2G. *Energy Policy* **2008**, *36*, 3578–3587. [CrossRef]
224. Popp, J.; Lakner, Z.; Harangi-Rákos, M.; Fári, M. The effect of bioenergy expansion: Food, energy, and environment. *Renew. Sustain. Energy Rev.* **2014**, *32*, 559–578. [CrossRef]
225. Bajracharya, S.; Srikanth, S.; Mohanakrishna, G.; Zacharia, R.; Strik, D.P.; Pant, D. Biotransformation of carbon dioxide in bioelectrochemical systems: State of the art and future prospects. *J. Power Source* **2017**, *356*, 256–273. [CrossRef]
226. Molina, M.G. Energy Storage and Power Electronics Technologies: A Strong Combination to Empower the Transformation to the Smart Grid. *Proc. IEEE* **2017**, *105*, 2191–2219. [CrossRef]
227. Chaudhary, D.K.; Kim, J. New insights into bioremediation strategies for oil-contaminated soil in cold environments. *Int. Biodeterior. Biodegrad.* **2019**, *142*, 58–72. [CrossRef]
228. Huang, C.H.; Tan, C.S. A review: CO_2 utilization. *Aerosol Air Qual. Res.* **2014**, *14*, 480–499. [CrossRef]
229. Singh Saharan, B.; Sharma, D.; Sahu, R.; Sahin, O.; Warren, A.; Biomonitoring, P.; Millipore, M.; Caddesi, K.; Ciftligi, K.; Kar, Y. Towards Algal Biofuel Production: A Concept of Green Bio Energy Development. *Innov. Rom. Food Biotechnol.* **2013**, *12*. Available online: http://www.bioaliment.ugal.ro/ejournal.htm (accessed on 8 November 2021).
230. Irfan, M.; Bai, Y.; Zhou, L.; Kazmi, M.; Yuan, S.; Maurice Mbadinga, S.; Yang, S.Z.; Liu, J.F.; Sand, W.; Gu, J.D.; et al. Direct microbial transformation of carbon dioxide to value-added chemicals: A comprehensive analysis and application potentials. *Bioresour. Technol.* **2019**, *288*, 121401. [CrossRef]

231. Johnson, M.C.; Heim, C.J.; Billingham, J.F.; Malczewski, M.L. Method for Analyzing Impurities in Carbon Dioxide 2006. Available online: https://patents.google.com/patent/US7064834B2 (accessed on 8 October 2021).
232. Lovley, D.R.; Nevin, K.P. Microbial Production of Multi-Carbon Chemicals and Fuels from Water and Carbon Dioxide Using Electric Current 2008. Available online: https://patents.google.com/patent/US9856449B2 (accessed on 8 October 2021).
233. Lajua Jo Eun-cheol. Cathode for Production of Biofuel, and Microbial Electrosynthesis System for Production of Biofuel Comprising the Same 2018. Available online: https://patents.google.com/patent/KR101892982B1 (accessed on 8 October 2021).
234. Deutzmann, J.S.; Spormann, A.M. Enhanced microbial electrosynthesis by using co-cultures. *ISME J.* **2016**. [CrossRef]
235. May, H.D.; Labelle, E.V. Bioelectrosynthesis of Organic Compounds 2019. Available online: https://patents.google.com/patent/US20190301029A1 (accessed on 8 October 2021).
236. Bergel, A.; Bernet, N.; Blanchet, E.; Bouchez, T.; Erable, B.; Etcheverry, L.; Huyard, A.; Quemener, E.L.; Mauricrace, P.; Moreau, S.; et al. Bioelectrochemical Reactor with Double Bioanode, Method for Anodic Regeneration and Use of the Reactor for Microbial Electrosynthesis 2020. Available online: https://patents.google.com/patent/WO2020053529A1 (accessed on 8 October 2021).
237. Li, W.; Menggen, L.; Yang, Y.; Ning, H.; Jiafang, Z. Microbial Cathode Catalytic Reduction CO2 Method for Electrosynthesis of Organic Matter 2020. Available online: https://patents.google.com/patent/CN111961691A (accessed on 8 October 2021).
238. Lovley, D.R.; Ueki, T.; LOVLEY, K.N. Microbial Strain for Electrosynthesis and Electrofermentation 2020. Available online: https://patents.google.com/patent/US10711318B2 (accessed on 8 October 2021).
239. Kun, G. Electrolytic Hydrogen Bubble Column Microbial Electrosynthesis Reactor and Use Method Thereof 2019. Available online: https://patents.google.com/patent/CN110528017B (accessed on 8 October 2021).
240. Prévoteau, A.; Carvajal-Arroyo, J.M.; Ganigué, R.; Rabaey, K. Microbial electrosynthesis from CO_2: Forever a promise? *Curr. Opin. Biotechnol.* **2020**, *62*, 48–57. [CrossRef] [PubMed]

 *fermentation*

Review

Recent Developments in Lignocellulosic Biofuels, a Renewable Source of Bioenergy

Ashutosh Kumar Rai [1,*], Naief Hamoud Al Makishah [2], Zhiqiang Wen [3], Govind Gupta [4], Soumya Pandit [5] and Ram Prasad [6,*]

1 Department of Biochemistry, College of Medicine, Imam Abdulrahman Bin Faisal University, Dammam 31441, Saudi Arabia
2 Environmental Sciences Department, Faculty of Meteorology, Environment and Arid Land Agriculture, King Abdulaziz University, Jeddah 21589, Saudi Arabia; nalmakishah@kau.edu.sa
3 School of Environmental and Biological Engineering, Nanjing University of Science & Technology, Nanjing 210094, China; zqwen@njust.edu.cn
4 School of Agriculture, Sanjeev Agrawal Global Educational University, Bhopal 462022, India; govind.g@sageuniversity.edu.in
5 Department of Life Sciences, School of Basic Sciences and Research, Sharda University, Greater Noida 201306, India; sounip@gmail.com
6 Department of Botany, Mahatma Gandhi Central University, Motihari 845401, India
* Correspondence: akraibiotech@gmail.com (A.K.R.); rpjnu2001@gmail.com (R.P.)

Abstract: Biofuel consists of non-fossil fuel derived from the organic biomass of renewable resources, including plants, animals, microorganisms, and waste. Energy derived from biofuel is known as bioenergy. The reserve of fossil fuels is now limited and continuing to decrease, while at the same time demand for energy is increasing. In order to overcome this scarcity, it is vital for human beings to transfer their dependency on fossil fuels to alternative types of fuel, including biofuels, which are effective methods of fulfilling present and future demands. The current review therefore focusses on second-generation lignocellulosic biofuels obtained from non-edible plant biomass (i.e., cellulose, lignin, hemi-celluloses, non-food material) in a more sustainable manner. The conversion of lignocellulosic feedstock is an important step during biofuel production. It is, however, important to note that, as a result of various technical restrictions, biofuel production is not presently cost efficient, thus leading to the need for improvement in the methods employed. There remain a number of challenges for the process of biofuel production, including cost effectiveness and the limitations of various technologies employed. This leads to a vital need for ongoing and enhanced research and development, to ensure market level availability of lignocellulosic biofuel.

Keywords: biomass; second-generation biofuel; bioenergy; bioethanol; biodiesel; non-fossil fuel

Citation: Rai, A.K.; Al Makishah, N.H.; Wen, Z.; Gupta, G.; Pandit, S.; Prasad, R. Recent Developments in Lignocellulosic Biofuels, a Renewable Source of Bioenergy. *Fermentation* **2022**, *8*, 161. https://doi.org/10.3390/fermentation8040161

Academic Editors: Alessia Tropea and Mohammad Taherzadeh

Received: 21 February 2022
Accepted: 16 March 2022
Published: 3 April 2022

1. Introduction

The term 'biofuel' is applied to fuel derived from renewable, living materials, e.g., plants and animals. Biofuels are energy carriers and non-fossil fuels that store the energy derived from the organic biomass of plants, animals, microorganisms, and waste. Energy derived from biofuels is known as bioenergy [1–5].

The development of renewable energy from biomass, solar, wind, water, and nuclear energy has now become an urgent issue as a result of the continued increase in demand for fossil fuel based petroleum products, along with their established role in global warming and climate change [1,6]. In addition, petroleum products have a limited reserve stock, leading to increased global attention being focused on studies of biomass based energy (i.e., biofuels) [5,7–20].

Biofuels can take the following forms: (1) liquid (i.e., ethanol and biodiesel); (2) solid (i.e., charcoal, wood pellets and fuelwood; and (3) gas (i.e., biogas). Biofuels have a renewable origin through the photosynthetic solar energy conversion to chemical energy, while petroleum products are derivatives of crude fossil fuel, obtained following its processing in oil refineries.

Based on its origin (i.e., biomass feedstock) and the technology used in biofuel production, biofuels are categorized between first- and fourth-generation biofuels [7,11–13,21–25] (Figure 1).

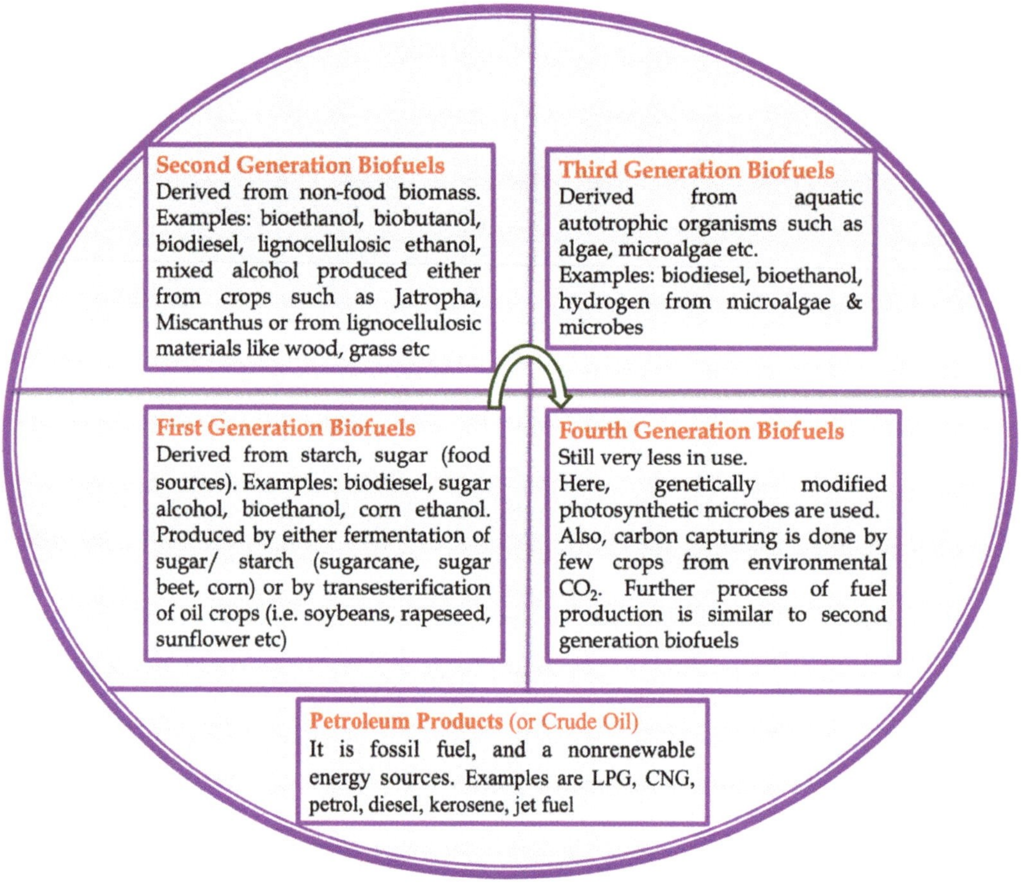

Figure 1. Comparison of first-, second-, third-, and fourth-generation biofuels, and petroleum fuels (adopted and modified from Naik et al. [13] and Dragone et al. [25]).

(i) First-generation biofuel is primarily derived from parts of edible plants (i.e., grains and oilseeds). These types of fuel have derived from sugar, starch, vegetable oil, and fats. Examples of most popular first-generation biofuels are biodiesel, ethanol, biofuel gasoline, biogas, etc. [7,13,24,26]. Presently, first-generation biofuel (biodiesel and bioethanol) is mainly produced by using agricultural feedstock such as sugarcane, corn, sugar beets, etc. [23]. Economic feasibility of biofuel production using crops (such as oilseed crops) as feedstock is not cost effective presently, therefore, a more efficient approach is needed to enhance the biofuel production and convert it to an economically feasible stage. Additionally, more research work is needed to increase the biodiesel production using first-generation feedstock such as oil [23].

(ii) Second-generation biofuel is a comparatively advanced biofuel which is derived from various non-food biomass of plant/or animal. Second-generation (lignocellulosic) biofuel is derived from non-edible plants or non-edible parts of the plants. It is well known that non-edible lignocellulosic biomass (such as vegetable grasses, forest residues, agricultural waste, etc.) is present abundantly in the natural ecosystem, therefore, it could be used as a feedstock for biofuel production. Examples of second-generation biofuels are lignocellulosic ethanol, butanol, mixed alcohols, etc. [4,13,24,27].
(iii) Third-generation biofuel is derived from photosynthetic microbes, e.g., microalgae. They derived from autotrophic organism. Here, carbon dioxide, light, and other nutrient sources are used in the synthesis of feedstock (biomass) which is further used in biofuel production [8,24,25,28]. Biofuels obtained from third-generation sources (such as microalgae) might be a better energy substitute as compared to previous generation biofuels, due to their short life cycle and less requirement of valuable agricultural land and resources for their growth [25]. Algae have rapid growth and higher rate of the photosynthesis compared to terrestrial plants used in first- and second-generation biofuel production. Due to their use in biofuel production, photosynthetic microbes (such as algae/microalgae) have recently received more attention from researchers worldwide [12].
(iv) Fourth-generation biofuel is not common and at an under developmental stage since a few years ago. Here, genetically altered photosynthetic microbes (such as cyanobacteria, algae, fungi) are used as feedstock. Photosynthetic microbes have the ability to convert atmospheric CO_2 to biofuel [24]. Some studies reported that carbon capturing is undertaken by some crops, taken from the atmosphere and further stored in their leaves, stems, etc., which is further converted into fuel using second-generation techniques [12]. Alalwan et al. [24] reported that, in the fourth-generation biofuels, genetically modified microorganisms are used to obtain more carbon (HC) yield and reduced carbon emissions [24].

Impact of Environmental Factors on Biofuel Production

Environmental factors also play a major role in the production of biofuels because growth of various crops and microbes employed in the biofuel production are directly influenced by environmental factors. The impacts of environmental factors on biofuel have been studied and discussed by several researchers [29–31]. Hosseinzadeh-Bandbafha et al. [30] described in detail the environmental impacts related to biodiesel. They used life cycle assessment method to discuss the biodiesel additives and their environmental impacts [30]. Sharma et al. [31] recently focused on the biofuel technologies used for sustainable environmental management. They explained, in detail, issues related to biofuel and its criteria of sustainability [31].

Since first-generation biofuel is obtained from crops such as sugar beet, grains, oil, and seeds, it has a number of limitations preventing it from attaining the targets demanded by the replacement of oil products, i.e., limited production and supply of the raw material. However, second-generation lignocellulosic biofuel is derived from non-edible parts of the biomass, therefore, it is more suitable for future applications. In this review, we are going to focus, in detail, on the second-generation lignocellulosic biofuels (Figure 2).

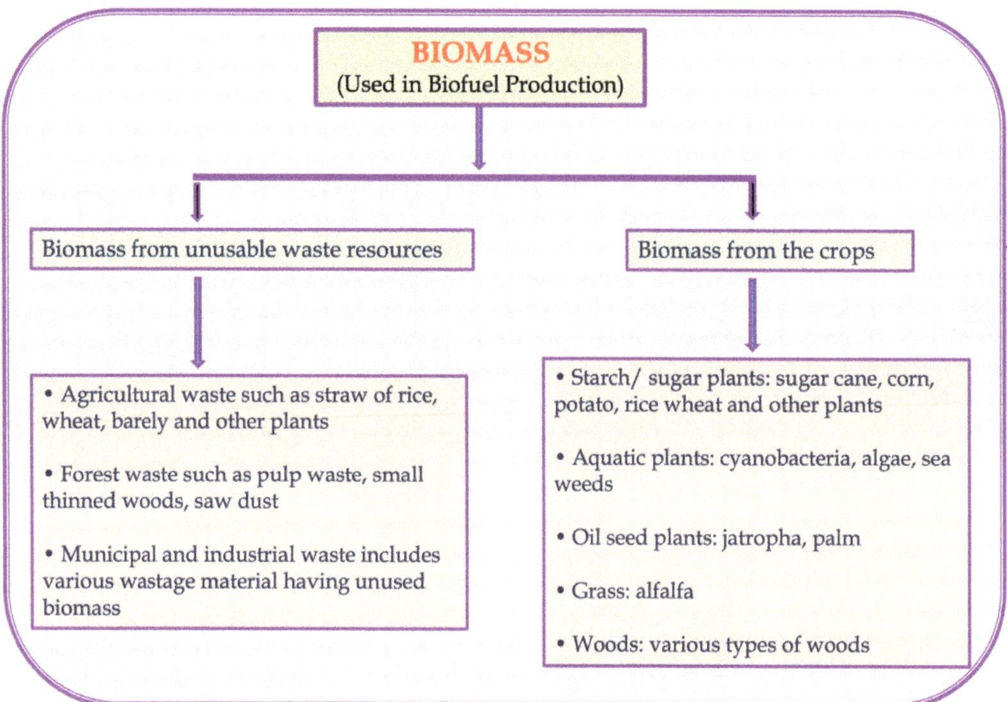

Figure 2. Biomass used as a renewable resource for biofuel production (adopted and modified from Naik et al. [13].)

2. Second-Generation Lignocellulosic Biofuels

Researchers and companies have now shifted their attention towards second generation biofuel production, in response to the limitations found in the production and supply of first-generation biofuels. Second-generation lignocellulosic biofuels are produced by employing non-edible biomass (e.g., cellulose, lignin, and hemi-celluloses) in a more sustainable manner rather than first-generation biofuel. Examples of second-generation biofuels are Fischer–Tropsch fuels and cellulosic ethanol. Such fuels are either carbon neutral or negative when it comes to CO_2 concentration [13,32–34].

Raw plant biomass material employed in the production of second-generation biofuels are generally referred to as lignocellulosic material and other non-food material of plants [4,20,35–38]. Such lignocellulosic raw material includes: (1) the by-products of plants (i.e., sugar cane bagasse, forest residues, and cereal straw); (2) the organic constituents of domestic waste; and (3) other forms of feedstock (i.e., crops, grasses, and short duration forests) (Figure 3).

Plant biomass is a widely and easily available biological resource for the raw materials for fuel [13,36,39,40]. There is considerable use of plant biomass in liquid biofuel production, due to these comprising cell walls composed largely of polysaccharides [13,40,41]. Badawy et al. [40] aimed to determine the most suitable biodiesel source among various sources such as Jatropha, rice straw, sugarcane, algae, etc. During their study, results showed that Jatropha was the most suitable biodiesel source [40]. Additionally, Arefin et al. [39] described biofuel production by floating aquatic plants, and discussed the methods related to biofuel production by aquatic plants (such as Azolla, duckweed, and water fern). Their observations showed that Azolla and water fern play a much better role in biofuel production as compared to other plants.

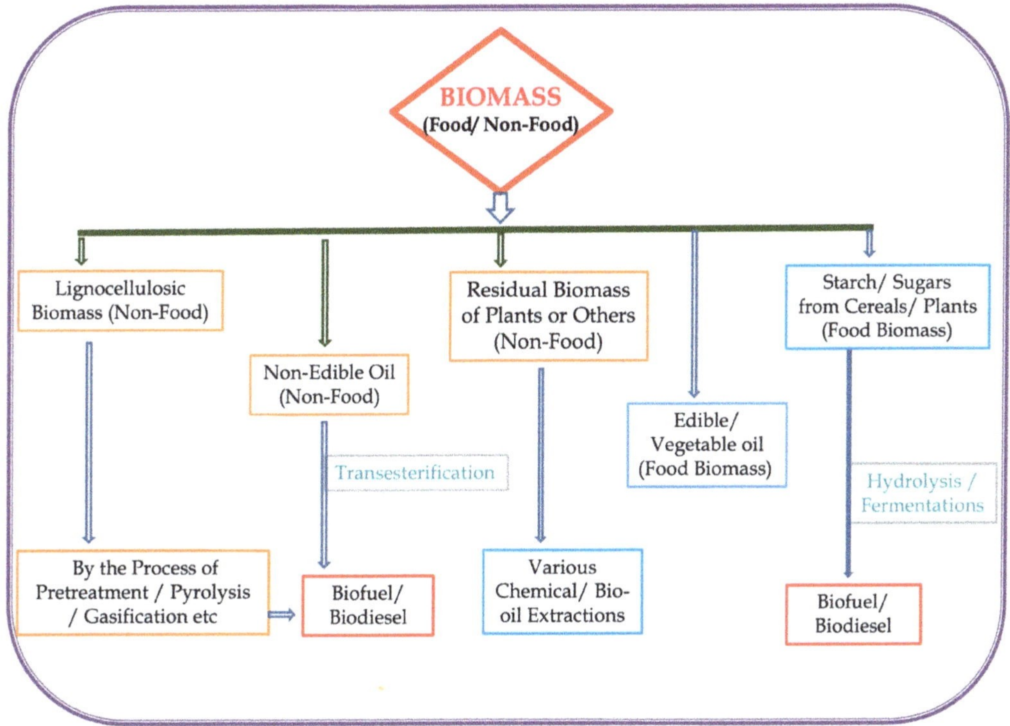

Figure 3. Schematic representations of second-generation biofuel (adopted and modified from Naik et al. [13]).

In addition to its direct use, second-generation biofuel can also be mixed with petroleum-based fuels in existing engines or used in slightly adapted vehicles with a compatible engine (e.g., vehicles for DME) [33].

It should be noted that the current production of such biofuels is not yet cost effective, due to various technical restrictions that require improvements in the methods employed [42].

3. Feedstock for Second-Generation Lignocellulosic Biofuels

Feedstock for second-generation lignocellulosic biofuels primarily consists of forms of biomass that are unfit for human consumption, hence, it does not compete with the production of food. Potential raw materials for second-generation biofuels consist of: (1) crop residue biomass; (2) non-food energy crops; (3) Jatropha; (4) wood residues; and (5) bacteria [7,22,34,43–46].

Second-generation biofuels production can also be enhanced by the growth of bio-energy crops in locations unsuitable for the farming of food crops, leading to maximum utilization of marginal land for second-generation biofuel production. Improving current methods will also enable the efficient creation of biofuels from the inedible parts of crops and forest trees. In addition, there is a potential for using waste-products for processing second-generation biofuels.

Firouzi et al. [44] used a hybrid Multi-Criteria Decision Making (MCDM) approach for the screening of biomass for the biofuel production. They noticed that wastes from municipal sewage, forest, and poultry were the most important resources for biofuel production. Narwane et al. [45] also discussed the integrated MCDM approach in the biofuel industry.

4. Lignocellulosic Feedstocks

The lignocellulosic biomass synthesized by photosynthesis on earth reaches more than 100 billion tons every year. However, only a small part is used by human activities, causing a great waste of resources. It is estimated that the lignocellulosic biomass that can be collected and utilized in the United States during a year is around 1.3 billion tons, and China has about 800 million tons [47]. In addition, although the use of fossil resources has brought great convenience to the development of human society, it has also brought a series of problems, such as environmental pollution and an energy crisis. Therefore, countries around the world are stepping up efforts to develop the conversion and utilization of lignocellulose, a renewable resource, to alleviate problems such as energy crises, environmental pollution, and sustainable development [48–52].

The major constituents of lignocellulosic feedstock consist of cellulose and hemicelluloses (i.e., about two thirds of the dry mass of plants), which are converted into sugars by means of thermochemical and biological processes. These lignocellulosic feedstocks are grouped into three types: (1) herbaceous and woody energy crops; (2) agricultural residues; and (3) forest residues. The cellulose present in the cell wall should split to form sugar, which can be further converted either to ethanol [53] or a fuel such as biodiesel or butanol. However, it should be noted that, due to its morphological characteristics, cell walls of the plant obstruct the cellulosic biofuel production.

For the synthesis of improved raw material, such as carbohydrates (which are further processed into biofuel), bioenergy crops should be grown on marginal land, employing the latest genomics and breeding technologies. Growing bioenergy crops on marginal land will lead to the production of sustainable biofuels.

There is a considerable variety of agricultural feedstock, which differ in their structural and chemical composition, leading to the production of a variety of biofuels, as discussed in the following sections.

5. Pretreatment of the Lignocellulosic Biomass

Pretreatment of raw material is an important step during biofuel production. It is applied during the process of biofuel and bioenergy production. It consists of mainly: (i) physical and chemical pretreatment and (ii) biological pretreatment.

(i) Physical and chemical pretreatment is widely used during biofuel production to improve quality of the substrate to be used for further digestion. Methods using heat, pressure, steam, hot water, ultrasonics, etc., are employed during the physical pretreatment process, while the oxidation, ozonization, acid or base pretreatment are used during chemical pretreatment methods [48]. These methods generally used in a combined way to obtain better results.

(ii) Biological pretreatment is mainly used for breaking lignin coatings and disrupting the cellulose structure so that it would be more susceptible for enzymatic or microbial digestion. During biological pretreatment methods, microorganisms play an important role and useful by-products are also produced [48].

Several researchers around the globe have discussed the importance of the pretreatment process used during biofuel production [2,48–52,54–60]. Wagner et al. [48] described the pretreatment methods to increase the production of biogas employing lignocellulosic biomass. Galbe and Wallberg [49] also described the common efficient pretreatment methods for lignocellulosic feedstock. Similarly, Sivamani et al. [54] studied acid pretreatment for the production of bioethanol.

Recently, Ab Rasid et al. [51] presented and discussed the lignocellulosic biomass pretreatment. They presented and focused on the green pretreatment strategies such as ionic liquids, ozonolysis, deep eutectic solvents, etc., for biomass pretreatment. Afolalu et al. [2] and Beig et al. [52] also discussed the different challenges related to the pretreatment of lignocellulosic biomass. Afolalu et al. [2] described the chemical, physical, and biological pretreatment processes.

Dionísio et al. [55] used dilute sulfuric acid for the pretreatment, which leads to 89.5% of hemicellulose solubilization. Lima et al. [56] discussed the ozone pretreatment of sugarcane for the ethanol and biogas integrated production. Morales-Martínez et al. [57] described the chemical pretreatment for ethanol production employing coffee husk waste. Mund et al. [58] discussed enzymatic hydrolysis and pretreatment of the leaf waste for biofuel production (Figure 4).

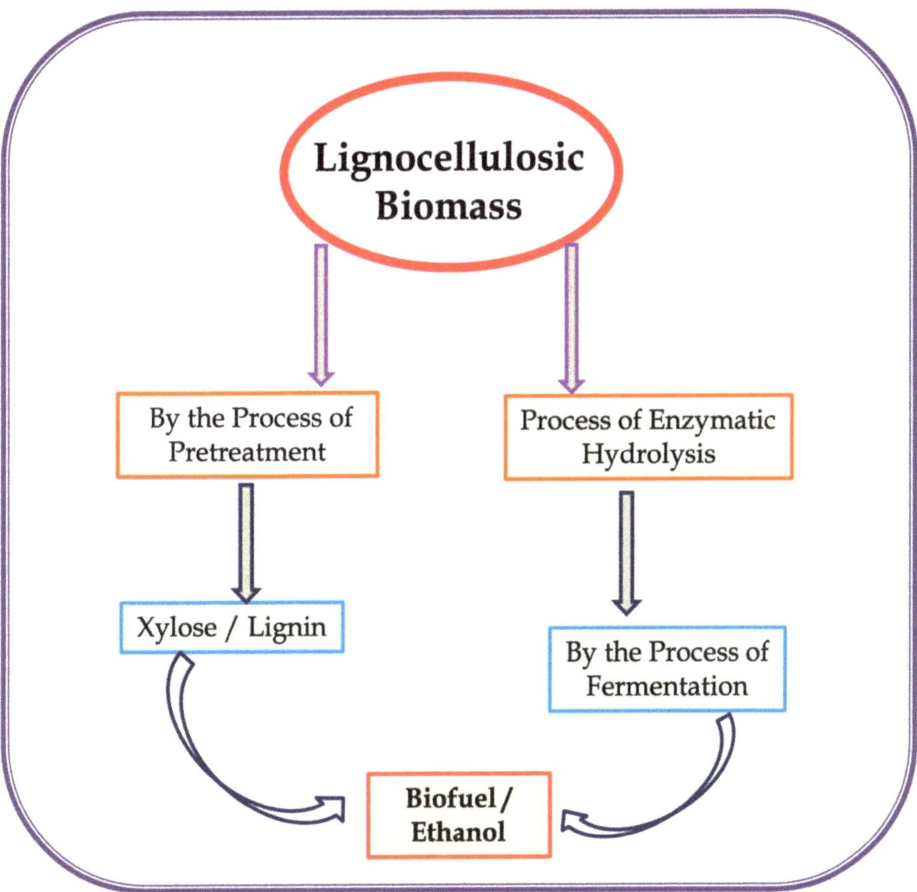

Figure 4. Lignocellulosic biomass conversion. A schematic view (adopted and modified from Naik et al. [13].

6. Conversion of Feedstock into Biofuels

Conversion of feedstock is an important step. There are primarily two processing routes followed for biofuel production from lignocellulosic raw materials:

(i) Biochemical route,
(ii) Thermochemical route.

6.1. Biochemical Route

Microorganisms and different enzymes are employed to convert various components of feedstocks (i.e., cellulose and hemicellulose) to sugars, followed by fermentation for ethanol production [27]. Generally, the reaction conditions of biochemical methods are relatively mild where pretreatment temperature is 60–220 °C, enzyme hydrolysis temperature is about 50 °C, and fermentation temperature is 20–60 °C [61].

This route employs various enzymes during conversion of feedstocks to biofuels, although it still requires a considerable amount of work to improve the characteristics of feedstocks and their cost effectiveness, i.e., lowering the production cost, and other related features, such as the efficacy of the enzymes and an improvement in the complete process of conversion [27,59,62–70].

It should be noted that, before the enzymatic hydrolysis, pre-treatment of the cellulosic and related feedstocks is a highly decisive step, which can be undertaken in either a physical manner, or chemically or biologically [71]. This reduces the cost of the overall process.

6.2. Thermochemical Route

Biomass changes into liquid fuel, employing gasification technologies and a number of different biofuels, i.e., ethanol, synthetic diesel, and aviation fuel. The Fischer–Tropsch conversion techniques may be used. Thermochemical methods include pyrolysis, liquefaction and gasification.

Pyrolysis is one of the important processes for biofuel production. The pyrolysis takes place at 300–1000 °C under oxygen-free conditions [72]. The pyrolysis products are synthesis gas (CO, CO_2, CH_4, and H_2, etc.), bio-oil, and bio-char. According to different conditions, the major rapid pyrolysis product is bio-oil, and bio-carbon is a major yield of slow pyrolysis. Various researchers around the globe have discussed and reported on the pyrolysis in biofuel production [60,72–74]. Djandja et al. [73] reported on the pyrolysis process and its use in the sewage sludge conversion in biofuel. They described the sewage sludge pyrolysis methods. Fombu and Ochonogor [74] designed a semi-batch pyrolysis reactor for enhanced biofuel production.

High temperature liquefaction is the biomass conversion into bio-oil at 250–370 °C under conditions containing more moisture [75].

Gasification is the conversion of biomass to CO, CO_2, H_2O, H_2, CH_4, etc., at 900–1200 °C, as well as the formation of by-products tar and coke [76].

This technique is also referred to as the biomass-to-liquids method. In this method, gasification/pyrolysis produces a gas ($CO + H_2$) that enables a broad range of carbon biofuels (i.e., aviation fuel and synthetic diesel) to be synthesized, employing the Fischer–Tropsch conversion. Recently, Cai et al. [77] studied the co-gasification methods for biomass and solid waste in the gasifier.

A comparison of both conversation methods reveals differing yields in terms of feedstock, while demonstrating a similarity in terms of energy. A comparison with different feedstocks reveals a complex picture. One major difference in biochemical and thermochemical methods is ethanol production in the biochemical method, while a range of higher hydrocarbons are synthesized by the thermochemical route (example: jet fuel). Doliente et al. [78] described in detail about the supply chain components of the bio-aviation fuel.

The biofuel costs of both pathways are not fixed, and vary between companies, thus leading to the potential for an alternative process to benefit the industry.

7. Other Approaches for Enhanced Biofuel Production

A change in the composition of cell walls may improve biofuel production. This may lead to improvement in biofuel production from lignocellulosic biomass through the employment of modern approaches of molecular biology, along with synthetic and systemic biology, to improve plant cell wall digestibility [31,79,80].

7.1. Synthetic Biology and System Biology

In recent years, vast development of synthetic and systems biology technology has provided a new perspective and tools for the research of lignocellulose biorefinery. The metagenomic, transcriptome, and metagenomics technologies developed in recent years can skip microbial pure-breeding and directly read the genome, transcriptome, and proteome information of the microbial community in the original environment to identify cellulase genes. Research related to expression level and mechanism of enzymes laid the foundation for the subsequent study of new cellulase gene extraction, heterologous protein expression, purification, and degradation mechanisms [81,82]. A number of research institutions such as the US Department of Energy directly extracted total DNA and total RNA from termite intestines and established a metagene library. By employing sequencing techniques, it become evident that there are a large variety of cellulose, hemicellulose hydrolysis related genes, which further enhances people's understanding of the richness of cellulase [83,84].

7.2. Microbial Community-Based Approaches

The genome, metabolome, flux group, and computational simulation techniques of systems biology and synthetic biology also provide rich tools for microbial community research [85]. At present, screening of microbial communities that degrade cellulose efficiently from nature, identifying their community structure, studying the fermentation kinetics, analyzing the mechanism of their efficient degradation and transformation, and then simulating the construction of similar systems or further strengthening their functions through transformation will provide us with new ideas for the establishment of a new cellulose degradation system [86–88].

Lipase producing or lipolytic bacteria are also our future hope since lipases used in the transesterification reaction further lead to biodiesel production [17]. Al Makishah et al. [17] isolated a bacterial stain (*Micrococcus luteus*) which has novel lipolytic transesterification activity.

7.3. Metabolic Engineering Techniques

Due to the complexity of sugar utilization and stress resistance traits, which involve multiple levels of genes, proteins, regulatory factors, and stress behaviors, it is difficult to achieve the desired result through simple genetic or metabolic engineering [89]. Adaptive evolutionary engineering based on metabolic engineering can allow microorganisms to quickly obtain excellent phenotypes, but there are problems such as unclear gene targets and negative mutation interference [90,91]. The multi-omics technology development has opened up a new perspective for evolutionary engineering, and also provided a reliable target for reverse metabolic engineering [92].

Systems and synthetic biology have improved people's knowledge about microbial physiology and metabolic processes, along with the complexity of interactions between metabolic pathways and their regulatory networks. Using accurate computer simulations of complex metabolic networks, the ability to optimize growth or produce a product under specific conditions can be obtained with minimal changes and printing [93].

7.4. Nanotechnology-Based Approaches

Nanotechnology has a very vast scope for the industries related to biofuel production. Nanotechnology along with its nanomaterials have emerged as an effective solution for the biofuel field in achieving cost-effective and efficient approaches to enhance biofuel production [94]. Worldwide, several researchers have reported and discussed the use of nanotechnology in the enhanced biofuel production [40,94–96]. Nizami and Rehan [94] discussed the use of nanotechnology and its tremendous ability to develop a cost-effective and efficient biofuel industry. Similarly, Sekoai et al. [95] discussed use of nanoparticles in the biofuel processes (such as biogas, biodiesel, bioethanol production), towards improving its process yields.

7.5. Integration of Various Approaches

A new strategy to systematically integrate microbiome data, gene expression profiling, proteomics, and metabolomics has enabled researchers to study cell metabolism in depth, so that they can know that they are rationally designing strains. Now, new tools provided by synthetic biology could introduce a wide range of genetic diversity into a microbial host.

Combined together with breeding, higher throughput technology, and adaptive evolution, a series of genetic transformations can be completed to optimize biological process project objectives [60,97–100]. Recently, Patel and Shah [100] discussed the integrated lignocellulosic biorefinery to obtain biofuel as well as more than 200 value added products. Sarkar and Sarkar [99] discussed the multi-stage smart system for the sustainable biofuel production, especially to generate purified biofuel using less energy and less carbon emissions. During their study, waste and consumed energy was reduced [99].

Advanced bio-refineries may prove beneficial for reducing second-generation biofuel costs, while the efficiency of the processing methods will be improved by using the entire biomass in advanced biorefineries. An additional significant method would be second-generation biofuel production from wood industries and pulp.

Apart from the above approaches, Rosson et al. [101] described the use of raw waste animal fats as bioliquids. It is also opening a new area for renewable energy. Al Hatrooshi et al. [102] used waste shark liver oil (WSLO) for making biodiesel as cost effective as possible, which is also a non-edible feedstock.

8. Challenges to Be Overcome

There are several constraints faced during biofuel production including production cost, environmental factors (loss of soil and land area), and others. Here, a number of challenges are presented.

The first major challenge is the production cost, i.e., high cost of the biofuel (bioethanol) production and its economic feasibility compared to the price of crude oil [29,38,42,59].

Mizik et al. [38] recently discussed the various constraints, specifically on the economical aspect, in detail. The authors raised various concerns associated to biofuel production and stated that higher generation biofuels are not price competitive due to their production costs and technology limitations [38].

The second major challenge is technology-based limitations, which need to improve in order to achieve cost effective and commercially suitable second-generation biofuels. The advancement in new techniques is proving a challenging task, in particular, when it comes to addressing the cost barriers linked with biofuel production [15,42].

Another major challenge may be the source of funding for continuous and enhanced research and development to raise the biofuel to a market level. This includes: (1) specific and enhanced support for higher yields of energy crops; (2) sustainable biomass production; (3) lowering the cost of the supply chain; and (4) improving the process of conversion.

9. Conclusions

During the present study we discussed the various aspects related to biofuels (especially second-generation lignocellulosic biofuels), concepts surrounding biofuels, and the challenges. Previously, the main focus was on the first-generation biofuels which have direct consequences/effects on various products obtained from agricultural resources, therefore, food prices might be affected. Additionally, poorer countries showed their resistance towards biofuels due to the lower cost effectiveness. Various research around the globe is going now to obtain solutions in the form of sustainable energy sources, i.e., second-generation lignocellulosic biofuels, which would be environmentally friendly and cost effective. Additionally, to avoid the negative impact of first-generation biofuel production on food supplies, use of agricultural waste residues and lignocellulosic feedstock (i.e., second-generation biofuel) might be a better option in a possible short-term period.

However, as first-generation biofuels include various challenges in their production and use, shifting to next generation lignocellulosic biofuel might become more economically feasible.

Current processes for production of these alternative fuels are still in development. It is expected that bio-refining plants based on derivatives of lignocellulose would be able to use a broader range of raw organic materials. This may lead to incorporation of the operational procedure and catalytic design, in order to increase the efficiency of biofuel production in a specific biofuel process. The prime objectives of any bio-refinery are to generate a variety of products using different biomass combinations. Finally, organic chemistry commitment requires the concepts of biological products and bioprospecting systems, thus forcing technological combination and chemical biological transformation of the materials.

Author Contributions: Idea/Plan of the manuscript—A.K.R. and N.H.A.M.; written manuscript's draft—A.K.R., N.H.A.M. and Z.W.; data analysis and critical revision of the manuscript—A.K.R., G.G., S.P. and R.P. All authors have read and agreed to the published version of the manuscript.

Funding: This research received no external funding.

Institutional Review Board Statement: Not applicable.

Informed Consent Statement: Not applicable.

Data Availability Statement: Not applicable.

Conflicts of Interest: The authors declare no conflict of interest.

References

1. Liu, Y.; Cruz-Morales, P.; Zargar, A.; Belcher, M.S.; Pang, B.; Englund, E.; Dan, Q.; Yin, K.; Keasling, J.D. Biofuels for a sustainable future. *Cell* **2021**, *184*, 1636–1647. [CrossRef]
2. Afolalu, S.A.; Yusuf, O.O.; Abioye, A.A.; Emetere, M.E.; Ongbali, S.O.; Samuel, O.D. Biofuel, a sustainable renewable source of energy—A review. *IOP Conf. Ser.* **2021**, *665*, 012040. [CrossRef]
3. Khan, M.A.H.; Bonifacio, S.; Clowes, J.; Foulds, A.; Holland, R.; Matthews, J.C.; Percival, C.J.; Shallcross, D.E. Investigation of Biofuel as a Potential Renewable Energy Source. *Atmosphere* **2021**, *12*, 1289. [CrossRef]
4. Perea-Moreno, M.A.; Samerón-Manzano, E.; Perea-Moreno, A.J. Biomass as renewable energy: Worldwide research trends. *Sustainability* **2019**, *11*, 863. [CrossRef]
5. Guo, M.; Song, W.; Buhain, J. Bioenergy and biofuels: History, status and prespective. *Renew. Sustain. Energy Rev.* **2015**, *42*, 712–725. [CrossRef]
6. Nazari, M.T.; Mazutti, J.; Basso, L.G.; Colla, L.M.; Brandli, L. Biofuels and their connections with the sustainable development goals: A bibliometric and systematic review. *Environ. Dev. Sustain.* **2021**, *23*, 11139–11156. [CrossRef]
7. Aro, E.-M. From first generation biofuels to advanced solar biofuels. *Ambio* **2016**, *45*, 24–31. [CrossRef]
8. Kumar, V.; Nanda, M.; Joshi, H.C.; Singh, A.; Sharma, S.; Verma, M. Production of biodiesel and bioethanol using algal biomass harvested from fresh water river. *Renew. Energy* **2018**, *116*, 606–612. [CrossRef]
9. Muhammad, U.L.; Shamsuddin, I.M.; Danjuma, A.; Musawa, R.S.; Dembo, U.H. Biofuels as the starring substitute to fossil fuels. *Pet. Sci. Eng.* **2018**, *2*, 44–49. [CrossRef]
10. Subramaniam, Y.; Masron, T.A.; Azman, N.H.N. Biofuels, environmental sustainability and food security: A review of 51 countries. *Energy Res. Soc. Sci.* **2020**, *68*, 101549. [CrossRef]
11. Ganguly, P.; Sarkhel, R.; Das, P. The second-and third-generation biofuel technologies: Comparative perspectives. In *Sustainable Fuel Technologies Handbook*; Academic Press: Cambridge, MA, USA, 2021; pp. 29–50. [CrossRef]
12. Dutta, K.; Daverey, A.; Lin, J.-G. Evolution retrospective for alternative fuels: First to fourth generation. *Renew. Energy* **2014**, *69*, 114–122. [CrossRef]
13. Naik, S.N.; Goud, V.V.; Rout, P.K.; Dalai, A.K. Production of first and second generation biofuels: A comprehensive review. *Renew. Sustain. Energy Rev.* **2010**, *14*, 578–597. [CrossRef]
14. Pandit, S.; Savla, N.; Sonawane, J.M.; Sani, A.M.; Gupta, P.K.; Mathuriya, A.S.; Rai, A.K.; Jadhav, D.A.; Jung, S.P.; Prasad, R. Agricultural waste and wastewater as feedstock for bioelectricity generation using microbial fuel cells: Recent advances. *Fermentation* **2021**, *7*, 169. [CrossRef]
15. Shahid, M.K.; Batool, A.; Kashif, A.; Nawaz, M.H.; Aslam, M.; Iqbal, N.; Choi, Y. Biofuels and biorefineries: Development, application and future perspectives emphasizing the environmental and economic aspects. *J. Environ. Manag.* **2021**, *297*, 113268. [CrossRef]
16. Sindhu, R.; Binod, P.; Pandey, A.; Ankaram, S.; Duan, Y.; Awasthi, M.K. Biofuel production from biomass: Toward sustainable development. In *Current Developments in Biotechnology and Bioengineering*; Kumar, S., Kumar, R., Pandey, A., Eds.; Elsevier: Amsterdam, The Netherlands, 2019; pp. 79–92. [CrossRef]
17. Al Makishah, N.H.; Rai, A.K.; Neamatallah, A.A.; Mabrouk, A.M. *Micrococcus luteus* 2030: A novel lipolytic bacterial strain isolated from local contaminated soil in Saudi Arabia. *Sylwan* **2019**, *163*, 132–152.
18. Sánchez, Ó.J.; Cardona, C.A. Trends in biotechnological production of fuel ethanol from different feedstocks. *Bioresour. Technol.* **2008**, *99*, 5270–5295. [CrossRef]
19. Escobar, J.C.; Lora, E.S.; Venturini, O.J.; Yánez, E.E.; Castillo, E.F.; Almazan, O. Biofuels: Environment, technology and food security. *Renew. Sustain. Energy Rev.* **2009**, *13*, 1275–1287. [CrossRef]

20. Jusakulvijit, P.; Bezama, A.; Thrän, D. The availability and assessment of potential agricultural residues for the regional development of second-generation bioethanol in Thailand. *Waste Biomass Valor.* **2021**, *12*, 6091–6118. [CrossRef]
21. EASAC. *The Current Status of Biofuels in the European Union, Their Environmental Impacts and Future Prospects*; German National Academy of Sciences Leopoldina: Haale, Germany, 2012; 47p, ISBN 978-3-8047-3118-9.
22. Oumer, A.N.; Hasan, M.M.; Baheta, A.T.; Mamat, R.; Abdullah, A.A. Bio-based liquid fuels as a source of renewable energy: A review. *Renew. Sust. Energ. Rev.* **2018**, *88*, 82–98. [CrossRef]
23. Hirani, A.H.; Javed, N.; Asif, M.; Basu, S.K.; Kumar, A. A review on first-and second-generation biofuel productions. In *Biofuels: Greenhouse Gas Mitigation and Global Warming*; Kumar, A., Ogita, S., Yau, Y.Y., Eds.; Springer: New Delhi, India, 2018; pp. 141–154. [CrossRef]
24. Alalwan, H.A.; Alminshid, A.H.; Aljaafari, H.A. Promising evolution of biofuel generations. Subject review. *Renew. Energy Focus.* **2019**, *28*, 127–139. [CrossRef]
25. Dragone, G.; Fernandes, B.; Vicente, A.A.; Teixeira, J.A. Third generation biofuels from microalgae. In *Current Research, Technology and Education Topics in Applied Microbiology and Microbial Biotechnology*; Mendez-Vilas, A., Ed.; FORMATEX: Badajoz, Spain, 2010; pp. 1355–1366.
26. Gumienna, M.; Szambelan, K.; Jelen, H.; Czarnecki, Z. Evaluation of ethanol fermentation parameters for bioethanol production from sugar beet pulp and juice. *J. Inst. Brew.* **2014**, *120*, 543–549. [CrossRef]
27. Sims, R.E.H.; Mabee, W.; Saddler, J.N.; Taylor, M. An overview of second generation biofuel technologies. *Bioresour. Technol.* **2010**, *101*, 1570–1580. [CrossRef] [PubMed]
28. Hussain, F.; Shah, S.Z.; Ahmad, H.; Abubshait, S.A.; Abubshait, H.A.; Laref, A.; Manikandan, A.; Kusuma, H.S.; Iqbal, M. Microalgae an ecofriendly and sustainable wastewater treatment option: Biomass application in biofuel and bio-fertilizer production. A review. *Renew. Sust. Energ. Rev.* **2021**, *137*, 110603. [CrossRef]
29. Popp, J.; Harangi-Rakos, M.; Gabnai, Z.; Balogh, P.; Antal, G.; Bai, A. Biofuels and their co-products as livestock feed: Global economic and environmental implications. *Molecules* **2016**, *21*, 285. [CrossRef]
30. Hosseinzadeh-Bandbafha, H.; Tabatabaei, M.; Aghbashlo, M.; Khanali, M.; Demirbas, A. A comprehensive review on the environmental impacts of diesel/biodiesel additives. *Energy Convers. Manag.* **2018**, *174*, 579–614. [CrossRef]
31. Sharma, S.; Kundu, A.; Basu, S.; Shetti, N.P.; Aminabhavi, T.M. Sustainable environmental management and related biofuel technologies. *J. Environ. Manag.* **2020**, *273*, 111096. [CrossRef]
32. Schenk, P.M.; Thomas-Hall, S.R.; Stephens, E.; Marx, U.C.; Mussgnug, J.H.; Posten, C.; Kruse, O.; Hankamer, B. Second Generation Biofuels: High-Efficiency Microalgae for Biodiesel Production. *Bioenerg. Res.* **2008**, *1*, 20–43. [CrossRef]
33. Eisentraut, A. *Sustainable Production of Second-Generation Biofuels: Potential and Perspectives in Major Economies and Developing Countries*; IEA Energy Papers, No. 2010/01; OECD Publishing: Paris, France, 2010; 221p. [CrossRef]
34. Scully, S.M.; Orlygsson, J. Recent advances in second generation ethanol production by thermophilic bacteria. *Energies* **2015**, *8*, 1–30. [CrossRef]
35. Gomez, L.D.; Clare, G.S.; McQueen-Mason, J. Sustainable liquid biofuels from biomass: The writing's on the walls. *New Phytol.* **2008**, *178*, 473–485. [CrossRef]
36. Zabaniotou, A.; Ioannidou, O.; Skoulou, V. Rapeseed residues utilization for energy and 2nd generation biofuels. *Fuel* **2008**, *87*, 1492–1502. [CrossRef]
37. Dhiman, S.; Mukherjee, G. Present scenario and future scope of food waste to biofuel production. *J. Food Process Eng.* **2021**, *44*, e13594. [CrossRef]
38. Mizik, T.; Gyarmati, G. Economic and sustainability of biodiesel production- A systematic literature review. *Clean Technol.* **2021**, *3*, 19–36. [CrossRef]
39. Arefin, M.A.; Rashid, F.; Islam, A. A review of biofuel production from floating aquatic plants: An emerging source of bio-renewable energy. *Biofuel. Bioprod. Biorefin.* **2021**, *15*, 574–591. [CrossRef]
40. Badawy, T.; Mansour, M.S.; Daabo, A.M.; Aziz, M.M.A.; Othman, A.A.; Barsoum, F.; Basouni, M.; Hussien, M.; Ghareeb, M.; Hamza, M.; et al. Selection of second-generation crop for biodiesel extraction and testing its impact with nano additives on diesel engine performance and emissions. *Energy* **2021**, *237*, 121605. [CrossRef]
41. Pauly, M.; Keegstra, K. Cell-wall carbohydrates and their modification as a resource for biofuels. *Plant J.* **2008**, *54*, 559–568. [CrossRef] [PubMed]
42. Raud, M.; Kikas, T.; Sippula, O.; Shurpali, N.J. Potentials and challenges in lignocellulosic biofuel production technology. *Renew. Sust. Energ. Rev.* **2019**, *111*, 44–56. [CrossRef]
43. Aderibigbe, F.A.; Shiru, S.; Saka, H.B.; Amosa, M.K.; Mustapha, S.I.; Alhassan, M.I.; Adejumo, A.L.; Abdulraheem, M.; Owolabi, R.U. Heterogeneous catalysis of second generation oil for biodiesel production: A review. *ChemBioEng Rev.* **2021**, *8*, 78–89. [CrossRef]
44. Firouzi, S.; Allahyari, M.S.; Isazadeh, M.; Nikkhah, A.; van Haute, S. Hybrid multi-criteria decision-making approach to select appropriate biomass resources for biofuel production. *Sci. Total Environ.* **2021**, *770*, 144449. [CrossRef]
45. Narwane, V.S.; Yadav, V.S.; Raut, R.D.; Narkhede, B.E.; Gardas, B.B. Sustainable development challenges of the biofuel industry in India based on integrated MCDM approach. *Renew. Energy* **2021**, *164*, 298–309. [CrossRef]
46. Sarangi, P.K.; Nayak, M.M. Agro-Waste for Second-Generation Biofuels. In *Liquid Biofuels: Fundamentals, Characterization, and Applications*; Shadangi, K.P., Ed.; Scrivener Publishing LLC: Beverly, MA, USA, 2021; pp. 697–709. [CrossRef]

47. Shi, Y. China's resources of biomass feedstock. *Eng. Sci.* **2011**, *13*, 16–23.
48. Wagner, A.O.; Lackner, N.; Mutschlechner, M.; Prem, E.M.; Markt, R.; Illmer, P. Biological pretreatment strategies for second-generation lignocellulosic resources to enhance biogas production. *Energies* **2018**, *11*, 1797. [CrossRef]
49. Galbe, M.; Wallberg, O. Pretreatment for biorefineries: A review of common methods for efficient utilisation of lignocellulosic materials. *Biotechnol. Biofuels* **2019**, *12*, 294. [CrossRef]
50. Machineni, L. Lignocellulosic biofuel production: Review of alternatives. *Biomass Convers. Biorefin.* **2020**, *10*, 779–791. [CrossRef]
51. Ab Rasid, N.S.; Shamjuddin, A.; Rahman, A.Z.A.; Amin, N.A.S. Recent advances in green pre-treatment methods of lignocellulosic biomass for enhanced biofuel production. *J. Clean. Prod.* **2021**, *321*, 129038. [CrossRef]
52. Beig, B.; Riaz, M.; Naqvi, S.R.; Hassan, M.; Zheng, Z.; Karimi, K.; Pugazhendhi, A.; Atabani, A.E.; Chi, N.T.L. Current challenges and innovative developments in pretreatment of lignocellulosic residues for biofuel production: A review. *Fuel* **2021**, *287*, 119670. [CrossRef]
53. Gnansounou, E.; Dauriat, A. Techno-economic analysis of lignocellulosic ethanol: A review. *Bioresour. Technol.* **2010**, *101*, 4980–4991. [CrossRef] [PubMed]
54. Sivamani, S.; Baskar, R.; Chandrasekaran, A.P. Response surface optimization of acid pretreatment of cassava stem for bioethanol production. *Environ. Prog. Sustain. Energy* **2020**, *39*, e13335. [CrossRef]
55. Dionísio, S.R.; Santoro, D.C.J.; Bonan, C.I.D.G.; Soares, L.B.; Biazi, L.E.; Rabelo, S.C.; Ienczak, J.L. Second-generation ethanol process for integral use of hemicellulosic and cellulosic hydrolysates from diluted sulfuric acid pretreatment of sugarcane bagasse. *Fuel* **2021**, *304*, 121290. [CrossRef]
56. Lima, D.R.S.; de Oliveira Paranhos, A.G.; Adarme, O.F.H.; Baêta, B.E.L.; Gurgel, L.V.A.; dos Santos, A.S.; de Queiroz Silva, S.; de Aquino, S.F. Integrated production of second-generation ethanol and biogas from sugarcane bagasse pretreated with ozone. *Biomass Convers. Biorefin.* **2022**, *12*, 809–825. [CrossRef]
57. Morales-Martínez, J.L.; Aguilar-Uscanga, M.G.; Bolaños-Reynoso, E.; López-Zamora, L. Optimization of chemical pretreatments using response surface methodology for second-generation ethanol production from coffee husk waste. *BioEnergy Res.* **2021**, *14*, 815–827. [CrossRef]
58. Mund, N.K.; Dash, D.; Mishra, P.; Nayak, N.R. Cellulose solvent–based pretreatment and enzymatic hydrolysis of pineapple leaf waste biomass for efficient release of glucose towards biofuel production. *Biomass Convers. Biorefin.* **2021**, 1–10. [CrossRef]
59. Sinitsyn, A.P.; Sinitsyna, O.A. Bioconversion of renewable plant biomass. Second-generation biofuels: Raw materials, biomass pretreatment, enzymes, processes, and cost analysis. *Biochemistry* **2021**, *86*, S166–S195. [CrossRef] [PubMed]
60. Talmadge, M.; Kinchin, C.; Chum, H.L.; de Rezende Pinho, A.; Biddy, M.; de Almeida, M.B.; Casavechia, L.C. Techno-economic analysis for co-processing fast pyrolysis liquid with vacuum gasoil in FCC units for second-generation biofuel production. *Fuel* **2021**, *293*, 119960. [CrossRef]
61. Lynd, L.R.; Weimer, P.J.; van Zyl, W.H.; Pretorius, I.S. Microbial Cellulose Utilization: Fundamental and Biotechnology. *Microbiol. Mol. Bio. Rev.* **2002**, *66*, 506–577. [CrossRef] [PubMed]
62. Gregg, D.J.; Boussaid, A.; Saddler, J.N. Techno-economic evaluations of a generic wood-to-ethanol process: Effect of increased cellulose yields and enzyme recycle. *Bioresour. Technol.* **1998**, *63*, 7–12. [CrossRef]
63. Weil, J.R.; Dien, B.; Bothast, R.; Hendrickson, R.; Mosier, N.S.; Ladisch, M.R. Removal of fermentation inhibitors formed during pretreatment of biomass by polymeric adsorbents. *Indust. Eng. Chem. Res.* **2002**, *41*, 6132–6138. [CrossRef]
64. Sheehan, J.; Aden, A.; Paustian, K.; Killian, K.; Brenner, J.; Walsh, M.; Nelson, R. Energy and environmental aspects of using corn stover for fuel ethanol. *J. Indust. Ecol.* **2004**, *7*, 117–146. [CrossRef]
65. Eggeman, T.; Elander, R.T. Process and economic analysis of pretreatment technologies. *Bioresour. Technol.* **2005**, *96*, 2019–2025. [CrossRef]
66. Galbe, M.; Liden, G.; Zacchi, G. Production of ethanol from biomass-research in Sweden. *J. Sci. Ind. Res.* **2005**, *64*, 905–919.
67. Mosier, N.; Wyman, C.; Dale, B.; Elander, R.; Lee, Y.Y.; Holtzapple, M.; Ladisch, M. Features of Promising Technologies for Pretreatment of Lignocellulosic Biomass. *Bioresour. Technol.* **2005**, *96*, 673–686. [CrossRef]
68. Rocha–Meneses, L.; Raud, M.; Orupõld, K.; Kikas, T. Second-generation bioethanol production: A review of strategies for waste valorization. *Agron. Res.* **2016**, *15*, 830–847.
69. Jutakridsada, P.; Saengprachatanarug, K.; Kasemsiri, P.; Hiziroglu, S.; Kamwilaisak, K.; Chindaprasirt, P. Bioconversion of *Saccharum officinarum* leaves for ethanol production using separate hydrolysis and fermentation processes. *Waste Biomass Valor.* **2019**, *10*, 817–825. [CrossRef]
70. Azam, M.; Jahromy, S.S.; Raza, W.; Raza, N.; Lee, S.S.; Kim, K.-H.; Winter, F. Status, characterization and potential utilization of municipal solid waste as renewable energy source: Lahore case study in Pakistan. *Environ. Int.* **2020**, *134*, 105291. [CrossRef]
71. Yang, B.; Wyman, C.E. Pre-treatment: The key to unlocking low cost cellulosic ethanol. *Biofuel Bioprod. Biorefin.* **2008**, *2*, 26–40. [CrossRef]
72. Roy, P.; Dias, G. Prospects for pyrolysis technologies in the bioenergy sector: A review. *Renew. Sustain. Energy Rev.* **2017**, *77*, 59–69. [CrossRef]
73. Djandja, O.S.; Wang, Z.C.; Wang, F.; Xu, Y.P.; Duan, P.G. Pyrolysis of municipal sewage sludge for biofuel production: A review. *Ind. Eng. Chem. Res.* **2020**, *59*, 16939–16956. [CrossRef]
74. Fombu, A.H., Ochonogor, A.E. Design and Construction of a Semi-batch Pyrolysis Reactor for the Production of Biofuel. In *IOP Conference Series: Earth and Environmental Science*; IOP Publishing Ltd.: Bristol, UK, 2021; p. 012041. [CrossRef]

75. Li, C.; Aston, J.E.; Lacey, J.A.; Thompson, V.S.; Thompson, D.N. Impact of feedstock quality and variation on biochemical and thermochemical conversion. *Renew. Sustain. Energy Rev.* **2016**, *65*, 525–536. [CrossRef]
76. Richardson, Y.; Joël, B.; Julbe, A. A short overview on purification and conditioning of syngas produced by biomass gasification: Catalytic strategies, process intensification and new concepts. *Prog. Energy Combust. Sci.* **2012**, *38*, 765–781. [CrossRef]
77. Cai, J.; Zeng, R.; Zheng, W.; Wang, S.; Han, J.; Li, K.; Luo, M.; Tang, X. Synergistic Effects of Co-Gasification of Municipal Solid Waste and Biomass in Fixed-Bed Gasifier. *Process. Saf. Environ. Prot.* **2021**, *148*, 1–12. [CrossRef]
78. Doliente, S.; Narayan, A.; Tapia, F.; Samsatli, N.J.; Zhao, Y.; Samsatli, S. Bio-aviation fuel: A comprehensive review and analysis of the supply chain components. *Front. Energy Res.* **2020**, *8*, 110. [CrossRef]
79. Furtado, A.; Lupo, J.S.; Hoang, N.V.; Healey, A.; Singh, S.; Simmons, B.A.; Henry, R.J. Modifying plants for biofuel and biomaterial production. *Plant Biotechnol. J.* **2014**, *12*, 1246–1258. [CrossRef] [PubMed]
80. Kalluri, U.C.; Yin, H.; Yang, X.; Davison, B.H. Systems and synthetic biology approaches to alter plant cell walls and reduce biomass recalcitrance. *Plant Biotechnol. J.* **2014**, *12*, 1207–1216. [CrossRef] [PubMed]
81. Ilmberger, N.; Streit, W.R. Screening for cellulase encoding clones in metagenomic libraries. In *Metagenomics. Methods in Molecular Biology*; Streit, W., Daniel, R., Eds.; Humana Press: New York, NY, USA, 2017; Volume 1539, pp. 205–217. [CrossRef]
82. Liu, G.; Qin, Y.; Li, Z.; Qu, Y. Development of highly efficient, low-cost lignocellulolytic enzyme systems in the post-genomic era. *Biotechnol. Adv.* **2013**, *31*, 962–975. [CrossRef] [PubMed]
83. Nimchua, T.; Thongaram, T.; Uengwetwanit, T.; Pongpattanakitshote, S.; Eurwilaichitr, L. Metagenomic analysis of novel lignocellulose-degrading enzymes from higher termite guts inhabiting microbes. *J. Microbiol. Biotechnol.* **2012**, *22*, 462–469. [CrossRef]
84. Rashamuse, K.; Tendai, W.S.; Mathiba, K.; Ngcobo, T.; Mtimka, S.; Brady, D. Metagenomic mining of glycoside hydrolases from the hindgut bacterial symbionts of a termite (*Trinervitermes trinervoides*) and the characterization of a multimodular beta-1,4-xylanase (GH11). *Biotechnol. Appl. Biochem.* **2017**, *64*, 174–186. [CrossRef]
85. Xie, S.; Syrenne, R.; Sun, S.; Yuan, J.S. Exploration of Natural Biomass Utilization Systems (NBUS) for advanced biofuel—From systems biology to synthetic design. *Curr. Opin. Biotechnol.* **2014**, *27*, 195–203. [CrossRef]
86. Liang, J.; Lin, Y.; Li, T.; Mo, F. Microbial consortium OEM1 cultivation for higher lignocellulose degradation and chlorophenol removal. *RSC Adv.* **2017**, *7*, 39011–39017. [CrossRef]
87. Wang, C.; Dong, D.; Wang, H.; Mueller, K.; Qin, Y.; Wang, H.; Wu, W. Metagenomic analysis of microbial consortia enriched from compost: New insights into the role of Actinobacteria in lignocellulose decomposition. *Biotechnol. Biofuels* **2016**, *9*, 22. [CrossRef]
88. Zhu, N.; Yang, J.; Ji, L.; Liu, J.; Yang, Y.; Yuan, H. Metagenomic and metaproteomic analyses of a corn stover-adapted microbial consortium EMSD5 reveal its taxonomic and enzymatic basis for degrading lignocellulose. *Biotechnol. Biofuels* **2016**, *9*, 243. [CrossRef]
89. Fu, Y.; Chen, L.; Zhang, W. Regulatory mechanisms related to biofuel tolerance in producing microbes. *J. Appl. Microbiol.* **2016**, *121*, 320–332. [CrossRef]
90. Dragosits, M.; Mattanovich, D. Adaptive laboratory evolution—Principles and applications for biotechnology. *Microb. Cell Fact.* **2013**, *12*, 64. [CrossRef] [PubMed]
91. Farwick, A.; Bruder, S.; Schadeweg, V.; Oreb, M.; Boles, E. Engineering of yeast hexose transporters to transport D-xylose without inhibition by D-glucose. *Proc. Natl. Acad. Sci. USA* **2014**, *111*, 5159–5164. [CrossRef] [PubMed]
92. dos Santos, S.C.; Sá-Correia, I. Yeast toxicogenomics: Lessons from a eukaryotic cell model and cell factory. *Curr. Opin. Biotechnol.* **2015**, *33*, 183–191. [CrossRef] [PubMed]
93. King, Z.A.; Lloyd, C.J.; Feist, A.M.; Palsson, B.O. Next-generation genome-scale models for metabolic engineering. *Curr. Opin. Biotechnol.* **2015**, *35*, 23–29. [CrossRef]
94. Nizami, A.S.; Rehan, M. Towards nanotechnology-based biofuel industry. *Biofuel Res. J.* **2018**, *5*, 798–799. [CrossRef]
95. Sekoai, P.T.; Ouma, C.N.M.; Du Preez, S.P.; Modisha, P.; Engelbrecht, N.; Bessarabov, D.G.; Ghimire, A. Application of nanoparticles in biofuels: An overview. *Fuel* **2019**, *237*, 380–397. [CrossRef]
96. Elumalai, P.V.; Nambiraj, M.; Parthasarathy, M.; Balasubramanian, D.; Hariharan, V.; Jayakar, J. Experimental investigation to reduce environmental pollutants using biofuel nano-water emulsion in thermal barrier coated engine. *Fuel* **2021**, *285*, 119200. [CrossRef]
97. Bassalo, M.C.; Liu, R.; Gill, R.T. Directed evolution and synthetic biology applications to microbial systems. *Curr. Opin. Biotechnol.* **2016**, *39*, 126–133. [CrossRef]
98. Ghim, C.-M.; Kim, T.; Mitchell, R.J.; Lee, S.K. Synthetic biology for biofuels: Building designer microbes from the scratch. *Biotechnol. Bioprocess Eng.* **2010**, *15*, 11–21. [CrossRef]
99. Sarkar, M.; Sarkar, B. How does an industry reduce waste and consumed energy within a multi-stage smart sustainable biofuel production system? *J. Clean. Prod.* **2020**, *262*, 121200. [CrossRef]
100. Patel, A.; Shah, A. Integrated lignocellulosic biorefinery: Gateway for production of second generation ethanol and value added products. *J. Bioresour. Bioprod.* **2021**, *6*, 108–128. [CrossRef]
101. Rosson, E.; Sgarbossa, P.; Pedrielli, F.; Mozzon, M.; Bertani, R. Bioliquids from raw waste animal fats: An alternative renewable energy source. *Biomass Convers. Biorefin.* **2021**, *11*, 1475–1490. [CrossRef]
102. Al Hatrooshi, A.S.; Eze, V.C.; Harvey, A.P. Production of biodiesel from waste shark liver oil for biofuel applications. *Renew. Energy* **2020**, *145*, 99–105. [CrossRef]

Article

Comparison of Water-Removal Efficiency of Molecular Sieves Vibrating by Rotary Shaking and Electromagnetic Stirring from Feedstock Oil for Biofuel Production

Cherng-Yuan Lin * and **Lei Ma**

Department of Marine Engineering, National Taiwan Ocean University, Keelung 202, Taiwan; awpcsawp@yahoo.com.tw
* Correspondence: Lin7108@ntou.edu.tw

Abstract: Adequate water-removal techniques are requisite to remain superior biofuel quality. The effects of vibrating types and operating time on the water-removal efficiency of molecular sieves were experimentally studied. Molecular sieves of 3 Å pore size own excellent hydrophilic characteristics and hardly absorb molecules other than water. Molecular sieves of 3 Å accompanied by two different vibrating types, rotary shaking and electromagnetic stirring, were used to remove initial water from the reactant mixture of feedstock oil in order to prevent excessive growth or breeding of microorganisms in the biofuel product. The physical structure of about 66% molecular sieves was significantly damaged due to shattered collision between the magnetic bar and molecular sieves during electromagnetic stirring for 1 h. The molecular sieves vibrated by the rotary shaker appeared to have relatively higher water-removal efficiency than those by the electromagnetic stirrer and by keeping the reactant mixture motionless by 6 and 5 wt.%, respectively. The structure of the molecular sieves vibrated by an electromagnetic stirrer and thereafter being dehydrated appeared much more irregular and damaged, and the weight loss accounted for as high as 19 wt.%. In contrast, the structure of the molecular sieves vibrated by a rotary shaker almost remained original ball-shaped, and the weight loss was much less after regenerative treatment for those molecular sieves. As a consequence, the water-removal process using molecular sieves vibrated by the rotary shaker is considered a competitive method during the biofuel production reaction to achieve a superior quality of biofuels.

Keywords: molecular sieve; water removal; rotary shaking; electromagnetic stirring; biofuel

Citation: Lin, C.-Y.; Ma, L. Comparison of Water-Removal Efficiency of Molecular Sieves Vibrating by Rotary Shaking and Electromagnetic Stirring from Feedstock Oil for Biofuel Production. *Fermentation* **2021**, *7*, 132. https://doi.org/10.3390/fermentation7030132

Academic Editor: Alessia Tropea

Received: 14 June 2021
Accepted: 23 July 2021
Published: 26 July 2021

1. Introduction

The mesocarp of palm accounting for 45–60 wt.% of whole palm fruit is the major contribution of palm oil. Palm oil is primarily composed of free fatty acids (FFA) and triglycerides consisting of glycerol and three fatty acids. Palm oil is the largest feedstock oil provider for food, biochemical, and biofuels due to its highest ratio of oil yield/production area and lowest production cost among terrestrial plants [1]. The oil yield of palm oil is 5.5 t/ha in comparison with 0.5 t/ha and 2 t/ha of soybean and rapeseed oils, respectively. The production cost of palm oil is 300 USD/t, which is much lower than 700 and 800 USD/t for rapeseed and soybean oils [2]. Global palm oil consumption is increasing rapidly from 61.6 million tons in 2016 to 71.5 million tons in 2020 [3]. Widespread use of various vegetable oils for biofuel production is considered one of the significant measures to mitigate the effects of greenhouse gases and acute climate change [4].

The preservation extent of vegetable oil quality partly depends on the water concentration in the oil. The existence of water in the palm oil might facilitate the oxidation rate, leading to the deterioration of oxidation stability and increase in peroxidation value. Further, fatty acids are hydrolyzed by water in lipids to form free fatty acids, which react with the alkaline compounds to cause saponification and produce more water in the compounds [5]. The rate of hydrolysis reaction is further accelerated with higher water concentration [6], leading to the deterioration of biofuel quality.

Water in vegetable oil frequently causes metallic corrosion of the oil storage tank and might even break down the oil feeding system, including the high-pressure injection pump. The water in vegetable oil also facilitates the growth and breeding of microorganisms such as bacteria and fungi ascribed to accelerating biodegradability characteristics of vegetable oils, resulting in worsening biofuel properties and damage to the flavors and nutrients in the food oils. Water content in biofuel might be increased with its intensity of microbial activity [7]. Hence, the lowest possible water content in biofuels is suggested [8].

Sebastian et al. [8] observed that the presence of water and free fatty acids in tallow feedstock oil inhibited the extent of transesterification reaction. Further, Ma and Hanna [9] suggested that the water content, acid value, and free fatty acids (FFA) are less than 0.06 wt.%, 1 mg KOH/g, and 0.5 wt.%, respectively, to prevent saponification reaction. Eze et al. [10] found that the conversion rate of transesterification was only 50 wt.% when the raw oil contained 5.6 wt.% FFA and 0.2 wt.% water. Hence, the separation of water from feedstock oil is considered a requisite to improve biofuel properties [11]. The frequently applied techniques of separating a liquid from adhered liquid include electrolytic dissociation, extractive distillation [12], vapor permeation [13], ion exchange adsorption [14], and absorbent selection [15].

Some water-separating processes involve heating liquid mixtures to reach their boiling points and thus require higher energy consumption. Here, the output energy to input energy ratio of the whole production process may be less than 1 and, therefore, is less economical in terms of energy requirement. The membrane separation method, which requires high vacuum pressure at the permeation side of the mixture and regular cleaning of its membrane structure, may not be an adequate choice for water separation during biofuel production. Distillation by heating is the most often applied water-removal method in the current transesterification with a strong alkaline catalyst [16]. Nevertheless, the fuel properties of biofuel products might deteriorate with continuous heating under high temperatures. Continuous or intermittent vacuum distillation was carried out by Lawrence and Jiang [17] for separating water from biofuel. The vacuum pressure and operating temperature were controlled under 2.6 kPa and 105~110 °C. Likewise, some other more efficient water-removal processes continue to be proposed and investigated. For example, aluminum silica materials have many tiny pores on the surface for absorbing compounds that are smaller than those pores' sizes, and those compounds can be stored within them [18]. The volume of the pore material may not change obviously with the stored compounds from the surrounding. The function of the pore material after absorbing water could be regenerated by heating under a high temperature and low humidity environment.

A few materials have been applied to those water-removal techniques, such as activated carbon, silica gel, activated aluminum oxide, and molecular sieve. Khalil et al. [19] used activated carbon to remove nitrate and phosphate from water. Hybrid membranes produced from activated carbon and whey protein fibrils were used to remove mercury and chromium in water [20]. In addition, solid silica gel (SiO_2) was applied to enhance the oxidative transformation of caffeine in water [21]. Among those, a molecular sieve composed of alkaline metal aluminosilicate is a constructive crystalline material possessing precise uni-pore size. A molecular sieve with a pore diameter less than 2 nm (i.e., 20 Å) is referred to as microporous material. Molecular sieves are frequently used as desiccants or catalytic applications, including fluid catalytic cracking or hydrocracking [22]. In particular, molecular sieves of precise pore size can be used to absorb materials of the corresponding size. Only sufficiently smaller compounds than the pore size can enter these pores and be adsorbed by the molecular sieves. Chemical compounds whose molecular diameter is larger than the corresponding pore size of the molecular sieves are hardly adsorbed [23]. Further, a molecular sieve has extremely high hydrophilic characteristics so that even under an ultra-low humid environment, its superior water absorption and storage properties are retained [24]. The characteristics of the molecular sieves also include excellent crushing resistance, convenient regeneration capability, and swift adsorption rate [25]. The pore sizes of commercialized molecular sieves are generally in the range of 3 and 10 Å. Zeolite molecular sieve beds of 4 Å pore size were observed to successfully absorb water from humid natural gas [26]. The carbon molecular sieve membrane was evaluated to be a highly cost-effective method for separating CH_4 from CO_2 to upgrade biogas characteristics [27]. A total of 4 Å zeolite molecular sieves synthesized from attapulgite were also applied to remove hydrogen sulfide (H_2S) in gaseous fuels. The removal rate of H_2S might reach nearly 100% [28]. The molecular sieve of 3 Å pore size has excellent water-absorption capacity and hardly absorbs molecules other than water [29]. Highly hygroscopic material such as molecular sieve might absorb a large amount of water several times of its own weight, while it is difficult to extrude out the absorbed water from the material.

The long-chained molecules of high-hygroscopicity material appear as a twisted and curled structure before they absorb water. After water absorption, the twisted structure of the molecular sieve swiftly expands to become cross-linked cubes full of water molecules [30]. Water content in feedstock oil might retard the transesterification reaction, leading to deteriorated fuel characteristics of the biofuel product [31]. The water-removal process is hence a necessity to enhance the reaction process for biofuel production. Although high water-absorbing materials have been widely applied in industrial practices, no study used those materials for water absorption during transesterification from feedstock oil or alcohol. Moreover, there is no study reporting the investigation of the effects of highly hygroscopic material accompanied by vibrating motion on the extent of water removal from raw oil or alcohol during transesterification. Hence, the effects of vibration modes and operating time of molecular sieves on the fraction of water removal from palm oil and ethanol and the extent of structural damage of the water-absorbing material after the process were experimentally investigated in this study. Molecular sieves accompanied by two different kinds of vibrating motions, including electromagnetic stirring and rotary shaking, were used to absorb water from the reactant mixture of transesterification. The efficiency and rate of water removal and appearance and weight loss of molecular sieves were observed and compared to find the optimum water-removal technique and thus improve the fuel characteristics of biofuel.

2. Materials and Methods

2.1. Experimental Materials

Molecular sieves were used to absorb water from palm oil or ethanol, which are the reactants of the transesterification for biofuel production. Palm oil was procured from Formosa Oilseed Processing Co., Ltd. in Taichung City, Taiwan. Molecular sieves of 3 Å pore size were provided by Eikme International Ltd. in Hsinchu County, Taiwan. The water-absorption capability of 3 Å pore size has been found to be superior among those commercialized molecular sieves of pore sizes in the range of 3 to 10 Å. One gram of the 3 Å molecular sieve is able to absorb as high as 0.83 g water [32]. The bulk density of the molecular sieve is 0.63 g/mL. The molecular sieves of 3 Å are also used for the desiccation of petroleum-cracking alkenes and gas [33].

2.2. Experimental Methods

Palm oil or ethanol has the disposition to absorb moisture from the environment [34]. Five grams of palm oil or ethanol was poured into a vial and sealed. The vial was placed within a constant-temperature tank (Model LE-509D, Yih Der Ltd. in Taichung City, Taiwan) at 25 °C. An internal circulating fan was operated to keep an even temperature distribution in the tank. The vial was vibrated by a rotary shaker (Model TS-520D, Yih Der Ltd. in Taichung City, Taiwan) within the tank for 5 min to mix the compound with 0.09 wt.% initial water uniformly in the vial. The vibration frequency of the rotary shaker was set at 140 rpm. A volumetric Karl Fischer Titrator (Model DL-31, Mettler Toledo Ltd. in Greifensee, Switzerland) was used to measure the water content of palm oil or ethanol stored in the vial. The experimental procedures are illustrated in Figure 1.

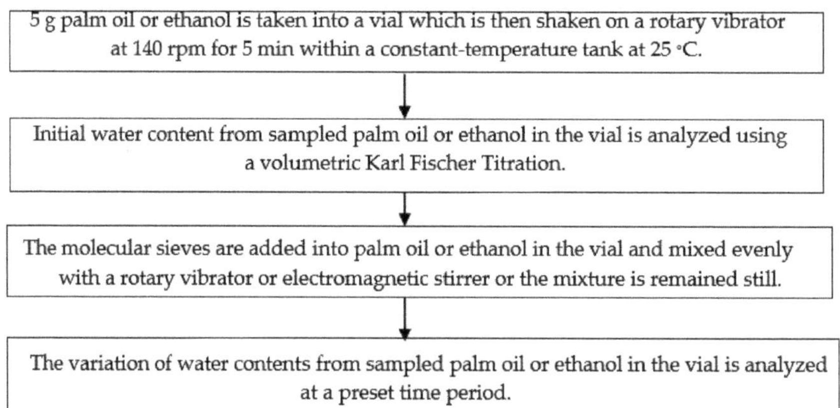

Figure 1. The experimental procedures for removing water from palm oil or ethanol by molecular sieves under different vibrating modes.

Molecular sieves of equal weight (5 g) to ethanol or palm oil were then added into the vial containing palm oil or ethanol. The vial at a constant temperature of 25 °C was then vibrated either by an electromagnetic stirrer or a rotary shaker for 1 h. The vial containing the oil or ethanol sample after removing water using the molecular sieves was then moved out from the constant-temperature tank to measure their water contents by a volumetric Karl Fischer Titration. After absorbing water from palm oil or ethanol under various operating conditions, the molecular sieves were collected and heated in a high-temperature furnace (Model DF 404, Deng Yng Ltd. in New Taipei City, Taiwan) at a constant temperature of 300 °C for 5 h for their dehydration and drying. The photographs of molecular sieves before and after dehydration processes were taken for comparison. The weights of the molecular sieves before and after dehydration were measured by a precise electronic balance (Model 210 g, Mettler Toledo Ltd. in Greifensee, Switzerland) to calculate weight loss of the molecular sieves during the water-removal process from palm oil or ethanol under different vibration modes.

At every consecutive 0.5 h of operating time period from 0 to 6 h for comparing the water-removal efficiencies between vibrating motions of electromagnetic stirring and rotary shaking, the water contents in the mixture of molecular sieves and palm oil or ethanol in the vial before and after the water-absorption processes were measured by a volumetric Karl Fischer Titration. When the water-removal efficiencies between ethanol and palm oil were compared, the water contents before and after the water-removal process at every consecutive 5 min of operating time from 0 to 60 min were measured. The water-removal efficiency from palm oil or ethanol by molecular sieves under rotary shaking and electromagnetic stirring was defined as shown in Equation (1):

$$\text{Water-removal efficiency (\%)} = (\text{water content prior to water removal (wt.\%)}) - \text{water content after water removal (wt.\%)})/(\text{water content prior to water removal (wt.\%)}) \times 100\% \quad (1)$$

The results of water-removal efficiencies between different vibrating motions or absorbents were thus be calculated and plotted. The water-removal rate, defined as the amount of water-removal per unit time, was also calculated for evaluating the effectiveness of potential water-removal techniques. The rate of water removal is formulated as shown in Equation (2):

$$\text{Rate of water-removal (wt.\%/min)} = (\text{water content prior to water removal (wt.\%)}) - \text{water content after water removal (wt.\%)})/\text{water-removal time (min)} \quad (2)$$

3. Results and Discussion

The effects of vibration modes and operating time on the water-removal efficiency were experimentally investigated in this study. At least three repetitions were carried out to obtain the mean values of the experimental data. The experimental uncertainties of the water-removal efficiency by electromagnetic stirring, rotary shaking, and being kept motionless were ±2.72%, ±4.36%, and ±1.84%, respectively; the uncertainty of the water-removal rate was ±3.35%. The experimental results were explained and discussed as follows.

3.1. Effect of Vibration Modes on Water-Removal Efficiency

The sample oil, molecular sieves, and a magnetic bar were kept in a vial to be vibrated by an electromagnetic stirrer. The magnetic bar began stirring the mixture of the sample oil and molecular sieves after triggering the electromagnetic stirrer. However, the magnetic bar was prone to collide with the molecular sieves, causing structural damage to the molecular sieves during the vibration of the vial. Part of the molecular sieves was even shattered to become powder after the vial's vibration, leading to fast deterioration of the water-absorption capability of the molecular sieves. The appearance of molecular sieves before and after the water-removal process vibrated by an electromagnetic stirrer is shown in Figure 2. Around 66% of the molecular sieves were subjected to structural damage ascribed to collision and friction between the molecular sieves and magnetic bar during electromagnetic stirring for 1 h. This caused the scaling-off of the surface material from the molecular sieves to mix with surrounding palm oil, resulting in a turbid palm oil liquid mixed with molecular sieve material. In contrast, under rotary shaking, most of the molecular sieves, after absorbing water content from palm oil or ethanol, appeared to be in original shape, with only a slight scale-off from the molecular sieves' surface. The comparative appearance of molecular sieves before and after water-removal processes under the vibration effect of the rotary shaker is shown in Figure 3. It was shown that after water removal by molecular sieves, palm oil appeared as only mildly turbid because of little damage to the molecular sieves by rotary shaking, and the shapes of molecular sieves remained almost integrated. Almost 100% structure of the molecular sieves after water-removal process by rotary shaking for 1 h remained but suffered only slight collision. Herold and Mokhatab [35] found that zeolite molecular sieves of 3 Å pore size are capable of selectively removing water from liquid aliphatic alcohol. The vibrating motion could help molecular sieves in promoting their rate of water adsorption from aliphatic alcohol.

(a) Before water absorption (b) After water absorption

Figure 2. A comparison of the appearance of molecular sieves (**a**) before water absorption and (**b**) after water absorption from palm oil accompanied by electromagnetic stirring.

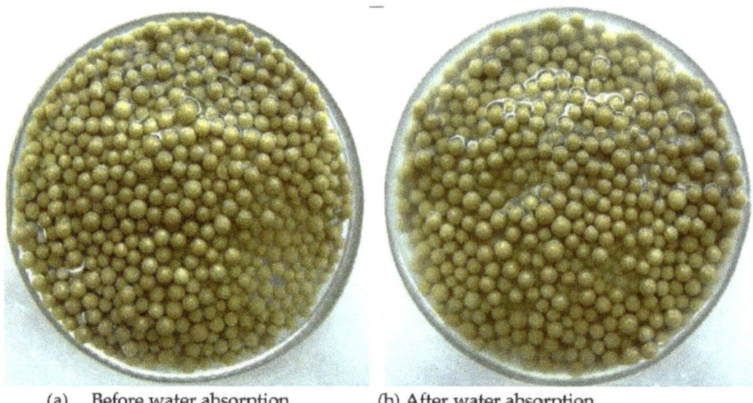

Figure 3. Comparison of the appearance of molecular sieves (**a**) before water absorption and (**b**) after water absorption from palm oil accompanied by rotary shaking.

The water-removal efficiency, as defined in Equation (1), was found to decrease with the operating time due to the broken structure and thus partial water-adsorption function loss of the molecular sieves after 3 h of absorbing water from ethanol accompanied by electromagnetic stirring in Figure 4. After 6 h of water absorption, part of the water absorbed by the molecular sieves accompanied by electromagnetic stirring released back to ethanol liquid, indicating that its water-removal efficiency decreased after 3 h of water absorption, as shown in Figure 4. In contrast, the water-removal efficiency steadily increased after 30 min of fast water absorption from ethanol by molecular sieves accompanied by the vibration effect of rotary shaking (Figure 4). The water-removal efficiency of the molecular sieves vibrated by rotary shaking for 6 h was observed to be higher than that of electromagnetic stirring by 6% for the same time, as shown in Figure 4. Molecular sieves are considered more efficient than other desiccants such as activated charcoal, silica gel, and alumina for dehydration in that they are small-pore adsorbents and with less possibility of co-adsorption of hydrocarbons [36]. Lad and Makkawi [37], after comparing the adsorption efficiency of methyl chloride among activated carbon, silica, and molecular sieves, found that the adsorption capability increased with the increase in adsorbent surface area. The molecular sieves vibrated by rotary shaking were found to have superior water-absorption efficiency due to its less damage by collision and higher complete adsorption surface area in turn in comparison with that by electromagnetic stirring.

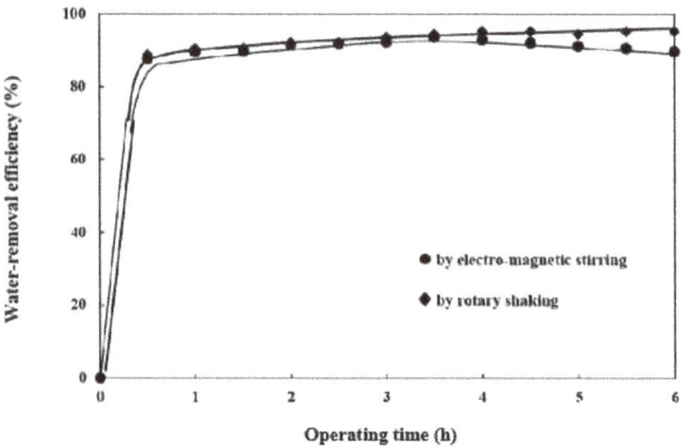

Figure 4. Effects of vibrating motion on water-removal efficiency of molecular sieves from ethanol.

The molecular sieves, after 6 h-water absorption under two different vibrating modes, were dehydrated, and their appearance was compared (Figure 5). The surface shapes of the dehydrated molecular sieves under electromagnetic stirring in Figure 5a appeared to be more irregular and broken than those under rotary shaking, as shown in Figure 5b, which were considerably glossier and mostly retained their original ball-shapes. The extent of damage caused to molecular sieves varied under different vibrating modes. When electromagnetic stirring was used to vibrate the mixture of molecular sieves and ethanol, a magnetic bar stirred the mixture under the effect of an electromagnetic field. Molecular sieves, being less hard than the magnetic bar, were prone to apparent damage due to the collision of the magnetic bar. In contrast, under rotary shaking action, direct collision among the molecular sieves with the same structure hardness rendered relatively less damage to the molecular sieves. The variation of total weight percentages of molecular sieves before and after water absorption under different vibrating modes are presented in Figure 6. Jemil et al. [38] found that increasing concentration of molecular sieves caused the gradual reduction in water in the reaction system. The vibration motion could decrease the concentration of molecular sieves required to reduce the same amount of water in the reaction system. Particularly, the molecular sieves suffered significantly larger weight loss under electromagnetic stirring treatment than rotary shaking action. The weight loss of molecular sieves under electromagnetic stirring accounted for as high as 19 wt.%, compared with less than 2 wt.% by rotary shaking after 6 h of water absorption. This indicated nearly 10-times higher weight loss after electromagnetic stirring than that after rotary shaking. This considerable weight loss of molecular sieves under electromagnetic stirring can also be judged by their indented and shrank surfaces, as shown in Figure 5a. Wang et al. [39] studied the effects of electromagnetic stirring (EMS) on mechanical properties and microstructure of Incoloy825 superalloy. They found that both the ultimate tensile strength and elongation rate of the superalloy increased with the application of electromagnetic stirring.

Based on the above-mentioned results, rotary shaking is suggested to be a more efficient vibrating mode to remove water from feedstock oil mixture during transesterification reaction. Further, rotary shaking causes considerably less structural damage to the adsorbent molecular sieves compared with electromagnetic stirring. The rotary shaking motion could also enhance the extent of fluidity [40] between the oil sample and adsorbent through the vibration of the rotary platform. The mixing extent of various reactants could also be raised by rotary shaking under an adequate rotary speed to facilitate a chemical reaction [41].

(a) Electromagnetic stirring (b) Rotary shaking

Figure 5. Surface of dehydrated molecular sieves after water-removal processes vibrated by (**a**) electromagnetic stirring and (**b**) rotary shaking.

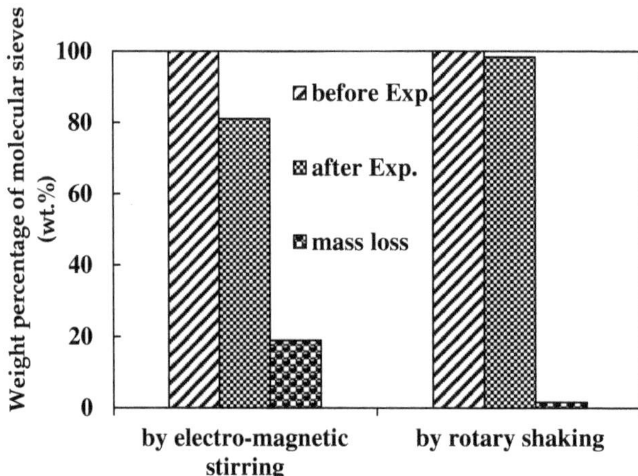

Figure 6. Comparison of weight percentages of molecular sieves before and after water-removal experiments vibrated by electromagnetic stirring and rotary shaking.

3.2. *Effects of Rotary Shaking and Motionless Treatment on Water Absorbency*

Rotary shaking was shown to have significantly higher water-removal efficiency than keeping the mixture of oil sample and molecular sieves motionless, as shown in Figure 7. After 30 min of water absorption from ethanol, the water-removal efficiency of rotary shaking was higher than that of the motionless mixture by 5%. The water-removal efficiencies of those two cases appeared to slightly increase with the operating time after 1 h of operation. This is attributed to the increase in the contact area and frequency of interaction between molecular sieves and ethanol under the rotary shaking motion at 140 rpm speed, resulting in superior water-absorption capacity of molecular sieves than keeping the mixture motionless. Tiadi et al. [42] found that higher adsorbent dosage, agitation speed, and contact time caused faster adsorption rates after comparing three different adsorbents. The increase in rotary shaking speed and reactant temperature was also observed to facilitate the activity of the chemical reaction and catalyst [43]. Raviadaran et al. [44] suggested that the increase in the absorption rate of contaminants by molecular sieves may be ascribed to more available adsorption sites provided by a higher adsorbent dosage or increased contact frequency between adsorbent and contaminants. Hence, the molecular sieves vibrated by rotary shaking were observed to remove more water from ethanol compared to other keeping motionless processes.

3.3. *Effects of Water-Absorption Time*

In the initial 5 min of water absorption by rotary shaking, the efficiency of water-removal from ethanol was considerably higher than that of palm oil (Figure 8). The water-removal efficiency of ethanol remained almost unchanged at nearly 85% after 10 min of water absorption. After 60 min of the absorption process, the water-removal efficiency of palm oil achieved that of ethanol, as shown in Figure 8. It is possible because ethanol is a polar solvent with low viscosity. Water molecules are thus prone to be absorbed and mixed with ethanol. The molecular structure of ethanol sustains its superior miscibility with both polar and nonpolar compounds. In particular, ethanol is rather hydroscopic to readily absorb water molecules from the environment [45]. Likewise, water can also be easily adsorbed out from ethanol by molecular sieves. In contrast, under rotary shaking, palm oil mixed with water tended to form a partial emulsion of water droplets-in-oil, resulting in the formation of many water droplets enveloped by the outer oil layer [46]. Palm oil is primarily composed of fatty acids, which are esterified with glycerol. The saturated fatty acids of palm oil could reach as high as 50 wt.% [47]. Hence, palm oil is considerably more viscous and heavier than ethanol, makes it relatively difficult to adsorb water out from the emulsion of palm oil by the molecular sieves. Hence, a milder increasing curve of water-removal efficiency with the operating time for palm oil appeared as, as observed in Figure 8.

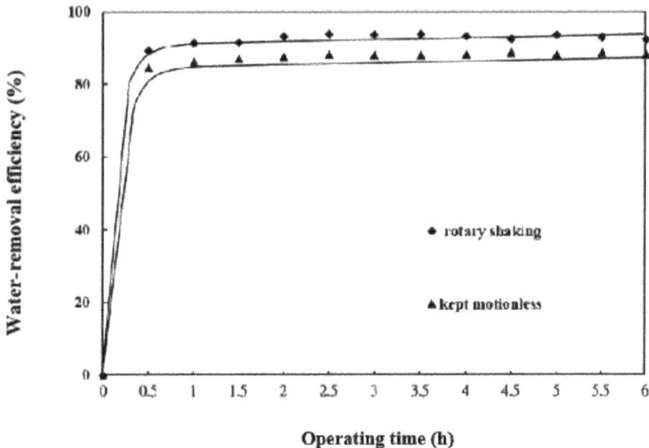

Figure 7. Comparison of water-removal efficiency of molecular sieves from ethanol under motionless and rotary shaking conditions.

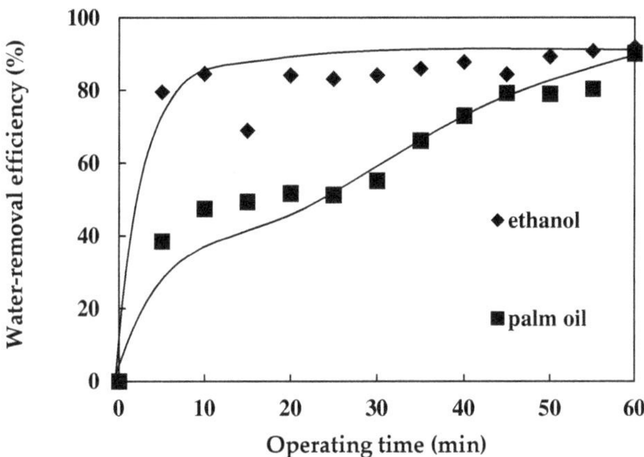

Figure 8. Comparison of the changes in water-removal efficiency with operating time between palm oil and ethanol by molecular sieves with rotary shaking.

Under rotary shaking motion, the water-removal efficiency of molecular sieves from ethanol soon reached 80% within 5 min of initiation. The water absorbency of molecular sieves for ethanol tended to reach saturation in the initial 10 min and thereafter appeared unvaried with the operating time while the water-removal efficiency for palm oil steadily increased and reached that of ethanol (at nearly 90%) after 60 min of operation. Water molecules were found to be more prone to be squeezed out from ethanol-water matrix under rotary shaking motion of the molecular sieves than from palm oil-water structure [48]. Superior water-absorbent materials such as molecular sieves provide high water-absorption function and excellent water retention performance [49]. Fluid and burning characteristics of biofuel products such as kinematic viscosity, cold filter plugging point (CFPP), and acid value are influenced by the water content in the reactant mixture [50]. The water removal rate in terms of wt.%/min as formulated in Equation (2) is defined as the amount of water removal (wt.%) per unit time (min). The water-removal rate of the molecular sieves for ethanol and palm oil under rotary shaking motion in the first 5 min was 1.005 and 0.476 wt.%/min, respectively. This indicates that the water-removal rate of ethanol was considerably faster than the palm oil by nearly two-fold. After 60 min of operation, the water-removal rate from both the analytes reached the same at 0.097 wt.%/min.

4. Conclusions

In this study, molecular sieves accompanied by two different vibration modes, including rotary shaking and electromagnetic stirring, were used to remove initial water from feedstock palm oil or ethanol. The water-removal efficiency and water-removal rate from ethanol or palm oil at various operating times were analyzed. The major results of this study are summarized as follows:

1. The magnetic bar was prone to collide with the molecular sieves in the vial to cause structural damage of the latter during electromagnetic stirring. The shape and structure of about 66% of molecular sieves were obviously damaged due to frequent collision and friction between the molecular sieves and magnetic bar under the effect of the electromagnetic field. It resulted in fast deterioration of water-absorption capability of the molecular sieves and a slight decrease in water-removal efficiency due to the release back of absorbed water from the molecular sieves;
2. The surface shapes of all the molecular sieves were nearly intact and only scaled off slightly from their surfaces after water removal from ethanol by rotary shaking motion for 6 h. The water-removing capability of the molecular sieves is almost sustained after the process accompanied by rotary shaking;
3. The water-removal efficiency of the molecular sieves vibrated by a rotary shaker was higher by 6% than that by an electromagnetic stirrer after 6 h water absorption from ethanol. The electromagnetic stirring motion caused an obvious loss of water-absorption capability of the molecular sieves due to severe structural damage during the water-removal process;
4. The shapes of molecular sieves were much more irregular and broken after being used for 6 h water-removal from ethanol under electromagnetic stirring than those by rotary shaking. In contrast, the molecular sieves under rotary shaking remained almost like original, ball-shaped, and much glossier. The extent of structural damage of the molecular sieves resulted in the accompanied loss of their water-adsorption capability;
5. The water-absorbing process by molecular sieves vibrated by electromagnetic stirring for 6 h caused significantly larger weight loss of the molecular sieves, which accounted for 19 wt.%, nearly 10 times, than that by rotary shaking, which was less than 2 wt.%. The rotary shaking motion is considered a much more adequate agitation method to increase contact frequency and area among the reactant mixtures of feedstock oil, water, and alcohol. This results in a higher reaction rate and faster water-removal efficiency;
6. The water-removal efficiency of molecular sieves vibrated by a rotary shaker is higher than that of the remaining motionless mixture of molecular sieves and ethanol by 5% after 30 min of the water-absorption process. The vibrating motion could facilitate the fluidity and mixing extents of the reactant mixture and thus accelerate the chemical reaction;
7. The water-removal efficiency from ethanol was considerably higher than that from palm oil by molecular sieves vibrated by a rotary shaker. The water-absorbing capability of the molecular sieves from ethanol reached saturation and steady-state after 10 min while the water-removal efficiency from palm oil increased with time and reached that of ethanol (90%) after 60 min of operation. Ethanol is highly hydroscopic and readily absorbs or desorbs water molecules than palm oil, composed of complex fatty acids and glycerol with much higher viscosity;

8. The water-removal rate of the molecular sieves from ethanol by rotary shaking motion in the first 5 min of the operation period was significantly higher and nearly twice that from palm oil in the same period of operation. The molecular structure of ethanol assures its superior miscibility with water molecules and higher water-removal rate.

Author Contributions: Conceptualization, C.-Y.L.; methodology, C.-Y.L.; validation, C.-Y.L.; formal analysis, C.-Y.L. and L.M.; investigation, C.-Y.L. and L.M.; resources, C.-Y.L.; data curation, L.M.; writing—original draft preparation, C.-Y.L. and L.M.; writing—review and editing, C.-Y.L.; visualization, L.M. supervision, C.-Y.L.; project administration, C.-Y.L.; funding acquisition, C.-Y.L. Both authors have read and agreed to the published version of the manuscript.

Funding: This research was funded by the Ministry of Science and Technology, Taiwan, ROC under grant number: MOST 109-2221-E-019-024 and 107-2221-E-019-056-MY2. The APC was funded by National Taiwan Ocean University, Taiwan, ROC.

Institutional Review Board Statement: Not applicable.

Informed Consent Statement: Not applicable.

Data Availability Statement: The data presented in this study are available in the article.

Acknowledgments: The authors are grateful for the financial support from the Ministry of Science and Technology, Taiwan, ROC.

Conflicts of Interest: The authors declare no conflict of interest.

References

1. Masudi, A.; Muraza, O. Vegetable oil to biolubricants: Review on advanced porous catalysts. *Energy Fuel* **2018**, *32*, 10295–10310. [CrossRef]
2. Zimmer, Y. Competitiveness of rapeseed, soybeans and palm oil. *J. Oilseed Brassica* **2016**, *1*, 84–90.
3. Statista Inc. Available online: https://www.statista.com/ (accessed on 11 May 2021).
4. Lin, C.Y.; Lu, C. Development perspectives of promising lignocellulose feedstocks for production of advanced generation biofuels: A review. *Renew. Sust. Energy Rev.* **2021**, *136*, 110445. [CrossRef]
5. Oliverira, E.D.C.; Silva, P.R.D.; Ramos, A.P.; Aranda, D.A.G.; Freire, D.M.G. Study of soybean oil hydrolysis catalyzed by Thermomyces Ianuginosus lipase and its application to biodiesel production via hydroesterification. *Enzym. Res.* **2011**, *2011*, 1–8. [CrossRef] [PubMed]
6. Porfyris, A.; Vasilakos, S.; Zotiadis, C.; Papaspyrides, C.; Moser, K.; Van Der Schueren, L.; Vouyiouka, S. Accelerated ageing and hydrolytic stabilization of poly (lactic acid) (PLA) under humidity and temperature conditioning. *Polym. Test.* **2018**, *68*, 315–332. [CrossRef]
7. Carrillo, L.; Bilason, L.; Loveres, S.F.; Paurillo, J.E.; Sacobo, C.J.; Sakay, C.; Espinosa, K.; Vergara, J. Utilization of vegetable oil refinery activated carbon-bleaching earth as an additive to the production of low-density facing bricks. *E&ES* **2020**, *463*, 012092.
8. Sebastian, J.; Muraleedharan, C.; Santhiagu, A. Enzyme catalyzed biodiesel production from rubber seed oil containing high free fatty acid. *Int. J. Green Energy* **2017**, *14*, 687–693. [CrossRef]
9. Ma, F.; Hanna, M.A. Biodiesel production: A review. *Bioresour. Technol.* **1999**, *70*, 1–15. [CrossRef]
10. Eze, V.C.; Phan, A.N.; Harvey, A.P. Intensified one-step biodiesel production from high water and free fatty acid waste cooking oils. *Fuel* **2018**, *220*, 567–574. [CrossRef]
11. Gil, I.D.; Uyazán, A.M.; Agular, J.L.; Rodríguez, G.; Caicedo, L.A. Separation of ethanol and water by extractive distillation with salt and solvent as entrainer: Process simulation. *Braz. J. Chem. Eng.* **2008**, *25*, 207–215. [CrossRef]
12. Bui, D.T.; Nida, A.; Ng, K.C.; Chua, K.J. Water vapor permeation and dehumidification performance of poly (vinyl alcohol)/lithium chloride composite membranes. *J. Membr. Sci.* **2016**, *498*, 254–262. [CrossRef]
13. Fang, Z.; He, C.; Li, Y.; Chung, K.H.; Xu, C.; Shi, Q. Fractionation and characterization of dissolved organic matter (DOM) in refinery wastewater by revised phase retention and ion-exchange adsorption solid phase extraction followed by ESI FT-ICR MS. *Talanta* **2017**, *162*, 466–473. [CrossRef] [PubMed]
14. Moreno, D.; Ferro, V.R.; De Riva, J.; Santiago, R.; Moya, C.; Larriba, M.; Palomar, J. Absorption refrigeration cycles based on ionic liquids: Refrigerant/absorbent selection by thermodynamic and process analysis. *Appl. Energy* **2018**, *213*, 179–194. [CrossRef]
15. Karmakar, A.; Karmakar, S.; Mukherjee, S. Properties of various plants and animals feedstocks for biodiesel production. *Bioresour. Technol.* **2010**, *101*, 7201–7210. [CrossRef] [PubMed]
16. Zhou, H.; Zuo, G. Discussion on the production technology of biodiesel reaching BD100 standard. *China Oil Fat* **2009**, *34*, 59–62.
17. Lawrence, M.; Jiang, Y. Porosity, pore size distribution, micro-structure. In *Bio-Aggregates Based Building Materials*; Springer: Dordrecht, The Netherlands, 2017; pp. 39–71.

18. Besnardiere, J.; Ma, B.; Torres-Pardo, A.; Wallez, G.; Kabbour, H.; González-Calbet, J.M.; Von Bardeleben, H.J.; Fleury, F.; Buissette, V.; Sanchez, C.; et al. Structure and electrochromism of two-dimensional octahedral molecular sieve h'-WO 3. *Nat. Commun.* **2019**, *10*, 1–9. [CrossRef]
19. Khalil, A.M.; Eljamal, O.; Amen, T.W.; Sugihara, Y.; Matsunaga, N. Optimized nano-scale zero-valent iron supported on treated activated carbon for enhanced nitrate and phosphate removal from water. *Chem. Eng. J.* **2017**, *309*, 349–365. [CrossRef]
20. Ramírez-Rodríguez, L.C.; Díaz Barrera, L.E.; Quintanilla-Carvajal, M.X.; Mendoza-Castillo, D.I.; Bonilla-Petriciolet, A.; Jiménez-Junca, C. Preparation of a hybrid membrane from whey protein fibrils and activated carbon to remove mercury and chromium from water. *Membranes* **2020**, *10*, 386. [CrossRef]
21. Manoli, K.; Nakhla, G.; Feng, M.; Sharma, V.K.; Ray, A.K. Silica gel-enhanced oxidation of caffeine by ferrate (VI). *Chem. Eng. J.* **2017**, *330*, 987–994. [CrossRef]
22. Li, X.; Chen, G.; Liu, C.; Ma, W.; Yan, B.; Zhang, J. Hydrodeoxygenation of lignin-derived bio-oil using molecular sieves supported metal catalysts: A critical review. *Renew. Sust. Energy Rev.* **2017**, *71*, 296–308. [CrossRef]
23. Atyaksheva, L.F.; Kasyanov, I.A.; Ivanova, I.I. Adsorptive Immobilization of proteins on mesoporous molecular sieves and zeolites. *Petrol. Chem.* **2019**, *59*, 327–337. [CrossRef]
24. Tasharrofi, S.; Golchoobi, A.; Fesahat, H.; Taghdisian, H.; Hosseinnia, A. Effects of water content on so2/n2 binary adsorption capacities of 13x and 5a molecular sieve, experiment, simulation, and modeling. *J. Petrol. Sci. Technol.* **2019**, *9*, 30–45.
25. Garcia, L.; Rodriguez, G.; Orjuela, A. Study of the pilot-scale pan granulation of zeolite-based molecular sieves. *Braz. J. Chem. Eng.* **2021**, *38*, 165–175. [CrossRef]
26. Santos, M.G.; Correia, L.M.; de Medeiros, J.L.; Ofélia de Queiroz, F.A. Natural gas dehydration by molecular sieve in offshore plants: Impact of increasing carbon dioxide content. *Energy Convers. Manag.* **2017**, *149*, 760–773. [CrossRef]
27. He, X.; Chu, Y.; Lindbråthen, A.; Hillestad, M.; Hägg, M.B. Carbon molecular sieve membranes for biogas upgrading: Techno-economic feasibility analysis. *J. Clean. Prod.* **2018**, *194*, 584–593. [CrossRef]
28. Liu, X.; Wang, R. Effective removal of hydrogen sulfide using 4A molecular sieve zeolite synthesized from attapulgite. *J. Hazard. Mater.* **2017**, *326*, 157–164. [CrossRef]
29. Xia, Z.; Ying, L.; Fang, J.; Du, Y.Y.; Zhang, W.M.; Guo, X.; Yin, J. Preparation of covalently cross-linked sulfonated polybenzimidazole membranes for vanadium redox flow battery applications. *J. Membr. Sci.* **2017**, *525*, 229–239. [CrossRef]
30. Lin, C.Y.; Ma, L. Influences of water content in feedstock oil on burning characteristics of fatty acid methyl esters. *Processes* **2020**, *8*, 1130. [CrossRef]
31. Pahl, C.; Pasel, C.; Luckas, M.; Bathen, D. Adsorptive water removal from organic solvents in the ppm-region. *Chem. Ing. Tech.* **2011**, *83*, 177–182. [CrossRef]
32. Xu, Y.M.; Tang, Y.P.; Chung, T.S.; Weber, M.; Maletzko, C. Polyarylether membranes for dehydration of ethanol and methanol via pervaporation. *Sep. Purif. Technol.* **2018**, *193*, 165–174. [CrossRef]
33. Salam, M.A.; Ahmed, K.; Hossain, T.; Habib, M.S.; Uddin, M.S.; Papri, N. Prospect of molecular sieves production using rice husk in bangladesh: A review. *Int. J. Chem. Math. Phys.* **2019**, *3*, 105–134. [CrossRef]
34. Gabruś, E.; Witkiewicz, K.; Nastaj, J. Modeling of regeneration stage of 3A and 4A zeolite molecular sieves in TSA process used for dewatering of aliphatic alcohols. *Chem. Eng. J.* **2018**, *337*, 416–427. [CrossRef]
35. Herold, R.H.M.; Mokhatab, S. Optimal Design and Operation of Molecular Sieve Gas Dehydration Units—Part 1. Gas Processing & LNG. 2017. Available online: http://www.gasprocessingnews.com/features/201708/optimal-design-and-operation-of-molecular-sieve-gas-dehydration-units%E2%80%94part-1.aspx (accessed on 25 July 2021).
36. Wang, H.; Jia, C.; Xia, X.; Karangwa, E.; Zhang, X. Enzymatic synthesis of phytosteryl lipoate and its antioxidant properties. *Food Chem.* **2018**, *240*, 736–742. [CrossRef]
37. Lad, J.B.; Makkawi, Y.T. Adsorption of methyl chloride on molecular sieves, silica gels, and activated carbon. *Chem. Eng. Technol.* **2020**, *43*, 436–446. [CrossRef]
38. Jemil, N.; Hmidet, N.; Ayed, H.B.; Nasri, M. Physicochemical characterization of Enterobacter cloacae C3 lipopeptides and their applications in enhancing diesel oil biodegradation. *Process. Saf. Environ.* **2018**, *117*, 399–407. [CrossRef]
39. Wang, F.; Wang, E.; Zhang, L.; Jia, P.; Wang, T. Influence of electromagnetic stirring (EMS) on the microstructure and mechanical property of Incoloy825 superalloy. *J. Manuf. Process.* **2017**, *26*, 364–371. [CrossRef]
40. Saremnia, B.; Esmaeili, A.; Sohrabi, M.R. Removal of total petroleum hydrocarbons from oil refinery waste using granulated NaA zeolite nanoparticles modified with hexadecyltrimethylammonium bromide. *Can. J. Chem.* **2016**, *94*, 163–169. [CrossRef]
41. Van Thuoc, D.; My, D.N.; Loan, T.T.; Sudesh, K. Utilization of waste fish oil and glycerol as carbon sources for polyhydroxyalkanoate production by Salinivibrio sp. M318. *Int. J. Biol. Macromol.* **2019**, *141*, 885–892. [CrossRef] [PubMed]
42. Tiadi, N.; Dash, R.R.; Mohanty, C.R.; Patel, A.M. Comparative studies of adsorption of chromium (VI) ions onto different industrial wastes. *J. Hazard Toxic Radioact. Waste* **2020**, *24*, 04020021. [CrossRef]
43. Dey, S.; Mehta, N.S. To optimized various parameters of Hopcalite catalysts in the synthetic processes for low temperature CO oxidation. *Appl. Energy Combust. Sci* **2021**, *6*, 100031.
44. Raviadaran, R.; Ng, M.H.; Manickam, S.; Chandran, D. Ultrasound-assisted water-in-palm oil nano-emulsion: Influence of polyglycerol polyricinoleate and NaCl on its stability. *Ultrason. Sonochem.* **2019**, *52*, 353–363. [CrossRef] [PubMed]
45. Davis, G.W. Addressing concerns related to the use of ethanol-blended fuels in marine vehicles. *J. Sustain. Dev. Energy Water Environ. Syst.* **2017**, *5*, 546–559. [CrossRef]

46. Anand, V.; Juvekar, V.A.; Thaokar, R.M. Coalescence, partial coalescence, and noncoalescence of an aqueous drop at an oil–water interface under an electric field. *Langmuir* **2020**, *36*, 6051–6060. [CrossRef] [PubMed]
47. Purnama, K.O.; Setyaningsih, D.; Hambali, E.; Taniwiryono, D. Processing, characteristics, and potential application of red palm oil—A review. *Int. J. Oil Palm* **2020**, *3*, 40–55. [CrossRef]
48. Zhu, Z.; Hu, D.; Liu, Y.; Xu, Y.; Zeng, G.; Wang, W.; Cui, F. Three-component mixed matrix organic/inorganic hybrid membranes for pervaporation separation of ethanol–water mixture. *J. Appl. Polym. Sci.* **2017**, *134*. [CrossRef]
49. Ai, F.; Yin, X.; Hu, R.; Ma, H.; Liu, W. Research into the super-absorbent polymers on agricultural water. *Agric. Water Manag.* **2021**, *245*, 106513. [CrossRef]
50. Lin, C.Y.; Ma, L. Fluid characteristics of biodiesel produced from palm oil with various initial water contents. *Processes* **2021**, *9*, 309. [CrossRef]

MDPI

Article

Enhanced Energy Recovery from Food Waste by Co-Production of Bioethanol and Biomethane Process

Teeraya Jarunglumlert [1], Akarasingh Bampenrat [1], Hussanai Sukkathanyawat [1] and Chattip Prommuak [2,*]

1 Faculty of Science, Energy and Environment, King Mongkut's University of Technology North Bangkok (Rayong Campus), Rayong 21120, Thailand; teeraya.j@sciee.kmutnb.ac.th (T.J.); akarasingh.b@sciee.kmutnb.ac.th (A.B.); hussanai.s@sciee.kmutnb.ac.th (H.S.)

2 Energy Research Institute, Chulalongkorn University, Bangkok 10330, Thailand

* Correspondence: chattip.p@chula.ac.th

Abstract: The primary objective of this research is to study ways to increase the potential of energy production from food waste by co-production of bioethanol and biomethane. In the first step, the food waste was hydrolysed with an enzyme at different concentrations. By increasing the concentration of enzyme, the amount of reducing sugar produced increased, reaching a maximum amount of 0.49 g/g food waste. After 120 h of fermentation with *Saccharomyces cerevisiae*, nearly all reducing sugars in the hydrolysate were converted to ethanol, yielding 0.43–0.50 g ethanol/g reducing sugar, or 84.3–99.6% of theoretical yield. The solid residue from fermentation was subsequently subjected to anaerobic digestion, allowing the production of biomethane, which reached a maximum yield of 264.53 ± 2.3 mL/g VS. This results in a gross energy output of 9.57 GJ, which is considered a nearly 58% increase in total energy obtained, compared to ethanol production alone. This study shows that food waste is a raw material with high energy production potential that could be further developed into a promising energy source. Not only does this benefit energy production, but it also lowers the cost of food waste disposal, reduces greenhouse gas emissions, and is a sustainable energy production approach.

Keywords: bioethanol; biomethane; food waste; co-production; biorefinery

Citation: Jarunglumlert, T.; Bampenrat, A.; Sukkathanyawat, H.; Prommuak, C. Enhanced Energy Recovery from Food Waste by Co-Production of Bioethanol and Biomethane Process. *Fermentation* **2021**, *7*, 265. https://doi.org/10.3390/fermentation7040265

Academic Editor: Alessia Tropea

Received: 28 October 2021
Accepted: 11 November 2021
Published: 16 November 2021

1. Introduction

With the outbreak of the COVID-19 virus since the beginning of 2020, the global economic growth rate has stalled and resulted in a sharp drop in energy demand. The International Energy Agency, IEA, forecasts that the severity of the epidemic will subside and economic growth will gradually recover in 2021, bringing back energy demand, which may increase by leaps and bounds to compensate for the contraction from the recent situation. In particular, despite the COVID-19 pandemic, demand for renewable energy continues to grow (approximately 3% in 2020) [1]. This implies that the world is becoming more conscious of the necessity of renewable energy use, and it pushes academics to investigate low-cost alternative and environmentally acceptable energy sources that contribute to sustainable development.

Bioethanol production from waste, such as organic fraction municipal waste, and agricultural waste, has consistently been one of the most popular alternative energy production pathways. In comparison to fossil fuels, bioethanol emits considerably lower greenhouse gases and thus receives widespread support as a vehicle fuel source. By mixing it in various proportions with gasoline, it transforms into gasohol, which can be used immediately in internal combustion engines without requiring further engine modifications.

Food waste (FW) is classified as a low-cost, high-potency second-generation feedstock due to its main constituents of biodegradable organic compounds (such as carbohydrates, proteins, and fats). It can be highly bioavailable for the production of various forms of bioenergy, such as biohydrogen [2,3], biogas [4,5], biodiesel [6–8], biobutanol [9], andbiohythane [10,11], as well as ethanol. Furthermore, converting FW into energy is a solution to the environmental crisis caused by the current amount of FW, which is steadily increasing as the economy and population grow. Globally, 931 million tons of FW were produced in 2019, with approximately 30% of food produced being discarded as waste [12]. The capacity to dispose of this FW is significantly less than the rate of production, resulting in a municipal waste overflow problem. FW contains a high level of moisture; therefore, it is difficult to transport, consumes more fuel when incinerated, and produces wastewater, odors, and greenhouse gases when disposed of in a landfill. As a result, eliminating FW is more challenging and more costly than other types of waste.

Many studies on the production of ethanol from FW and other organic waste have been conducted in a variety of areas. In particular, the effects of the composition of FW, effects of pretreatment prior to the fermentation process, types of enzymes, types of microorganisms used in ethanol production, and suitable operational conditions were investigated with the main goal of increasing ethanol productivity [13–19]. Even so, commercial production of bioethanol from second-generation feedstocks is being questioned for its cost-effectiveness in terms of both economic feasibility and energy efficiency. Previous studies have reported that the cost-effectiveness of ethanol production from second-generation biomass has a very low Energy Return on Investment (EROI) (approximately 0.8–1.6) compared to fossil fuels (18–45) [20]. As a result, there has been an increase in a number of studies conducted on the process of co-production in which multiple products are produced. In particular, the co-production process in this context refers to the production of other fuels or valuable substances as a by-product of ethanol production, which can be sold to increase revenue or used as fuel to reduce energy costs in ethanol production. As a result, the economic competitiveness of second-generation bioethanol production is enhanced [21]. In this regard, previous research has demonstrated the production of bioethanol coupled with biogas from red oak [22], sugarcane bagasse [23], wheat straw [24], corn stover [25], spruce wood [26], and switchgrass [27]. However, research on co-production processes that utilize FW as a feedstock remains limited.

This research examines the production of bioethanol and biomethane from FW, with an emphasis on a simple approach in which the smallest amounts of chemicals and energy are used. This began with a mechanical pretreatment of FW to reduce their size without the use of heat or chemicals. After hydrolysis with enzyme, the liquid fraction of the hydrolysate was separated to produce ethanol, while the solid fraction was subjected to anaerobic digestion to produce biogas (Scenario 1). The energy yield was then compared to that obtained by first producing ethanol from the entire fraction of hydrolysate and then biogas (Scenario 2). Eventually, a production scenario that achieved the highest productivity and gross energy output was suggested.

2. Materials and Methods

2.1. Raw Material

FW used in this research was collected from the cafeteria of King Mongkut's University of Technology North Bangkok, Rayong campus. To mitigate variability due to differences in starting raw material, FW was collected continuously for two weeks, allowing for the use of a single lot of samples throughout the trial. After, non-biodegradable components such as fishbones, chicken bones, packaging fragments, and so forth were removed. The remainder of the sample was reduced in size using a food processor and then stored at −20 °C for further use. The total solids (TS) and volatile solids (VS) of FW were determined using a method proposed by Sluiter et al. (2008) [28].

2.2. Enzymatic Hydrolysis

In the enzymatic hydrolysis stage, FW was digested using α-amylase from *Aspergillus oryzae* with a specific activity of 30 U/mg (Sigma–Aldrich, St. Louis, MI, USA) in 500-mL Erlenmeyer flasks with 200 mL working volumes. The concentration of the original FW was 10% (*w*/*v*). To determine the optimal enzyme concentration and duration of hydrolysis that led to the highest reducing sugar (RS) content, the enzyme concentration was varied at 1, 3, 4, and 5% *w*/*w* (g enzyme/g dry FW), and hydrolysis was carried out for 1, 2, 3, 4, 5, and 9 h at 60 °C with a stirring rate of 150 rpm. As the pH of the FW was in the optimal range for enzymatic activity (4.0–6.5), pH adjustment was negligible.

2.3. Bioethanol and Biomethane Production

The main objective of this research is to study the energy production methods from FW that provide the highest gross energy output from bioethanol and biomethane production. Following the hydrolysis, the experiment is divided into two scenarios (Figure 1):

- Scenario 1: In the ethanol fermentation, only the liquid fractions from the hydrolysis were used, and the solid residues were separated for use in the production of biomethane.
- Scenario 2: The entire hydrolysate was used in ethanol fermentation, followed by the extraction of fermented solid residues for additional anaerobic digestion.

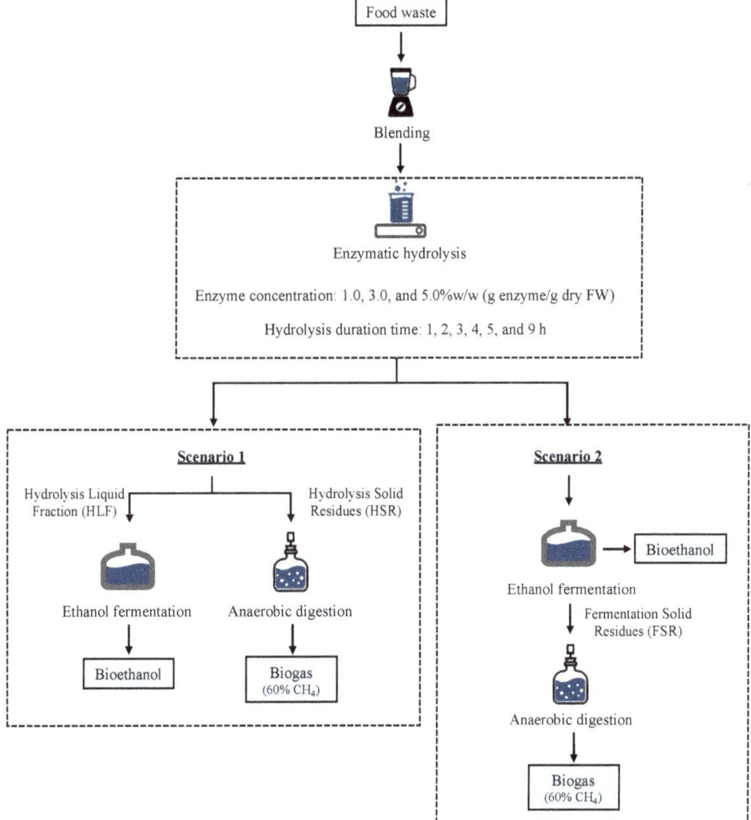

Figure 1. Schematic diagram of two process configurations for co-production of bioethanol and biomethane from FW.

The process of producing bioethanol began with the preparation of yeast. *Saccharomyces cerevisiae* (commercial dry baker's yeast purchased at a local store) was dissolved in sterile water at a concentration of 10 g/L and stored at 4 °C prior to use without cultivation [29]. The prepared solution was placed in 500-mL Erlenmeyer flasks (with working volumes of 200 mL), the pH was adjusted to 5.5, and *S. cerevisiae* yeast solution was added at the concentration of 5.0, 10.0, and 15.0% (v/v). Fermentation took place at 35 °C for 120 h with a stirring rate of 150 rpm. At the end of fermentation, the liquid was separated by filtration. The clear solution was then analysed for ethanol content by gas chromatography (Flame Ionization (FID), Bruker Scion 456-GC) following the method by Cutaia (1984) [30].

In the anaerobic digestion process to produce biomethane, the experiment was carried out in batches using 5 L volumetric flasks with working volumes of 2 L. The substrate was prepared by mixing the fermentation solid residues (FSR), or hydrolysis solid residues (HSR) with 500 mL DI water to a concentration of 20% (w/v). Then, inoculum (biogas digester sludge from swine farm wastewater) was added at a substrate to inoculum ratio of 1:3, followed by a 5-min purge with N_2 before closing the lid to allow anaerobic conditions. During the 80-day digestion at 37 °C, the produced biomethane was measured with a mass flow meter (F-111B-100-RAD-22-K, Bronkhorst, The Netherlands) and the biogas composition was determined using a biogas analyser (Biogas 5000, Geotech, England).

3. Results and Discussion

3.1. Food Waste Characteristic

The properties of the raw materials to be fed into the production process are essential prerequisites in determining the optimum conditions and estimating the expected yield. The composition of FW varies according to eating culture. Asian foods are similar in that they are primarily composed of starch. The FW used in the study was collected from university cafeterias, almost all of which sell local Asian food based on starchy ingredients such as rice and noodles, which contributed up to 51.03% of the total carbon content, 2.11% of nitrogen, and a very high moisture content of 89.01% (Table 1). This composition is similar to FW collected from canteens in Korea [31–33], China [34,35], and Japanese households [36], with total carbon content ranging between 40% and 54% (dry basis) and moisture content ranging between 70% and 90%. According to recent research, the amount of ethanol produced is proportional to the carbon content of the raw material. That is, the higher the carbon content of the raw material, the greater the productivity, and the higher the moisture content, the less water is required during the process [37]. Additionally, the FW collected for this study possessed a VS of 10.59% and an ash content of 0.4%. The VS and ash contents could be used to estimate the amount of biofuel to be produced. This high VS and low ash raw material demonstrate the presence of a significant amount of organic matter that microorganisms can consume during the biological process. Based on the initial composition, it was certain that the collected FW was a potential feedstock for biofuel production due to the abundant nutrients required by microbes to convert to valuable chemicals, particularly bioethanol and biogas.

Table 1. Characteristics of the raw material.

Component of FW	Fraction
Moisture content (%)	89.01 ± 0.61
Total solids [1], TS (%)	10.99 ± 0.61
Volatile solids [1], VS (%)	10.59 ± 0.58
Ash [1] (%)	0.40 ± 0.03
Total carbon [2] (%)	51.03 ± 0.75
Total nitrogen [2] (%)	2.11 ± 0.21

[1] Calculated on wet basis. [2] Calculated on dry basis.

3.2. Enzymatic Hydrolysis

Hydrolysis is a process that converts carbohydrates in raw materials into sugars, which are then used as food by microorganisms to produce ethanol in the following step. Thus, this first step is critical as it determines the overall efficiency of the biofuel production process. In past studies of bioethanol production from FW, thermochemical pretreatment with acid or base, followed by enzymatic hydrolysis, is often used to ensure the complete conversion of carbohydrates to sugars. However, research indicates that severe conditions are not always beneficial. For instance, high-temperature treatments cause sugars to undergo partial degradation [38] and the formation of microbial inhibitors can occur as a result of acid treatments [39]. Enzymatic hydrolysis is thus the most widely used method as it performs under mild conditions, does not produce inhibitors that interfere with the subsequent fermentation [40], and allows for fully biological processes. In this study, FW was hydrolysed with α-amylase without pretreatment other than size reduction. The amount of RS obtained from FW degradation by the α-amylase enzyme is depicted in Figure 2, which clearly indicates the increase in RS yield with the increasing enzyme dosage and the duration of the hydrolysis. After 9 h of hydrolysis, the highest RS yields were 17.90, 27.08, and 49.45 g/L, corresponding to 0.18, 0.27, and 0.49 g/g FW, when 1.0%, 3.0%, and 5.0% of enzyme was introduced, respectively. This range of RS yields was similar to that reported by Kim et al. (2011) [41], who obtained RS yields of 0.436 and 0.627 g/g TS from FW digestion with glucoamylase and carbohydrase, respectively, and by Han et al. (2020) [42], who obtained 0.784 g RS/g FW from waste hamburger hydrolysis with α-amylase. It is noted that the amount of RS produced rapidly increased during the initial stages of hydrolysis and gradually increased over time until it reached a stable state. This is consistent with the findings of Han et al. (2019) [43], who found a slight increase in the amount of RS obtained from waste cake after 80 min of hydrolysis with α-amylase. Likewise, a study conducted by Hong et al. (2011) [31] on the enzymatic hydrolysis of FW demonstrated that the glucose content remained stable after 10 h. Additionally, increasing the enzyme concentration resulted in an increase in glucose content, which reached its maximum at 600 g/kg FW when 4 mL of enzyme was introduced. However, in this study, the enzyme dose of 0.5 mL/100 g FW was chosen as the optimal condition for further processing. This was because increasing the enzyme dose from 0.5 mL to 1, 2, and 4 mL increased the amount of glucose obtained only slightly. In other words, almost all carbohydrate molecules had already been digested and thus increasing the amount of enzyme further from this point did not result in a significant increase in glucose. This is consistent with the findings of Kim et al. (2011) [41], who found that when the enzyme content was increased from 5 to 10%, the amount of glucose produced remained similar.

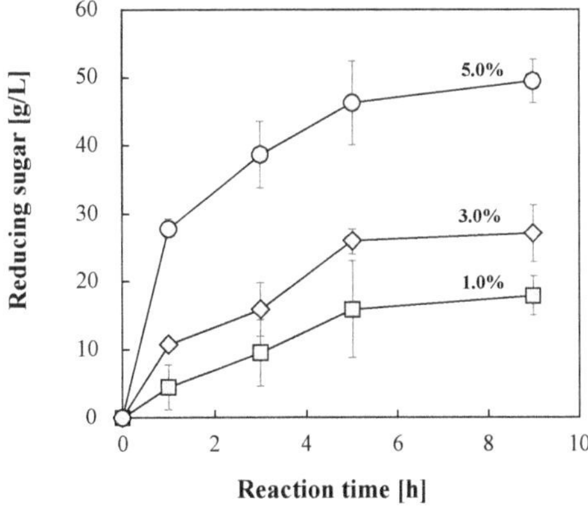

Figure 2. RS concentration from the enzymatic hydrolysis of FW.

3.3. Ethanol Production from Food Waste Hydrolysate

The hydrolysate was processed in two ways after being hydrolysed by the enzyme. Scenario 1 divided FW hydrolysate into two fractions: HLF for ethanol fermentation and HSR for biomethane production via anaerobic digestion. In Scenario 2, the hydrolysate was completely fermented to produce ethanol. After the ethanol was separated, the remainder was used to produce biomethane. In both scenarios, the fermentation was set to take place under the same conditions. That was, at 35 °C for 120 h with various concentrations of *S. cerevisiae*. It was found from the experiment that the amount of ethanol produced from HLF (Scenario 1) ranged from 21.26–23.24 g/L (Figure 3), which only slightly increased as the yeast content increased from 5 to 10 and 15% (*v*/*v*). This is because most of the RS had already been converted to ethanol, considered based on the theory where 100 g of glucose can be converted to 51 g of ethanol and 49 g of CO_2 by a fermentation process [44]. The amount of ethanol produced from HLF at yeast concentrations of 5, 10, and 15% (*v*/*v*) resulted in ethanol yields of 0.43, 0.46, and 0.45 g/g RS, accounting for 84.3, 90.2, and 88.2% of theoretical yield, respectively. Unlike Scenario 2, the ethanol produced from whole hydrolysate increased from 20.77 to 24.77, and 26.23 g/L as yeast dosage increased from 5 to 10, and 15% (*v*/*v*), respectively. The reason why the ethanol content produced from whole hydrolysate was higher than that produced from HLF may be due to the greater amount of RS as there was no loss during the filtration process as in Scenario 1, as well as remaining sugars in the solid fraction. According to recent ethanol production studies, FW ethanol yields range between 0.4 and 0.5 g/g RS [41,45,46]. Moreno et al. (2021) [47] reported that the ethanol yield from unpretreated organic fraction municipal waste was highest at 80% of the theoretical yield. Kiran et al. (2015) [46] produced 0.5 g/g glucose from waste cake ethanol, accounting for 98% of the theoretical yield after 32 h-fermentation. These are consistent with Han et al. (2019) [43], where waste cake was used as a substrate in ethanol production, yielding as high as 1.13 g ethanol/g RS as the waste cake was readily biodegradable and contained other constituents that promote yeast activity. When the amount of bioethanol produced from FW is considered, this research produced ethanol with a maximum yield of 0.22 and 0.25 g ethanol/g dry FW from the liquid fraction (Scenario 1) and from whole hydrolysate (Scenario 2), respectively. At this point, it can be seen that the yields of ethanol produced from untreated FW are comparable to those produced from pretreated FW. Considering the costs that can be saved by eliminating pretreatment or detoxification procedures, such as costs of chemicals, energy, and investment, this approach can bring the overall cost of ethanol production in line with conventional processes.

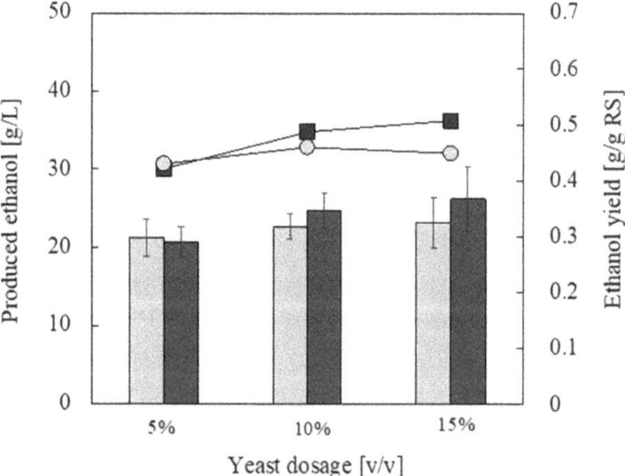

Figure 3. Ethanol production from Scenario 1 (light grey) and Scenario 2 (dark grey), where stacks represent the ethanol concentrations in g/L and marks represent ethanol yields in g/g RS.

3.4. Biomethane Production

The anaerobic digestion process was chosen for the production of biomethane from the solid residues from enzymatic hydrolysis and fermentation of FW. This is because the process has been shown to be one of the most efficient methods for extraction and conversion of remaining vital substances into energy. In this study, the anaerobic digestion process was carried out at 35 °C using sludge from an anaerobic digester from a swine farm as an inoculum. Figure 4 shows the cumulative methane content and production rate over an 80-day period. Methane produced from FSR (Scenario 2) was the highest, 264.53 ± 2.3 mL/g VS, followed by HSR (Scenario 1), 224.29 ± 1.8 mL/g VS and raw FW (RFW), 215 ± 3.2 mL/g VS. The greater biomethane production potential of this FSR may be due to the fact that FW had undergone two stages of degradation, enzymatic hydrolysis, and ethanol fermentation, leaving the remaining organic matter with higher solubility which was prompt to be converted to biomethane. Furthermore, considering the methane production rate, the FSR showed a relatively higher methane production rate than the HSR and RFW, especially during the first 30 days where more than 90% of the methane was produced. As seen in the graph of the FSR's daily methane production rate, methane began to accumulate in the first week of the anaerobic digestion process at a rate of approximately 100–300 mL/d with no lag phase. This is in contrast to HSR and RFW, where methane production began in the second and fifth weeks, respectively. The results of this study are consistent with research by Wu et al. (2015) [44] reporting that pretreatment by fermenting FW to ethanol with alcohol active dry yeast at 35 °C for 24 h prior to anaerobic digestion resulted in a 71.7% increase in methane production yield. Due to the FW conversion to ethanol rather than volatile fatty acids (VFA) during the normal acidification process of anaerobic digestion, the pH of the system remained constant until the methanogens' activity was inhibited. As a result, biogas production was increased and the lag phase was shortened. This is similar to research conducted by Zhao et al. (2016) [48] who studied the effect of ethanol pre-fermentation on pre-anaerobic digestion pretreatment and found that methane production was increased by 49.6% compared to untreated FW. Furthermore, Refai et al. (2014) [49] investigated the effects of various volatile fatty acids, including acetate and ethanol, on methane formation during anaerobic digestion processes. It was found that by adding acetate and ethanol to the system, methane formation rates increased by 35–126% as a result of the metabolic activity of aceticlastic methanogens being promoted. Similarly, Prasertsan et al. (2021) [50] reported that the addition of approximately 5% of ethanol to the palm oil mill effluent resulted in an increase in bioassimilation of the process. This not only enhanced the amount of methane produced by 2.7 times, but also improved COD removal efficiency.

3.5. Gross Energy Output

Bioethanol and biomethane co-production can increase the total energy yield of FW energy production through bioprocessing. The energy values obtained from the co-production in this study were calculated from the low heating value (LHV) of 26.7 MJ/kg and 35.8 MJ/m^3, for bioethanol and biomethane, respectively [24]. Figure 5 illustrates the yield and gross energy output of 1 ton dry FW. In this study, a ton of FW could be converted to 0.49 tons of RS under optimal enzymatic hydrolysis conditions. When the liquid fraction was separated to produce bioethanol and the solid fraction was used to produce biomethane (Scenario 1), the yields were 282 L and 68 m^3, respectively, corresponding to a gross energy output of 8.37 GJ, or 262 L of gasoline equivalent. Meanwhile, continuous production (Scenario 2) where all hydrolysate was processed into ethanol fermentation and then fermentation solid residues were used to produce biomethane produced higher total bioethanol and biomethane of 318 L and 80 m^3, respectively, corresponding to gross energy output of 9.57 GJ, equivalent to 299 L of gasoline. The gross energy outputs produced in both scenarios are slightly higher than outputs reported in previous research by Karimi and Karimi (2018) [51], where co-production of ethanol and biogas from kitchen and garden wastes yielded a maximum gasoline equivalent of 162.1 L/ton waste. However, these gross energy outputs were comparable to those reported by Papa et al. (2015) [52] in which the energy recovery obtained from the co-production of bioethanol and biomethane from corn stover and switchgrass pretreated with mild ionic liquid was higher than that pretreated with pressurized hot water, ranging from 8.8–10.9 GJ/ton biomass. Similarly, Bondesson et al. (2013) [53] yielded total energy output of 9.2–9.8 GJ/ton of corn stover which was pretreated with 0.2% H_2SO_4. Additionally, when considering the production of ethanol or biomethane alone, the energy obtained from ethanol produced by Scenarios 1 and 2 was 5.94 and 6.70 GJ/ton, respectively. Meanwhile, biomethane production from food waste without ethanol production yielded 215 + 3.2 mL/g VS (Figure 4), equivalent to 2.34 GJ/ton of energy. This result clearly demonstrates the potential for bioethanol and biomethane co-production to increase gross energy output by approximately 1.4 times when compared to bioethanol alone and approximately 4 times when compared to biomethane production alone. This is consistent with a study by Moshi et al. (2015) [54], the co-production of ethanol and methane from cassava peels resulted in a 1.2–1.3-fold increase in energy yield compared to methane-only production, and a 3–4-fold increase compared to ethanol production alone. The results are also in line with the study by Wu et al., 2021 [55], who found that the co-production process of *Pennisetum purpureum* increased the energy recovery by 98.9% and 53.6% compared to ethanol production and biomethane alone, respectively. However, with the increase in gross energy output, it is still questionable whether the additional steps and associated extra capital and operating costs are worthwhile. In this respect, further research is required.

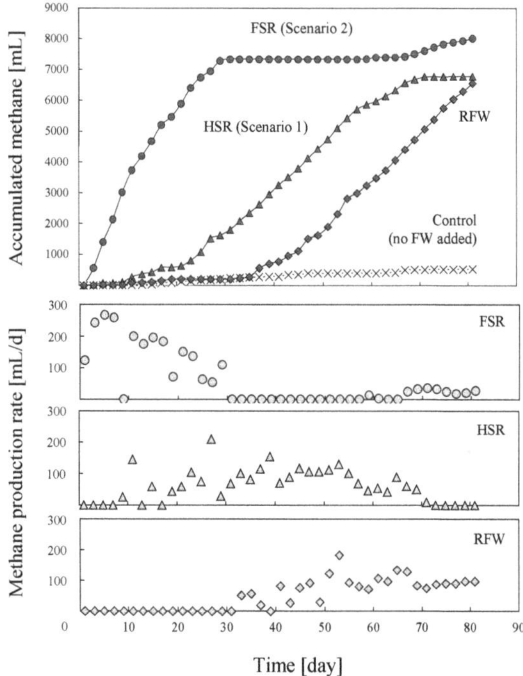

Figure 4. Biomethane production from fermentation solid residues (FSR), hydrolysis solid residues (HSR), raw food waste (RFW), and control (no FW added).

While co-production of bioethanol and biomethane from FW has resulted in high yields, the majority of research has still been carried out on a laboratory scale. As a result, economics, cost effectiveness, and life cycle assessment remain unexplored areas for researchers to investigate further. According to previous research, the price of ethanol produced from FW varies significantly. Sondhi et al. (2020) [56] determined the minimum ethanol selling price (MESP) of bioethanol produced from microwave-treated kitchen waste at a power level of 90 W for 30 min. The ethanol yield of 0.32 g/g biomass resulted in a MESP of approximately 0.14 USD/L, which is very low compared to the market price of 0.59 USD/L for ethanol. This is significantly different than the estimate of 0.64 USD/L of ethanol by Intan Shafinas Muhammad and Rosentrater (2020) [57]. Additionally, while FW is classified as a zero-cost raw material for energy production, there are hidden costs associated with management, collection, and sorting, which are major impediments to industrial waste utilization. As FW is typically disposed of as municipal waste, which contains both organic and inorganic materials, sorting this waste at the waste disposal site is nearly impossible. Proper management, which includes source sorting, collection planning, public awareness, and government promotion, are therefore critical factors in determining the future success of energy production from FW.

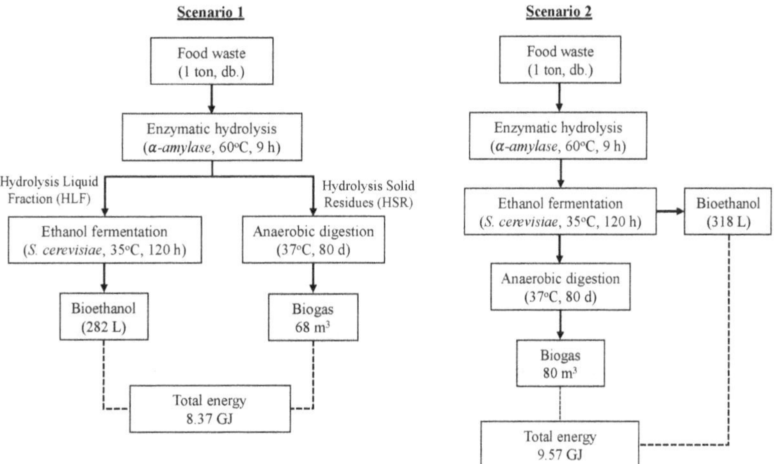

Figure 5. Gross energy output from each ton of dry FW obtained from two scenarios of co-production of bioethanol and biomethane.

4. Conclusions

FW is primarily composed of carbohydrates, possessing the potential and suitability to be used as a raw material for biofuels. Without any pretreatment other than size reduction, FW in this study was hydrolysed with α-amylase to decompose the primary constituent carbohydrates into fermentable sugars. This, under optimal conditions, yielded the highest amount of RS of 0.49 g/g FW. The study was then divided into two scenarios: (1) the hydrolysis liquid fraction was used for ethanol production and the hydrolysis solid residues were used for biomethane production; and (2) the entire hydrolysate was used for ethanol production followed by biomethane production from fermentation solid residues. The study found that the ethanol production yields obtained from both scenarios were in the range of 0.43–0.5 g/g RS when fermented with *S. cerevisiae* for 120 h at 35 °C. The maximum ethanol content of 0.25 g/g dry FW was obtained from fermentation of the entire hydrolysate (Scenario 2). Additionally, the fermentation solid residues in Scenario 2 resulted in a greater and faster potential for biomethane production than the hydrolysis solid residues in Scenario 1. However, both Scenarios 1 and 2 show high potential for biofuel production from FW, with gross energy output of 8.37 and 9.57 GJ/ton dry FW, equivalent to 262 and 299 L of gasoline, respectively.

This study demonstrates that co-production of bioethanol and biomethane from food waste is an extremely efficient method of increasing gross energy output when compared to producing either product alone. Additionally, it is a truly sustainable and environmentally friendly method of energy production that reduces GHG emissions in two ways: by reducing food waste and by reducing the use of fossil fuels. The findings of this study serve as an important starting point for demonstrating the feasibility of converting food waste to energy, potentially paving the way for industrial scale production in the future. However, additional research on the economics, investment costs, operation costs, and energy consumption of the entire process is required. This includes the collection and sorting of food waste, as well as the sorting and purification of all products from the subsequent manufacturing process.

Author Contributions: Conceptualization, C.P. and T.J.; validation, C.P.; formal analysis, T.J., A.B. and H.S.; writing—original draft preparation, C.P., T.J., A.B. and H.S.; writing—review and editing, C.P.; supervision, C.P. All authors have read and agreed to the published version of the manuscript.

Funding: This research was funded by King Mongkut's University of Technology North Bangkok. Contract no. KMUTNB-63-NEW-18.

Institutional Review Board Statement: Not applicable.

Informed Consent Statement: Not applicable.

Data Availability Statement: Not applicable.

Conflicts of Interest: The authors declare no conflict of interest.

References

1. IEA. Global Energy Review 2021. 2021. Available online: https://www.iea.org/reports/global-energy-review-2021 (accessed on 1 September 2021).
2. Silva, F.M.; Oliveira, L.B.; Mahler, C.F.; Bassin, J. Hydrogen production through anaerobic co-digestion of food waste and crude glycerol at mesophilic conditions. *Int. J. Hydrogen Energy* **2017**, *42*, 22720–22729. [CrossRef]
3. Jarunglumlert, T.; Prommuak, C.; Putmai, N.; Pavasant, P. Scaling-up bio-hydrogen production from food waste: Feasibilities and challenges. *Int. J. Hydrogen Energy* **2018**, *43*, 634–648. [CrossRef]
4. Westerholm, M.; Liu, T.; Schnürer, A. Comparative study of industrial-scale high-solid biogas production from food waste: Process operation and microbiology. *Bioresour. Technol.* **2020**, *304*, 122981. [CrossRef]
5. Shamurad, B.; Sallis, P.; Petropoulos, E.; Tabraiz, S.; Ospina, C.; Leary, P.; Dolfing, J.; Gray, N. Stable biogas production from single-stage anaerobic digestion of food waste. *Appl. Energy* **2020**, *263*, 114609. [CrossRef]
6. Papanikolaou, S.; Dimou, A.; Fakas, S.; Diamantopoulou, P.; Philippoussis, A.; Galiotou-Panayotou, M.; Aggelis, G. Biotechnological conversion of waste cooking olive oil into lipid-rich biomass using *Aspergillus* and *Penicillium* strains. *J. Appl. Microbiol.* **2011**, *110*, 1138–1150. [CrossRef] [PubMed]
7. Omar, W.N.N.W.; Amin, N.A.S. Optimization of heterogeneous biodiesel production from waste cooking palm oil via response surface methodology. *Biomass Bioenergy* **2011**, *35*, 1329–1338. [CrossRef]
8. Sunthitikawinsakul, A.; Sangatith, N. Study on the quantitative fatty acids correlation of fried vegetable oil for biodiesel with heating value. *Procedia Eng.* **2012**, *32*, 219–224. [CrossRef]
9. Huang, H.; Singh, V.P.; Qureshi, N. Butanol production from food waste: A novel process for producing sustainable energy and reducing environmental pollution. *Biotechnol. Biofuels* **2015**, *8*, 1–12. [CrossRef]
10. Yeshanew, M.M.; Frunzo, L.; Pirozzi, F.; Lens, P.N.L.; Esposito, G. Production of biohythane from food waste via an integrated system of continuously stirred tank and anaerobic fixed bed reactors. *Bioresour. Technol.* **2016**, *220*, 312–322. [CrossRef] [PubMed]
11. Sarkar, O.; Mohan, S.V. Pre-aeration of food waste to augment acidogenic process at higher organic load: Valorizing biohydrogen, volatile fatty acids and biohythane. *Bioresour. Technol.* **2017**, *242*, 68–76. [CrossRef]
12. UNEP. Food Waste Index Report 2021. 2021. Available online: https://www.unep.org/resources/report/unep-food-waste-index-report-2021 (accessed on 10 September 2021).
13. Mahmoodi, P.; Karimi, K.; Taherzadeh, M.J. Hydrothermal processing as pretreatment for efficient production of ethanol and biogas from municipal solid waste. *Bioresour. Technol.* **2018**, *261*, 166–175. [CrossRef] [PubMed]
14. Maurya, D.P.; Singla, A.; Negi, S. An overview of key pretreatment processes for biological conversion of lignocellulosic biomass to bioethanol. *3 Biotech* **2015**, *5*, 597–609. [CrossRef] [PubMed]
15. Matsakas, L.; Christakopoulos, P. Ethanol production from enzymatically treated dried food waste using enzymes produced on-site. *Sustainability* **2015**, *7*, 1446–1458. [CrossRef]
16. Ntaikou, I.; Antonopoulou, G.; Lyberatos, G. Sustainable second-generation bioethanol production from enzymatically hydrolyzed domestic food waste using pichia anomala as biocatalyst. *Sustainabiltiy* **2020**, *13*, 259. [CrossRef]
17. Mihajlovski, K.; Radovanović, Ž.; Carević, M.; Dimitrijević-Branković, S. Valorization of damaged rice grains: Optimization of bioethanol production by waste brewer's yeast using an amylolytic potential from the *Paenibacillus chitinolyticus* CKS1. *Fuel* **2018**, *224*, 591–599. [CrossRef]
18. Bibra, M.; Rathinam, N.K.; Johnson, G.R.; Sani, R.K. Single pot biovalorization of food waste to ethanol by *Geobacillus* and *Thermoanaerobacter* spp. *Renew. Energy* **2020**, *155*, 1032–1041. [CrossRef]
19. Dhiman, S.S.; David, A.; Shrestha, N.; Johnson, G.R.; Benjamin, K.M.; Gadhamshetty, V.; Sani, R.K. Simultaneous hydrolysis and fermentation of unprocessed food waste into ethanol using thermophilic anaerobic bacteria. *Bioresour. Technol.* **2017**, *244*, 733–740. [CrossRef] [PubMed]

20. Rocha-Meneses, L.; Raud, M.; Orupõld, K.; Kikas, T. Second-generation bioethanol production: A review of strategies for waste valorisation. *Agron. Res.* **2017**, *15*, 830–847.
21. Jarunglumlert, T.; Prommuak, C. Net energy analysis and techno-economic assessment of co-production of bioethanol and biogas from cellulosic biomass. *Fermentation* **2021**, *7*, 229. [CrossRef]
22. Li, W.; Ghosh, A.; Bbosa, D.; Brown, R.C.; Wright, M.M. Comparative techno-economic, uncertainty and life cycle analysis of lignocellulosic biomass solvent liquefaction and sugar fermentation to ethanol. *ACS Sustain. Chem. Eng.* **2018**, *6*, 16515–16524. [CrossRef]
23. Gubicza, K.; Nieves, I.U.; Sagues, W.; Barta, Z.; Shanmugam, K.; Ingram, L.O. Techno-economic analysis of ethanol production from sugarcane bagasse using a liquefaction plus simultaneous saccharification and co-fermentation process. *Bioresour. Technol.* **2016**, *208*, 42–48. [CrossRef] [PubMed]
24. Joelsson, E.; Galbe, M.; Wallberg, O.; Dienes, D.; Kovacs, K. Combined production of biogas and ethanol at high solids loading from wheat straw impregnated with acetic acid: Experimental study and techno-economic evaluation. *Sustain. Chem. Process.* **2016**, *4*, 22. [CrossRef]
25. Zhao, L.; Zhang, X.; Xu, J.; Ou, X.; Chang, S.; Wu, M. Techno-economic analysis of bioethanol production from lignocellulosic biomass in China: Dilute-acid pretreatment and enzymatic hydrolysis of corn stover. *Energies* **2015**, *8*, 4096–4117. [CrossRef]
26. Shafiei, M.; Karimi, K.; Taherzadeh, M. Techno-economical study of ethanol and biogas from spruce wood by NMMO-pretreatment and rapid fermentation and digestion. *Bioresour. Technol.* **2011**, *102*, 7879–7886. [CrossRef] [PubMed]
27. Tao, L.; Aden, A.; Elander, R.T.; Pallapolu, V.R.; Lee, Y.; Garlock, R.J.; Balan, V.; Dale, B.E.; Kim, Y.; Mosier, N.S.; et al. Process and technoeconomic analysis of leading pretreatment technologies for lignocellulosic ethanol production using switchgrass. *Bioresour. Technol.* **2011**, *102*, 11105–11114. [CrossRef]
28. Sluiter, A.; Hames, B.; Ruiz, R.; Scarlata, C.; Sluiter, J.; Templeton, D.; Crocker, D.L.A.P. *Determination of Structural Carbohydrates and Lignin in Biomass*; NREL: Golden, CO, USA, 2008.
29. Chiang, L.-C.; Gong, C.-S.; Chen, L.-F.; Tsao, G.T. D-xylulose fermentation to ethanol by *Saccharomyces cerevisiae*. *Appl. Environ. Microbiol.* **1981**, *42*, 284–289. [CrossRef]
30. Cutaia, A.J. Malt beverages and brewing materials: Gas chromatographic determination of ethanol in beer. *J. Assoc. Off. Anal. Chem.* **1984**, *67*, 192–193. [CrossRef]
31. Hong, Y.S.; Yoon, H.H. Ethanol production from food residues. *Biomass Bioenergy* **2011**, *35*, 3271–3275. [CrossRef]
32. Kim, Y.S.; Jang, J.Y.; Park, S.J.; Um, B.H. Dilute sulfuric acid fractionation of Korean food waste for ethanol and lactic acid production by yeast. *Waste Manag.* **2018**, *74*, 231–240. [CrossRef] [PubMed]
33. Kim, J.K.; Han, G.H.; Oh, B.R.; Chun, Y.N.; Eom, C.-Y.; Kim, S.W. Volumetric scale-up of a three stage fermentation system for food waste treatment. *Bioresour. Technol.* **2008**, *99*, 4394–4399. [CrossRef] [PubMed]
34. Saeed, M.A.; Wang, Q.; Jin, Y.; Yue, S.; Ma, H. Assessment of bioethanol fermentation performance using different recycled waters of an integrated system based on food waste. *BioResources* **2019**, *14*, 3717–3730.
35. Yu, M.; Wu, C.; Wang, Q.; Sun, X.; Ren, Y.; Li, Y.-Y. Ethanol prefermentation of food waste in sequencing batch methane fermentation for improved buffering capacity and microbial community analysis. *Bioresour. Technol.* **2018**, *248*, 187–193. [CrossRef]
36. Nagao, N.; Tajima, N.; Kawai, M.; Niwa, C.; Kurosawa, N.; Matsuyama, T.; Yusoff, F.M.; Toda, T. Maximum organic loading rate for the single-stage wet anaerobic digestion of food waste. *Bioresour. Technol.* **2012**, *118*, 210–218. [CrossRef]
37. Taheri, M.E.; Salimi, E.; Saragas, K.; Novakovic, J.; Barampouti, E.M.; Mai, S.; Malamis, D.; Moustakas, K.; Loizidou, M. Effect of pretreatment techniques on enzymatic hydrolysis of food waste. *Biomass Convers. Biorefinery* **2021**, *11*, 219–226. [CrossRef]
38. Cekmecelioglu, D.; Uncu, O.N. Kinetic modeling of enzymatic hydrolysis of pretreated kitchen wastes for enhancing bioethanol production. *Waste Manag.* **2013**, *33*, 735–739. [CrossRef] [PubMed]
39. Hossain, S.; Theodoropoulos, C.; Yousuf, A. Techno-economic evaluation of heat integrated second generation bioethanol and furfural coproduction. *Biochem. Eng. J.* **2019**, *144*, 89–103. [CrossRef]
40. Saeed, M.A.; Ma, H.; Yue, S.; Wang, Q.; Tu, M. Concise review on ethanol production from food waste: Development and sustainability. *Environ. Sci. Pollut. Res.* **2018**, *25*, 28851–28863. [CrossRef] [PubMed]
41. Kim, J.H.; Lee, J.C.; Pak, D. Feasibility of producing ethanol from food waste. *Waste Manag.* **2011**, *31*, 2121–2125. [CrossRef] [PubMed]
42. Han, W.; Liu, Y.; Xu, X.; Huang, J.; He, H.; Chen, L.; Qiu, S.; Tang, J.; Hou, P. Bioethanol production from waste hamburger by enzymatic hydrolysis and fermentation. *J. Clean. Prod.* **2020**, *264*, 121658. [CrossRef]
43. Han, W.; Xu, X.; Gao, Y.; He, H.; Chen, L.; Tian, X.; Hou, P. Utilization of waste cake for fermentative ethanol production. *Sci. Total Environ.* **2019**, *673*, 378–383. [CrossRef]
44. Wu, C.; Wang, Q.; Xiang, J.; Yu, M.; Chang, Q.; Gao, M.; Sonomoto, K. Enhanced productions and recoveries of ethanol and methane from food waste by a three-stage process. *Energy Fuels* **2015**, *29*, 6494–6500. [CrossRef]
45. Uncu, O.N.; Cekmecelioglu, D. Cost-effective approach to ethanol production and optimization by response surface methodology. *Waste Manag.* **2011**, *31*, 636–643. [CrossRef] [PubMed]
46. Kiran, E.U.; Liu, Y. Bioethanol production from mixed food waste by an effective enzymatic pretreatment. *Fuel* **2015**, *159*, 463–469. [CrossRef]

47. Moreno, A.D.; Magdalena, J.A.; Oliva, J.M.; Greses, S.; Lozano, C.C.; Latorre-Sánchez, M.; Negro, M.J.; Susmozas, A.; Iglesias, R.; Llamas, M.; et al. Sequential bioethanol and methane production from municipal solid waste: An integrated biorefinery strategy towards cost-effectiveness. *Process. Saf. Environ. Prot.* **2021**, *146*, 424–431. [CrossRef]
48. Zhao, N.; Yu, M.; Wang, Q.; Song, N.; Che, S.; Wu, C.; Sun, X. Effect of ethanol and lactic acid pre-fermentation on putrefactive bacteria suppression, hydrolysis, and methanogenesis of food waste. *Energy Fuels* **2016**, *30*, 2982–2989. [CrossRef]
49. Refai, S.; Wassmann, K.; Deppenmeier, U. Short-term effect of acetate and ethanol on methane formation in biogas sludge. *Appl. Microbiol. Biotechnol.* **2014**, *98*, 7271–7280. [CrossRef] [PubMed]
50. Prasertsan, P.; Leamdum, C.; Chantong, S.; Mamimin, C.; Kongjan, P.; O-Thong, S. Enhanced biogas production by co-digestion of crude glycerol and ethanol with palm oil mill effluent and microbial community analysis. *Biomass Bioenergy* **2021**, *148*, 106037. [CrossRef]
51. Karimi, S.; Karimi, K. Efficient ethanol production from kitchen and garden wastes and biogas from the residues. *J. Clean. Prod.* **2018**, *187*, 37–45. [CrossRef]
52. Papa, G.; Rodriguez, S.; George, A.; Schievano, A.; Orzi, V.; Sale, K.; Singh, S.; Adani, F.; Simmons, B. Comparison of different pretreatments for the production of bioethanol and biomethane from corn stover and switchgrass. *Bioresour. Technol.* **2015**, *183*, 101–110. [CrossRef]
53. Bondesson, P.-M.; Galbe, M.; Zacchi, G. Ethanol and biogas production after steam pretreatment of corn stover with or without the addition of sulphuric acid. *Biotechnol. Biofuels* **2013**, *6*, 11. [CrossRef]
54. Moshi, A.P.; Crespo, C.F.; Badshah, M.; Hosea, K.M.; Mshandete, A.M.; Elisante, E.; Mattiasson, B. Characterisation and evaluation of a novel feedstock, *Manihot glaziovii, Muell. Arg*, for production of bioenergy carriers: Bioethanol and biogas. *Bioresour. Technol.* **2014**, *172*, 58–67. [CrossRef] [PubMed]
55. Wu, P.; Kang, X.; Wang, W.; Yang, G.; He, L.; Fan, Y.; Cheng, X.; Sun, Y.; Li, L. Assessment of coproduction of ethanol and methane from pennisetum purpureum: Effects of pretreatment, process performance, and mass balance. *ACS Sustain. Chem. Eng.* **2021**, *9*, 10771–10784. [CrossRef]
56. Sondhi, S.; Kaur, P.S.; Kaur, M. Techno-economic analysis of bioethanol production from microwave pretreated kitchen waste. *SN Appl. Sci.* **2020**, *2*, 1–13. [CrossRef]
57. Muhammad, N.I.S.; Rosentrater, K.A. Economic assessment of bioethanol recovery using membrane distillation for food waste fermentation. *Bioengineering* **2020**, *7*, 15. [CrossRef] [PubMed]

MDPI

Article

Pineapple Waste Cell Wall Sugar Fermentation by *Saccharomyces cerevisiae* for Second Generation Bioethanol Production

Fabio Salafia [1], Antonio Ferracane [2,*] and Alessia Tropea [3,*]

1 Food Chemistry, Safety and Sensoromic Laboratory (FoCuSS Lab), Department of Agriculture, University "Mediterranea" of Reggio Calabria, Via dell'Università, 25, 89124 Reggio Calabria, Italy; fabio.salafia@unirc.it
2 Department of Chemical, Biological, Pharmaceutical and Environmental Sciences, University of Messina, Polo Annunziata, Viale Annunziata, 98166 Messina, Italy
3 Department of Research and Internationalization, University of Messina, Via Consolato del Mare, 41, 98100 Messina, Italy
* Correspondence: aferracane@unime.it (A.F.); atropea@unime.it (A.T.)

Abstract: Agricultural food waste is rich in cellulosic and non-cellulosic fermentable substance. In this study, we investigated the bioconversion of pineapple waste cell wall sugars into bioethanol by simultaneous saccharification and fermentation using *Saccharomyces cerevisiae* ATCC 4126. Soluble and insoluble cell wall sugars were investigated during the fermentation process. Moreover, the fermentation medium was investigated for protein, moisture, ash, lignin and glycerol determinations with a particular focus on the increase in single cell protein due to yeast growth, allowing a total valorization of the resulting fermentation medium, with no further waste production, with respect to environmental sustainability. Soluble and insoluble sugars in the starting material were 32.12% and 26.33% respectively. The main insoluble sugars resulting from the cell wall hydrolysis detected at the beginning of the fermentation, were glucose, xylose and uronic acid. Glucose and mannose were the most prevalent sugars in the soluble sugars fraction. The ethanol theoretical yield, calculated according to dry matter lost, reached up to 85% (3.9% EtOH). The final fermentation substrate was mainly represented by pentose sugars. The protein content increased from 4.45% up to 20.1% during the process.

Keywords: ethanol; simultaneous saccharification and fermentation; *Saccharomyces cerevisiae*; single cell protein; food waste; pineapple waste; cell wall sugar; fermentation

Citation: Salafia, F.; Ferracane, A.; Tropea, A. Pineapple Waste Cell Wall Sugar Fermentation by *Saccharomyces cerevisiae* for Second Generation Bioethanol Production. *Fermentation* **2022**, *8*, 100. https://doi.org/10.3390/fermentation8030100

Academic Editor: Gunnar Lidén

Received: 11 February 2022
Accepted: 25 February 2022
Published: 27 February 2022

1. Introduction

Waste disposal is one of the major problems facing most food processing plants [1,2]. According to Campos et al. [3], there is an increasing interest in the valorization of the wastes generated by the food industry, including waste generated as a consequence of the new developments in process engineering and the resulting byproducts [4].

Waste utilization in the fruit and vegetable processing industry is an important challenge that governments must address in order to promote sustainability [3,5]. Additionally, these substrates have a high potential, due to their micro and macro composition [6,7], as a low-cost high-potency second-generation feed-stock that can easily undergo biodegradation [8].

Among agricultural food waste, pineapple industrialization is known to generate a significant amount of solid residues, and values between 75–80% have been reported [5,9]. In the past, pineapple wastes were utilized as sources for bromelain extraction, wine and vinegar production, yeast cultivation for food/feed proteins, or also for organic acid production. They can also be a source for other bioactive compounds, such as antioxidants [10–14].

Pineapple wastes, such as the fruit peel and crown, are comprised of lignin, hemicellulose and cellulose. For this reason, they are considered to be lignocellulosic materials that can be used in the production of second-generation bioethanol, after pre-treatment and hydrolysis, in order to provide fermentable sugars for the subsequent fermentation [15,16].

Hydrolyzation is the main step for lignocellulosic biomass fermentation; in fact, the polysaccharides are tightly packed in plant cell walls and are often surrounded by lignin, forming highly recalcitrant structures resistant to direct enzymatic attack [17,18]. Enzymatic hydrolysis is regarded today as the most promising approach for liberating fermentable sugars in an energy-efficient way from the carbohydrates found in lignocelluloses in order to produce bioethanol via fermentation [19,20].

According to Pereira et al. [21], among the different microbes used for bioethanol production, the yeast *Saccharomyces cerevisiae* is the most commonly used organism because of its good fermentative capacity, high tolerance to ethanol and other inhibitors (either formed during raw material pre-treatments or produced during fermentation), and its capacity to grow rapidly under anaerobic conditions, as are typically established in large-scale vessels [22].

Ethanol production is mainly dependent on glucose concentration (the theoretical alcohol yield is about 0.5 g of ethanol per g of glucose), but nutrient supplementation is also an important parameter to take into consideration, since an adequate amount of specific nutrients, such as trace elements, vitamins and nitrogen, often poor in agricultural waste, can significantly improve yeast viability and resistance to the medium, stimulating ethanol production performances [15,23].

Several related studies about bioethanol production from pineapple wastes report different fermentation approaches, such as direct fermentation (DF), separate hydrolysis and fermentation (SHF) and simultaneous saccharification and fermentation (SSF) [24–26]. Among these fermentation processes, SSF has the advantage of preventing the buildup of hydrolysis, such as cellobiose and glucose, which can reduce the rate of further substrate hydrolysis. However, it has to be carried out at temperatures that suit the fermenting organism. In the case of yeast, the temperature is generally below 40 °C, which is below the optimum temperature for enzymatic hydrolysis (50 °C) [27].

The present research is focused on the evaluation of pineapple waste cell wall sugars as an alternative source of second-generation bioethanol. This study utilizes *Saccharomyces cerevisiae* ATCC 4126 to carry out an SSF process using a supplemented medium, by the addition of a specific nitrogen source, salts, and vitamins, which are required by the yeast in order to improve its ability to use the substrate both for alcohol production and for its own growth. The high amount of cell wall sugars in pineapple waste prompted us to utilize it as a raw material for bioethanol production and as a cheap medium. Moreover, the initial and final fermentation mediums were investigated with a particular focus on the increase of single cell protein due to yeast growth, making the resulting fermented substrate suitable as animal feed. This allows a total valorization of the resulting fermentation medium, with no further waste production, with respect to environmental sustainability.

2. Materials and Methods

2.1. Substrate

Pineapples were purchased from a local market in Messina, IT. The pineapples were manually cleaned by removing the crown and the pulp. For analytical purposes only, the waste represented by pineapple peel and core (the inner part) have been used as a fermentative substrate. Wastes were cut into small pieces and homogenized in a fruit blender for 5 min.

2.2. Microorganism

Saccharomyces cerevisiae ATCC 4126 was maintained on yeast medium (YM) agar (yeast extract 3 g/L, malt extract 3 g/L, peptone 5 g/L, glucose 10 g/L, agar 20 g/L, Oxoid, Basingstoke, UK) at 4 °C. To carry out the tests, *S. cerevisiae* was cultured overnight at 30 °C on a rotary shaker (INNOVA 44, Incubator Shaker Series, New Brunswick Scientific, Edison, NJ, USA) at 250 rpm, in 20 mL YM medium tubes [28].

After overnight incubation, the cell suspensions were aseptically harvested by centrifugation (3000 rpm, 5 min, Centrifuge 5810 R, Eppendorf UK Ltd., Stevenage, UK), the supernatant (YM media) was discarded, and the yeast cells were washed twice in 5 mL 0.9% (*w/v*) NaCl to minimize nutrient transfer from seed culture to fermentation medium [28].

The total viable yeast cells were measured by using a cell count reader (Nucleocounter® YC 100™, Chemo Metec, Allerød, Denmark). The standard yeast culture contained 10^8 cells per mL of *S. cerevisiae* ATCC 4126 [28].

2.3. Experimental Setup

Fermentation tests were carried out in a 5 L batch fermenter (Biostat Biotech B, Sartorius Stedim Biotech, Goettingen, Germany). The fermenter was equipped with one four-bladed Rushton turbine and the usual control systems as follows: temperature, pH, pO_2, pCO_2 and a foam detector.

Pineapple waste, comprising fruit skin and core, were homogenized in a fruit blender for 5 min. The resulting homogenate, with a dry matter content of 14% (*w/w*), was diluted with water to a 9% dry matter, in a working volume of 3.5 L and immediately treated at 100 °C for 10 min under continuous mixing to inactivate endogenous enzymes and reduce microbial spoilage. No further sterilization procedures were adopted [4].

SSF fermentation was carried out by adding a 2% (*v/v*) inoculum of *S. cerevisiae* (10^8 cells per mL) and the enzymes (20 µL/g dry matter of Depol™ 740 L and 250 µL/g dry matter of Accellerase® 1500 enzymes) to the substrate. Both of the enzymes were added to the medium according to Tropea et al. [25].

According to Tropea et al. [23], the fermentation medium was supplemented with urea phosphate salt 2.3 g/L, KCl 0.2 g/L, $MgSO_4{\cdot}7H_2O$ 3.8 g/L, Ca-pantothenate 0.0833 mg/L and biotin 0.0833 mg/L.

Fermentation parameters were 30 °C, pH 5 and constant stirring at 200 rpm. The pH value was previously adjusted from 3.8 up to 5, using 2 M NaOH.

CO_2 evolution was measured during all fermentation tests using a BioPAT® Xgas 1 analyser for BIOSTAT® B-DCU II system (Sartorius Stedim Biotech, Goettingen, Germany) and duplicate broth samples were withdrawn from the reaction vessel using a 20 mL syringe. Samples for ethanol analysis were immediately frozen at −18 °C until analysis, whereas samples for the other determinations were heated at 100 °C for 10 min, to inactivate the enzymes and stop any further fermentation, and then frozen at −18 °C until analyzed. All fermentations were carried out until no further CO_2 fluctuations were observed. The pH was not controlled by the addition of an alkali during fermentation [4].

2.4. Chemicals

Chemicals were purchased from Sigma Aldrich (Bellefonte, PA, USA), except for galacturonic acid and glucose, which were purchased from Fluka Biochemical (Buchs, Switzerland); glycerol, KCl, $MgSO_4{\cdot}7H_2O$, and Ca-pantothenate, which were provided by Fisher Scientific (UK Ltd., Loughborough, UK); and biotin, which was provided by Calbiochem.

Commercially available enzyme solutions DepolTM 740 L (ferulic acid esterase), provided by Biocatalysts Ltd., Cefn Coed, Wales, U.K and Accellerase® 1500 (endoglucanase), provided by Genencor (Rochester, NY, USA), were used.

2.5. Protein, Moisture, Ash and Lignin Determinations

Representative samples were drained off for protein content testing using the method suggested by the AOAC [29]. The protein percentage was calculated considering a conversion factor of 6.25. The increase in protein was quantified by the Büchi Kjeldahal (Büchi, Switzerland) instrument, equipped with the Büchi Distillation Unit B-324 (Büchi, Switzerland), Digestion Unit K-424 and Scrubber B-414 (Büchi, Switzerland), used for crude protein determination as total N, multiplying the results by the conversion factor.

The dry weights were calculated as steady weights after 2 h at 110 °C using a Mettler PM 200 equipped with a Mettler LP16 IR balance (Mettler-Toledo GmbH, Laboratory & Weighing Technologies, Greifensee, Switzerland).

Ash determination was carried out according to the AOAC method [29]. Klason lignin was quantified gravimetrically according to Carrier et al. [30]. All samples were analyzed in triplicate.

2.6. Alcohol-Insoluble Residues (AIR)

AIR samples were prepared prior to analysis for cell wall sugars. Wet fermented pineapple waste samples, after defrosting, were homogenized for 1 min at maximum speed in a Janke & Kunnel, Ika-Werk Ultra-Turrax homogenizer at room temperature and then poured into boiling ethanol for obtaining a final mixture that had an EtOH concentration of 85% (v/v). Sample particles from the homogenizer were collected using 50 mL of 70% EtOH. The insoluble residue was recovered by vacuum filtration using a 5 μm nylon filter NYBOLT by a Buchner funnel. After two further sequential extractions in boiling 85% ethanol (v/v) the residue was extracted in boiling absolute ethanol and then washed with cold absolute ethanol. The final filtrate was dried by a rotating evaporator (Büchi, Switzerland) at 40 °C, recovered in water and tested for residual soluble sugars. The insoluble residue was washed with two volumes of acetone and after removal by suction, dried to a constant weight at 40 °C [31,32] and analyzed for insoluble sugars determination.

2.7. Sugar Analysis

Insoluble sugars were released from AIR samples by hydrolysis and analyzed by gas chromatography-flame ionization detection (GC-FID) after conversion to their alditol acetates. As an internal standard, 2-deoxyglucose was used [33]. Monosaccharides were released from polysaccharides with pre-hydrolysis of the samples using 0.2 mL of 72% (w/w) H_2SO_4 for 3 h at room temperature, followed by 2.5 h of hydrolysis in 1 M H_2SO_4 at 100 °C. A total of 0.5 mL was collected for uronic acid determination after 1 h of hydrolysis. Hydrolysis was followed by the reduction and acetylation of the monosaccharides, and the alditol acetates were analyzed by Shimadzu Gas Chromatograph GC-2010, equipped with a Flame Ionization Detector (GC/FID) (Kyoto, Japan), by using a capillary column DB-225 (30 m length, 0.25 mm ID and 0.15 μm d_f, (50%-Cyanopropylphenyl)-dimethylpolysiloxane) [34].

The same protocol, starting from hydrolysis in 1 M H_2SO_4, was carried out for the determination of the residual soluble sugar in the supernatant fraction. The oven temperature program was as follows: 200 °C to 220 °C at a rate of 40 °C/min (7 min), increasing to 230 °C at a rate of 20 °C/min (1 min). The temperature of the injector was 220 °C and the detector was 230 °C. The carrier gas used was hydrogen, at a flow rate of 1.7 mL/min. The free sugars were identified and quantified based on their retention times, and response factors obtained by the injection of standards. Uronic acid content was determined by the m-phenylphenol colorimetric method [35], modified according to Rae et al. [36], and the galacturonic acid was used as the standard. To the 0.5 mL of diluted hydrolyzed sample (1:4), 3 mL of boric acid 50 mM H_2SO_4 98% (w/w) was added. After shaking, the test tubes were heated at 100 °C for 10 min. A quantity of 100 μL of m-phenylphenol was added after cooling, reacting for 30 min in the dark, and the absorbance was measured at 520 nm. All samples were analyzed in triplicate.

2.8. Alcohols Determination

Ethanol and glycerol were quantified by HPLC. A total of 500 μL of supernatant sample from fermented pineapple waste were centrifuged for 10 min at 500 rpm and 20 °C in a 96-deep well plate using an Eppendorf Centrifuge 5810 R, then filtered through AcroPrep^{TM} 0.2 μm GHP Membrane 96-Well Filter Plates into a 96-deep well collection plate for a further 10 min at the same speed. After centrifugation, plates were covered by a rubber lid and loaded directly onto a Shimadzu HPLC system (Kyoto, Japan), equipped with an autosampler SIL-20A HT, a degasser DGU-20A3, a pump LC-20AD, a column oven CTO-20A and a Refractive Index Detector model: RID-10A. Analyses were carried out using an Aminex HPX-87P 300 × 7.8 mm carbohydrate analysis column (Bio-Rad Laboratories Ltd., Hemel Hempstead, UK. Resin ionic form: lead. Support: sulfonated divinyl benzene-styrene copolymer. Particle size: 9 µm.) with matching guard columns (BIO-RAD, MicroGuard® Carbo-P, Hercules, CA, USA), operating at 65 °C with ultrapure water at a flow rate of 0.6 mL/min as the mobile phase, in isocratic mode. The sample injection volume was 20 µL. Two injections were performed for each sample. Standard curves of anhydrous sugars were produced and myo-inositol (cyclohexane-1,2,3,4,5,6-hexol) was used as the internal standard. The total analysis time was 42 min. All samples were analyzed in triplicate [37].

3. Results and Discussion

3.1. Protein, Moisture, Ash and Lignin

As shown in Figure 1, protein increased following the same trend observed for ethanol and glycerol production, increasing from an initial 4.45% to 7.3% at $t = 9$, and reaching the highest concentration (21.3%) at $t = 21$. According to Aruna et al. [38] and Aruna [39], this trend can be ascribed to the yeast cell growth, which can be also referred as single cells proteins (SCP). Moreover, the protein percentage reached in the present study is in line with previous findings, making the final fermented substrate suitable as animal feed. The last fermentation phase was characterized by a 1.2% protein decrease (Figure 1), due, of course, to the natural yeast cell autolysis during the decline phase of the growth curve [40,41].

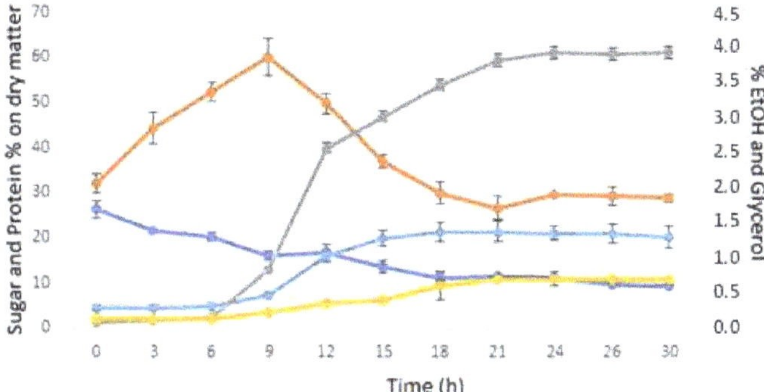

Figure 1. Trend of the main parameters evaluated during the SSF process. Light blue: % of protein, blue: % of insoluble sugars, orange: % of soluble sugars, grey: % EtOH, yellow: % of glycerol.

During the SSF processes, the dry matter dropped down from 9% to 2.5%, leaving around 30% of the dry matter in the substrate unused. This could be ascribed to the pH falling during the fermentation period. In fact, the pH value dropped down from 5.0 ± 0.3 to 3.3 ± 0.2. The pH drop was probably caused both by yeast catabolite production and D-galacturonic acid release from pectin [42]. The observed pH decrease could hamper the enzymatic activity with a consequent arrest of fiber saccharification.

Lignin and ash, whose percentages are shown in Table 1, were 3.89% and 0.56%, respectively, in the starting material. Whereas, at the end of the fermentation process, the percentages detected in the fermented material were 6.54% and 0.58%, respectively. The increase in lignin can be explained by remembering that lignin is not involved in alcoholic fermentation [43], and so, according to the literature, the increase in lignin dry matter is typically due to enzymatic fiber hydrolysis [44].

Table 1. Fermentation medium composition [a].

	Starting Material	Fermented Material
Soluble sugar	32.12 ± 2.05	28.7 ± 0.80
Insoluble sugar	26.33 ± 1.83	9.36 ± 0.39
Protein	4.45 ± 0.6	20.1 ± 2.5
Lignin	3.89 ± 0.3	6.54 ± 0.1
Ash	0.56 ± 0.01	0.58 ± 0.01
Dry matter	9 ± 0.5	2.5 ± 0.4

[a] Composition reported as percentage of dry matter. Results are means ± Standard Deviation of triplicate analyses.

The ash percentage was stable around 0.6% during the whole process. A previous study, where there was no supplementation with minerals and vitamins to the medium, reported an ash percentage decrease, due to a partial ash utilization by the yeast as a source of minerals [45]. In this study, according to Tropea et al. [23], the supplementation with salts and vitamins was followed by a minor ash utilization by the yeast.

3.2. Cell Wall Insoluble and Soluble Sugars

Initial soluble and insoluble sugars in pineapple waste processed by SSF were 32.12% and 26.33%, respectively (Table 1).

Table 2 shows the percentage of the insoluble monosaccharides detected during the whole fermentation process. The main sugars resulting from the cell wall hydrolysis of AIR pineapple waste residues detected at the beginning of the fermentation, were 9.84% glucose (Glc), 8.16% xylose (Xyl) and 3.18% uronic acid (UA), followed by 2.46% arabinose (Ara) and 1.58% galactose (Gal), with smaller amounts of mannose (Man), rhamnose (Rha) and fucose (Fuc).

Table 2. Cell wall insoluble monosaccharide composition [a].

Hours	Residue	Totals	Rhamnose	Fucose	Arabinose	Xylose	Mannose	Galactose	Glucose	UA
0	3.7	26.33 ± 1.83	0.06 ± 0.01	0.06 ± 0.01	2.46 ± 0.18	8.16 ± 0.64	0.99 ± 0.02	1.58 ± 0.12	9.84 ± 1.07	3.18 ± 0.17
3	2.5	21.43 ± 0.62	0.07 ± 0.01	0.03 ± 0.01	1.88 ± 0.07	7.63 ± 0.66	0.68 ± 0.05	1.51 ± 0.15	7.30 ± 1.14	2.33 ± 0.14
6	1.9	20.25 ± 1.11	0.07 ± 0.01	0.03 ± 0.01	2.21 ± 0.20	6.48 ± 0.76	0.59 ± 0.08	1.29 ± 0.07	6.71 ± 0.71	2.85 ± 0.25
9	1.5	16.04 ± 0.95	0.04 ± 0.01	0.02 ± 0.01	1.19 ± 0.02	6.59 ± 1.42	0.35 ± 0.03	0.74 ± 0.07	5.98 ± 0.85	1.12 ± 0.36
12	1.4	16.58 ± 1.89	0.04 ± 0.01	0.03 ± 0.01	1.49 ± 0.19	6.71 ± 1.94	0.39 ± 0.02	0.80 ± 0.14	6.00 ± 0.23	1.13 ± 0.33
15	1.3	13.48 ± 1.50	0.04 ± 0.01	0.02 ± 0.0	1.25 ± 0.11	5.45 ± 1.11	0.39 ± 0.01	0.69 ± 0.04	4.80 ± 0.57	0.82 ± 0.03
18	1.0	11.13 ± 1.08	0.03 ± 0.0	0.02 ± 0.0	0.69 ± 0.09	5.00 ± 0.68	0.48 ± 0.01	0.55 ± 0.02	3.43 ± 0.42	0.94 ± 0.10
21	0.9	11.41 ± 0.27	0.03 ± 0.0	0.02 ± 0.0	0.89 ± 0.02	4.45 ± 0.47	0.55 ± 0.03	0.52 ± 0.04	3.73 ± 0.25	1.22 ± 0.41
24	0.8	11.00 ± 1.60	0.03 ± 0.01	0.01 ± 0.0	1.25 ± 0.08	3.91 ± 0.94	0.52 ± 0.01	0.61 ± 0.07	3.47 ± 0.33	1.19 ± 0.21
26	0.8	9.51 ± 0.17	0.04 ± 0.01	0.02 ± 0.01	1.09 ± 0.09	3.23 ± 0.58	0.58 ± 0.01	0.67 ± 0.12	2.83 ± 0.39	1.05 ± 0.09
30	0.8	9.36 ± 0.39	0.04 ± 0.01	0.02 ± 0.0	0.99 ± 0.10	3.19 ± 0.72	0.57 ± 0.02	0.66 ± 0.11	2.80 ± 0.29	1.09 ± 0.15

[a] Expressed as percentage of insoluble sugar on dry matter calculated in AIR mass basis. Results are shown as means of triplicate analysis ± Standard Deviation; residue (%) = proportion of biomass recovered as alcohol insoluble residue (AIR); UA = uronic acid

In Table 3, the sugars in the soluble fraction detected in alcohol-soluble residue samples (ASR) are reported. The main sugars detected at the beginning of the SSF process were represented by Glc and Man, reaching up to a percentage of 26.33% and 4.36%, respectively. This starting material sugar composition was in accordance with the results obtained by Abdullah and Mat [9] and Huang et al. [46].

Figure 1 shows the time course of ethanol production and the corresponding levels of soluble and fiber-bound sugars. As it can be observed, the substrate was hydrolyzed in the early phases of the process, as a consequence of the enzyme addition. In fact, a decrease in the insoluble fraction was recorded by $t = 3$ in contraposition with an increase in the concentration in soluble sugar.

In all the samples of digested materials, the insoluble sugar decrease was followed by an increase in the concentration of soluble Glc, Man, Xyl, Ara and UA.

Table 3. Cell wall soluble monosaccharide composition [a].

Hours	Totals	Rhamnose	Fucose	Arabinose	Xylose	Mannose	Galactose	Glucose	UA
0	32.12 ± 2.05	0.01 ± 0.00	0.01 ± 0.00	0.04 ± 0.01	0.03 ± 0.01	4.36 ± 0.24	0.19 ± 0.00	26.63 ± 1.80	0.84 ± 0.03
3	44.05 ± 3.48	0.01 ± 0.01	0.01 ± 0.00	0.57 ± 0.02	1.96 ± 0.08	5.61 ± 0.18	0.40 ± 0.02	34.06 ± 3.06	1.43 ± 0.31
6	52.39 ± 2.10	0.01 ± 0.00	0.03 ± 0.00	1.46 ± 0.08	5.57 ± 0.52	4.35 ± 0.48	0.77 ± 0.08	37.07 ± 2.14	3.13 ± 0.41
9	59.94 ± 4.05	0.02 ± 0.00	0.04 ± 0.01	2.22 ± 0.25	9.15 ± 0.88	2.89 ± 0.18	1.26 ± 0.08	40.91 ± 2.07	3.45 ± 0.77
12	4962 ± 2.26	0.02 ± 0.00	0.04 ± 0.00	2.86 ± 0.45	10.88 ± 0.34	2.91 ± 0.51	2.33 ± 0.29	26.30 ± 1.90	4.29 ± 0.37
15	36.98 ± 1.35	0.01 ± 0.01	0.10 ± 0.02	3.38 ± 0.81	10.40 ± 1.62	1.31 ± 0.57	0.81 ± 0.17	16.58 ± 0.79	4.39 ± 0.76
18	29.91 ± 2.50	0.04 ± 0.00	0.02 ± 0.00	3.23 ± 0.82	12.58 ± 0.81	0.96 ± 0.27	0.53 ± 0.09	8.31 ± 1.39	4.24 ± 0.85
21	26.50 ± 2.55	0.03 ± 0.01	0.07 ± 0.01	3.59 ± 0.32	12.71 ± 1.50	0.47 ± 0.14	0.95 ± 0.19	3.73 ± 0.41	4.95 ± 0.81
24	29.45 ± 0.63	0.04 ± 0.00	0.09 ± 0.01	3.94 ± 0.05	15.25 ± 1.26	0.57 ± 0.09	0.86 ± 0.13	3.68 ± 0.38	5.02 ± 0.88
26	29.30 ± 2.10	0.03 ± 0.00	0.07 ± 0.01	3.97 ± 0.55	15.56 ± 1.04	0.64 ± 0.26	1.08 ± 0.21	2.94 ± 0.66	5.01 ± 0.74
30	28.70 ± 0.80	0.03 ± 0.00	0.07 ± 0.01	3.89 ± 0.35	15.16 ± 0.80	0.61 ± 0.21	1.05 ± 0.18	2.92 ± 0.43	4.97 ± 0.12

[a] Expressed as a percentage of soluble sugar on dry matter. Results are shown as means of triplicate analysis ± Standard Deviation. UA = uronic acid.

The highest concentration of soluble sugars was reached at $t = 9$ (Figure 1) when the Glc concentration detected was 40.91%, followed by 9.15% Xyl, 3.45% UA, 2.89% Man, 2.22% Ara and 1.26% Gal. The soluble sugar increase was the result of the insoluble sugar percentage decreasing, as can be observed in Figure 1. In fact, at that stage, the total insoluble sugar decreased from 26.33% down to 16.04%. This decrease was mainly due to the same monosaccharides increasing in the soluble fraction, as described above (Table 2).

At $t = 18$, the substrate utilization reached a plateau; in fact, both the insoluble and the soluble sugar compositions were stable (Figure 1). The main insoluble sugars that could be detected during the last steps of the fermentation process were Xyl, Glu, Ara and UA; whereas Xyl, Ara, Glu and UA could be detected in the soluble fraction.

While the hexoses were used by *S. cerevisiae* for growth and ethanol production, this yeast species was unable to use the pentoses [23,25]. This behavior explains the progressive concentration increase of xylose and arabinose throughout the fermentation process [47].

The decrease in fiber, of course, was due to the enzymatic saccharification of pineapple cell walls. The pentose increase, due to hemicellulose hydrolysis, was caused mostly by Depol™ 740L [48]. At the same time, Depol's 740 L activity probably enhanced the Accellerase® 1500 activity, considering the presence of ferulic acid, esterified to glucuronoarabinoxylans, in pineapple cell walls [49].

At the end of fermentation, the total insoluble sugars percentage, calculated on dry matter, dropped down to 9.36%; whereas, the total soluble sugars percentage, calculated on dry matter base, decreased from 32.12% to 28.70%. This value was observed mainly due to the percentage of the unused Xyl remaining in the substrate, which increased from 0.03% up to 15.16% during fermentation. On the contrary, the soluble Glc percentage dropped from 26.63% down to 2.92%.

3.3. Ethanol and Glycerol Production

Ethanol production, as well as glycerol production (Figure 1), started at $t = 9$, reaching a concentration of 3.45% and 0.68%, respectively, at $t = 15$. While glycerol concentration was not followed by a further increase, ethanol production went up until $t = 24$, reaching the highest concentration recorded in this process at 3.9% (30.77 g/L). This represents 85% of the theoretical yield (TY), calculated as the maximum ethanol yield in relation to dry matter loss (0.511 g alcohol per 1.0 g dry matter). In comparison with previous studies, where the fermentation substrate was not supplemented with a nitrogen source, vitamins and salts [18,50–52], the highest ethanol production ranged from around 6 g/L to 10 g/L, reached between 24 and 72 h. Whereas, in this study, the ethanol production at the end of the SSF process was higher and it was reached within 24 h. This increase in ethanol production could be ascribable to the nutrient supplementation, which enhanced the ethanol production by *S. cerevisiae*, according to Tropea et al. [23]. The last fermentation phase was characterized by no further ethanol production.

Figure 1 reports the glycerol production during the fermentation time. Glycerol is the main by-product of alcoholic fermentation [53,54] and its synthesis represents an undesirable loss of carbon source, if the aim is to maximize ethanol production. Previous studies reported a glycerol percentage of around 1% [55]. The lower percentage recorded in this study could be ascribable to the addition of salts and vitamin during the fermentation process, as they could promote the NADH re-oxidation by supporting different cellular metabolisms [56–59], resulting in a higher sugar availability for ethanol production.

4. Conclusions

The amount of cell wall sugars detected in pineapple waste after enzymatic hydrolysis makes this substrate an interesting resource for bioethanol production. The TY, calculated on dry matter loss, was 85%, making pineapple waste an excellent raw material for ethanol production by *S. cerevisiae* ATTC 4126. The enzymatic release of xylose and arabinose, sugars not fermented by wild *Saccharomyces* spp., suggest the use of mixed cultures and/or recombinant yeasts, or to the development of robust strains that could ferment hexoses and pentoses simultaneously, with high ethanol production. This would lead to the improvement of the final ethanol concentration and productivity, since, after fermentation, an amount of pentoses was left unutilized in the medium. A further TY improvement could be finally achieved by carrying out further tests with a strict pH control during the process, because this could improve the dry matter utilization and, consequently, also the ethanol production. This study pointed out the possibility of using the supplemented pineapple waste cell wall sugar as a fermentation medium for producing second-generation bioethanol, representing the partial valorization of this food industry residue. However, an integrated approach requires producing more value-added products. In this case, the resulting fermentation substrate was enriched in SCP, and was consequently suitable as animal feed, thus replacing expensive conventional sources of protein, like fishmeal and soymeal, and preventing the production of further waste by the end of the fermentation process.

Author Contributions: Conceptualization, F.S., A.F. and A.T.; methodology, F.S., A.F. and A.T.; validation, F.S., A.F. and A.T.; formal analysis, F.S., A.F. and A.T.; investigation, F.S., A.F. and A.T.; data curation, F.S., A.F. and A.T; writing—original draft preparation, F.S., A.F. and A.T.; writing—review and editing, F.S., A.F. and A.T.; supervision, A.T. All authors have read and agreed to the published version of the manuscript.

Funding: This research received no external funding.

Institutional Review Board Statement: Not applicable.

Informed Consent Statement: Not applicable.

Data Availability Statement: Not applicable.

Conflicts of Interest: The authors declare no conflict of interest.

References

1. Ferracane, A.; Tropea, A.; Salafia, F. Production and maturation of soaps with non-edible fermented olive oil and comparison with classic olive oil soaps. *Fermentation* **2021**, *7*, 245. [CrossRef]
2. Goula, A.M.; Lazarides, H.N. Integrated processes can turn industrial foodwaste into valuable food by-products and/or ingredients: The cases of olive milland pomegranate wastes. *J. Food Eng.* **2015**, *167*, 45–50. [CrossRef]
3. Campos, D.A.; Gómez-García, R.; Vilas-Boas, A.A.; Madureira, A.R.; Pintado, M.M. Management of fruit industrial by-products—A case study on circular economy approach. *Molecules* **2020**, *25*, 320. [CrossRef] [PubMed]
4. Tropea, A.; Potortì, A.G.; Lo Turco, V.; Russo, E.; Vadalà, R.; Rand, R.; Di Bella, G. Aquafeed production from fermented fish waste and lemon peel. *Fermentation* **2021**, *7*, 272. [CrossRef]
5. Roda, A.; de Faveri, D.M.; Giacosa, S.; Dordori, R.; Lambri, M. Effect of pretreatments on the saccharification of pineapple waste as a potential source for vinegar production. *J. Clean. Prod.* **2016**, *112*, 4477–4484. [CrossRef]
6. Lo Turco, V.; Potortì, A.G.; Tropea, A.; Dugo, G.; Di Bella, G. Element analysis of dried figs (*Ficus carica* L.) from the Mediterranean areas. *J. Food Compos. Anal.* **2020**, *90*, 103503. [CrossRef]
7. Jarunglumlert, T.; Bampenrat, A.; Sukkathanyawat, H.; Prommuak, C. Enhanced energy recovery from food waste by co-production 2 of bioethanol and biomethane process. *Fermentation* **2021**, *7*, 265. [CrossRef]
8. Pandit, S.; Savla, N.; Sonawane, J.M.; Sani, A.M.; Gupta, P.K.; Mathuriya, A.S.; Rai, A.K.; Jadhav, D.A.; Jung, S.P.; Prasad, R. Agricultural waste and wastewater as feedstock for bioelectricity generation using microbial fuel cells: Recent advances. *Fermentation* **2021**, *7*, 169. [CrossRef]

9. Abdullah, M.B.; Mat, H. Characterisation of solid and liquid pineapple waste. *Reaktor* **2008**, *12*, 48–52. [CrossRef]
10. Busairi, M.A. Conversion of pineapple juice waste into lactic acid in batch and fed—Batch fermentation systems. *Reaktor* **2008**, *12*, 98–101. [CrossRef]
11. Jamal, P.; Fahrurrazi, T.M.; Zahangir, A.M. Optimization of media composition for the production of bioprotein from pineapple skins by liquid-state bioconversion. *J. Appl. Sci.* **2009**, *9*, 3104–3109. [CrossRef]
12. Raji, Y.O.; Jibril, M.; Misau, I.M.; Danjuma, B.Y. Production of vinegar from pineapple peel. *Int. J. Adv. Sci. Res. Technol.* **2012**, *3*, 656–666.
13. Imandi, S.B.; Bandaru, V.V.R.; Somalanka, S.R.; Bandaru, S.R.; Garapati, H.R. Application of statistical experimental designs for the optimization of medium constituents for the production of citric acid from pineapple waste. *Bioresour. Technol.* **2008**, *99*, 4445–4450. [CrossRef] [PubMed]
14. Gil, L.S.; Maupoey, P.F. An integrated approach for pineapple waste valorisation. Bioethanol production and bromelain extraction from pineapple residues. *J. Clean. Prod.* **2018**, *172*, 1224–1231. [CrossRef]
15. Beigbeder, J.-B.; de Medeiros Dantas, J.M.; Lavoie, J.-M. Optimization of yeast, sugar and nutrient concentrations for high ethanol production rate using industrial sugar beet molasses and response surface methodology. *Fermentation* **2021**, *7*, 86. [CrossRef]
16. Asimakopoulou, G.; Karnaouri, A.; Staikos, S.; Stefanidis, S.D.; Kalogiannis, K.G.; Lappas, A.A.; Topakas, E. Production of omega-3 fatty acids from the microalga *Crypthecodinium cohnii* by utilizing both pentose and hexose sugars from agricultural residues. *Fermentation* **2021**, *7*, 219. [CrossRef]
17. Sun, Y.; Cheng, J. Hydrolysis of lignocellulosic materials for ethanol production: A review. *Bioresour. Technol.* **2002**, *83*, 1–11. [CrossRef]
18. Himmel, M.E.; Ding, S.Y.; Johnson, D.K.; Adney, W.S.; Nimlos, M.R.; Brady, J.W.; Foust, T.D. Biomass recalcitrance: Engineering plants and enzymes for biofuels production. *Science* **2007**, *315*, 804–807. [CrossRef]
19. Galbe, M.; Zacchi, G. Pretreatment of lignocellulosic materials for efficient bioethanol production. *Adv. Biochem. Eng. Biotechnol.* **2007**, *108*, 41–65. [CrossRef]
20. Prasad, R.K.; Chatterjee, S.; Mazumder, P.B.; Gupta, S.K.; Sharma, S.; Vairale, M.G.; Datta, S.; Dwivedi, S.K.; Gupta, D.K. Bioethanol production from waste lignocelluloses: A review on microbial degradation potential. *Chemosphere* **2019**, *231*, 588–606. [CrossRef]
21. Pereira, F.B.; Guimaraes, P.M.; Teixeira, J.A.; Domingues, L. Robust industrial *Saccharomyces cerevisiae* strains for very high gravity bio-ethanol fermentations. *J. Biosci. Bioeng.* **2011**, *112*, 130–136. [CrossRef] [PubMed]
22. Mussatto, S.I.; Dragone, G.; Guimarães, P.M.R.; Silva, J.P.A.; Carneiro, L.M.; Roberto, I.C.; Vicente, A.; Domingue, L.; Teixeira, J.A. Technological trends, global market, and challenges of bio-ethanol production. *Biotechnol. Adv.* **2010**, *28*, 817–830. [CrossRef] [PubMed]
23. Tropea, A.; Wilson, D.; Cicero, N.; Potortì, A.G.; La Torre, G.L.; Dugo, G.; Richardson, D.; Waldron, K.W. Development of minimal fermentation media supplementation for ethanol production using two *Saccharomyces cerevisiae* strains. *Nat. Prod. Res.* **2016**, *30*, 1009–1016. [CrossRef] [PubMed]
24. Dahnum, D.; Tasum, S.O.; Triwahyuni, E.; Nurdin, M.; Abimanyu, H. Comparison of SHF and SSF processes using enzyme and dry yeast for optimization of bioethanol production from empty fruit bunch. *Energy Procedia* **2015**, *68*, 107–116. [CrossRef]
25. Tropea, A.; Wilson, D.; Lo Curto, R.B.; Dugo, G.; Saugman, P.; Troy-Davies, P.; Waldron, K.W. Simultaneous saccharification and fermentation of lignocellulosic waste material for second generation ethanol production. *J. Biol. Res.* **2015**, *88*, 142–143.
26. Chintagunta, A.D.; Ray, S.; Banerjee, R. An integrated bioprocess for bioethanol and biomanure production from pineapple leaf waste. *J. Clean. Prod.* **2017**, *165*, 1508–1516. [CrossRef]
27. Tengborg, C.; Galbe, M.; Zacchi, G. Reduced inhibition of enzymatic hydrolysis of steam pretreated softwood. *Enzym. Microb. Technol.* **2001**, *28*, 835–844. [CrossRef]
28. Tropea, A.; Ferracane, A.; Albergamo, A.; Potortì, A.G.; Lo Turco, V.; Di Bella, G. Single cell protein production through multi food-waste substrate fermentation. *Fermentation* **2022**, *8*, 91. [CrossRef]
29. AOAC. *Official Methods of Analysis*, 18th ed.; Association of Official Analytical Chemists Arlington: Gaithersburg, MD, USA, 2012.
30. Carrier, M.; Loppinet Serani, A.; Denux, D.; Lasnier, J.; Ham Pichavant, F.; Cansell, F.; Aymonier, C. Thermogravimetric analysis as a new method to determine the lignocellulosic composition of biomass. *Biomass Bioenergy* **2011**, *35*, 298–307. [CrossRef]
31. Waldron, K.W.; Selvendran, R.R. Composition of the cell walls of different asparagus (*Asparagus officinalis*) tissues. *Physiol. Plant.* **1990**, *80*, 568–575. [CrossRef]
32. Mandalari, G.; Faulds, C.B.; Sancho, A.I.; Saija, A.; Bisignano, G.; Lo Curto, R.; Waldron, K.W. Fractionation and characterisation of arabinoxylans from brewers spent grain and wheat bran. *J. Cereal Sci.* **2005**, *42*, 205–212. [CrossRef]
33. Blakeney, A.B.; Harris, P.J.; Henry, R.J.; Stone, B.A. A simple and rapid preparation of alditol acetates for monosaccharide analysis. *Carbohydr. Res.* **1983**, *113*, 291–299. [CrossRef]
34. Bastos, R.; Coelho, E.; Coimbra, M.A. Modifications of *Saccharomyces pastorianus* cell wall polysaccharides with brewing process. *Carbohydr. Polym.* **2015**, *124*, 322–330. [CrossRef] [PubMed]
35. Blumenkrantz, N.; Asboe-Hansen, G. New method for quantitative determination of uronic acids. *Anal. Biochem.* **1973**, *54*, 484–489. [CrossRef]
36. Rae, A.L.; Harris, P.J.; Bacic, A.; Clarke, A.E. Composition of the cell walls of *Nicotiana alata* Link et Otto pollen tubes. *Planta* **1985**, *166*, 128–133. [CrossRef]

37. Eliston, A.; Collins, A.; Wilson, D.; Robert, N.; Waldron, K.W. High concentrations of cellulosic ethanol achieved by fed batch semi simultaneous saccharification and fermentation of waste-paper. *Bioresour. Technol.* **2013**, *134*, 117–126. [CrossRef]
38. Aruna, T.E.; Aworh, O.C.; Raji, A.O.; Olagunju, A.I. Protein enrichment of yam peels by fermentation with *Saccharomyces cerevisiae* (BY4743). *Ann. Agric. Sci.* **2017**, *62*, 33–37. [CrossRef]
39. Aruna, T.E. Production of value-added product from pineapple peels using solid state fermentation. *Innov. Food Sci. Emerg. Technol.* **2019**, *57*, 102193. [CrossRef]
40. Alexandre, H.; Guilloux-Benatier, M. Yeast autolysis in sparkling wine. *Aust. J. Grape Wine Res.* **2006**, *12*, 119–127. [CrossRef]
41. Schiavone, M.; Sieczkowski, N.; Castex, M.; Dague, E.; François, J.M. Effects of the strain background and autolysis process on the composition and biophysical properties of the cell wall from two different industrial yeasts. *FEMS Yeast Res.* **2015**, *15*, fou012. [CrossRef]
42. Filippov, M.P.; Shkolenko, G.A.; Kohn, R. Determination of the esterification degree of the pectin of different origin and composition by the method of infrared spectroscopy. *Chem. Zvesti.* **1978**, *32*, 218–222.
43. Zaldivar, J.; Nielsen, J.; Olsson, L. Fuel ethanol production from lignocellulose: A challenge for metabolic engineering and process integration. *Appl. Microbiol. Biotechnol.* **2001**, *56*, 17–34. [CrossRef] [PubMed]
44. Forssell, P.; Kontkanen, H.; Schols, H.A.; Hinz, S.; Eijsink, V.G.H.; Treimo, J.; Robertson, J.A.; Waldron, K.W.; Faulds, C.B.; Buchert, J. Hydrolysis of brewers' spent grain by carbohydrate degrading enzymes. *J. Inst. Brew.* **2008**, *4*, 114–120. [CrossRef]
45. Araya-Cloutier, C.; Rojas-Garbanzo, C.; Velàzquez-Carrillo, C. Effect of initial sugar concentration on the production of L(+) lactic acid by simultaneous enzymatic hydrolysis and fermentation of an agro-industrial waste product of pineapple (*Ananas comosus*) using *Lactobacillus casei* subspeies *rhamnosus*. *Int. J. Biotechnol. Wellness Ind.* **2012**, *1*, 91–100. [CrossRef]
46. Huang, Y.L.; Chow, C.J.; Fang, Y.J. Preparation and physicochemical properties of fiber-rich fraction from pineapple peels as a potential ingredient. *J. Food Drug Anal.* **2011**, *19*, 318–323. [CrossRef]
47. Van Maris, A.J.A.; Abbott, D.A.; Bellissimi, E.; van den Brink, J.; Kuyper, M.; Luttik, M.A.H.; Wisselink, H.W.; Scheffers, W.A.; van Dijken, J.P.; Pronk, J.T. Alcoholic fermentation of carbon sources in biomass hydrolysates by *Saccharomyces cerevisiae*: Current status. *Antonie Van Leeuwenhoek* **2006**, *90*, 391–418. [CrossRef]
48. Treimo, J.; Westereng, B.; Horn, S.J.; Forssell, P.; Robertson, J.A.; Faulds, C.B.; Waldron, K.W.; Buchert, J.; Eijsink, V.G.H. Enzymatic solubilization of brewers' spent grain by combined action of carbohydrases and peptidases. *J. Agric. Food Chem.* **2009**, *57*, 3316–3324. [CrossRef]
49. Smith, B.G.; Harris, P.J. Ferulic acid is esterified to glucuronoarabinoxylans in pineapple cell walls. *Phytochemistry* **2001**, *56*, 513–519. [CrossRef]
50. Bhatia, L.; Johri, S. Biovalorization potential of peels of *Ananas cosmosus* (L.) Merr. for ethanol production by *Pichia stipitis* NCIM 3498 & *Pachysolen tannophilus* MTCC 1077. *Indian J. Exp. Biol.* **2015**, *53*, 819–827.
51. Choonut, A.; Saejong, M.; Sangkharak, K. The production of ethanol and hydrogen from pineapple peel by *Saccharomyces cerevisiae* and *Enterobacter aerogenes*. *Energy Procedia* **2014**, *52*, 242–249. [CrossRef]
52. Casabar, J.T.; Unpaprom, Y.; Ramaraj, R. Fermentation of pineapple fruit peel wastes for bioethanol production. *Biomass Convers. Biorefinery* **2019**, *9*, 761–765. [CrossRef]
53. Choi, W.J. Glycerol-based biorefinery for fuels and chemicals. *Recent Pat. Biotechnol.* **2008**, *2*, 173–180. [CrossRef] [PubMed]
54. Da Silva, G.P.; Mack, M.; Contiero, J. Glycerol: A promising and abundant carbon source for industrial microbiology. *Biotechnol. Adv.* **2009**, *27*, 30–39. [CrossRef] [PubMed]
55. Bai, F.W.; Anderson, W.A.; Moo-Young, M. Ethanol fermentation technologies from sugar and starch feedstocks. *Biotechnol. Adv.* **2008**, *26*, 89–105. [CrossRef]
56. Nissen, T.L.; Kielland-Brandt, M.C.; Nielsen, J.; Villadsen, J. Optimization of ethanol production in *Saccharomyces cerevisiae* by metabolic engineering of the ammonium assimilation. *Metab. Eng.* **2000**, *2*, 69–77. [CrossRef]
57. Nissen, T.L.; Hamann, C.W.; Kielland-Brandt, M.C.; Nielsen, J.; Villadsen, J. Anaerobic and aerobic batch cultivations of *Saccharomyces cerevisiae* mutants impaired in glycerol synthesis. *Yeast* **2000**, *16*, 463–474. [CrossRef]
58. Hou, J.; Lages, N.F.; Oldiges, M.; Vemuri, G.N. Metabolic impact of redox cofactor perturbations in *Saccharomyces cerevisiae*. *Metab. Eng.* **2009**, *11*, 253–261. [CrossRef]
59. Guadalupe, M.V.; Almering, M.J.; van Maris, A.J.; Pronk, J.T. Elimination of glycerol production in anaerobic culture sofa *Saccharomyces cerevisiae* strain engineered to use acetic acid as an electron acceptor. *Appl. Environ. Microbiol.* **2010**, *76*, 190–195. [CrossRef]

Article

Bioethanol Production from Spent Sugar Beet Pulp—Process Modeling and Cost Analysis

Damjan Vučurović [1], Bojana Bajić [1,*], Vesna Vučurović [1], Rada Jevtić-Mučibabić [2] and Siniša Dodić [1]

[1] Department of Biotechnology and Pharmaceutical Engineering, Faculty of Technology Novi Sad, University of Novi Sad, Bulevar cara Lazara 1, 21000 Novi Sad, Serbia; dvdamjan@uns.ac.rs (D.V.); vvvesna@uns.ac.rs (V.V.); dod@uns.ac.rs (S.D.)

[2] Institute of Food Technology in Novi Sad, University of Novi Sad, Bulevar cara Lazara 1, 21000 Novi Sad, Serbia; rada.jevtic@fins.uns.ac.rs

* Correspondence: baj@uns.ac.rs; Tel.: +381-214853620

Abstract: Global economic development has led to the widespread use of fossil fuels, and their extensive use has resulted in increased environmental pollution. As a result, significantly more attention is being paid to environmental issues and alternative renewable energy sources. Bioethanol production from agro-industrial byproducts, residues, and wastes is one example of sustainable energy production. This research aims to develop a process and cost model of bioethanol production from spent sugar beet pulp. The model was developed using SuperPro Designer® v.11 (Intelligen Inc., Scotch Plains, NJ, USA) software, and determines the capital and production costs for a bioethanol-producing plant processing about 17,000 tons of spent sugar beet pulp per year. In addition, the developed model predicts the process and economic indicators of the analyzed biotechnological process, determines the share of major components in bioethanol production costs, and compares different model scenarios for process co-products. Based on the obtained results, the proposed model is viable and represents a base case for further bioprocess development.

Keywords: bioethanol; spent sugar beet pulp; model; economics

Citation: Vučurović, D.; Bajić, B.; Vučurović, V.; Jevtić-Mučibabić, R.; Dodić, S. Bioethanol Production from Spent Sugar Beet Pulp—Process Modeling and Cost Analysis. *Fermentation* **2022**, *8*, 114. https://doi.org/10.3390/fermentation8030114

Academic Editor: Alessia Tropea

Received: 22 February 2022
Accepted: 3 March 2022
Published: 6 March 2022

1. Introduction

In the past few decades, sustainability has become a key consideration due to the depletion of fossil fuels and other natural resources, increased environmental awareness, and the social benefits of reducing environmental pollution [1–3]. Fossil fuels are the main contributors to climate change; therefore, in order to meet the increasing demand for energy production, it is necessary to utilize a valuable and eco-friendly alternative to non-renewable fuels, such as bioethanol produced from renewable feedstock. Due to growing concerns regarding the global food supply, second-generation bioethanol (from lignocellulosic non-edible biomass) and third-generation bioethanol (from algal sources) are becoming increasingly attractive [4,5]. Additionally, in accordance with the Renewable Energy Directive (Directive EU 2018/2001) [6], a common framework was established for the promotion of energy from renewable sources in the EU, setting a binding target for the EU's gross final consumption, being that the overall share of energy from renewable sources should be 32% by 2030. This legislation also promotes the use of non-food crops for biofuel production, and has limited the amount of biofuels and bioliquids produced from food or feed crops. According to the Energy Development Strategy [7], the Republic of Serbia will require further sustainable energy development until 2030, based on activities that include intensive use of renewable energy sources.

Lignocellulosic biomass is considered to be a key renewable resource of the future, while agro-industrial byproducts, residues, and wastes have enormous potential to generate sustainable bioproducts and bioenergy [8,9]. One such agro-industrial byproduct is sugar beet pulp, which is obtained in the sugar-processing industry after sucrose extraction from sugar beet, and which represents a very attractive raw material for bioethanol production due to its composition. This sugar processing industry byproduct is typically used as animal feed; hence, it is significant to investigate the possibility of obtaining greater economic and environmental benefits by using a given raw material to produce a value-added product, such as bioethanol, along with the valorization of other process byproducts to achieve a sustainable bioprocess [10,11].

Significant research has been carried out on bioethanol production from different lignocellulosic raw materials [12–14]. For bioprocess design and optimization, simulations are of great importance for reducing costs and the number of required experiments, as well as predicting different potential scenarios. The application of bioprocess optimization, modeling, and simulation is of enormous importance in the development of each bioprocess [15,16]. An economic analysis of the whole process on a commercial scale can be performed using an in-depth process model which includes all unit operations from biomass handling to bioethanol distillation. Tradeoffs in energy and water use in the process, as well as capital costs, can be understood using such models. The data (emissions, energy and utilities requirements) generated by these models can be utilized for the analysis of the environmental impact of the process [17]. Computer simulation process models have been used by various researchers to study bioethanol production from different agro-industrial byproducts, residues, and wastes, such as grass straw [17], triticale grain and straw [18], sugarcane and blue agave bagasse [19], oil palm frond [20], sweet potato [21], and sugar beet raw juice [22].

According to data from the Statistical Office of the Republic of Serbia, sugar beet represents one of the most important crops in our country. More specifically, it is the third most produced crop in recent years, with a production of 2,018,215 tons in the year 2020 [23]. After processing this amount of sugar beet, about 500,000 tons of wet-pressed spent sugar beet pulp (water content approximately 75–80%) remain, which can be converted into dry spent sugar beet pulp (about 10% water content) [24]. Due to its availability and low price, spent sugar beet pulp could have great potential for bioethanol production in Serbia [25]. Furthermore, spent sugar beet pulp is especially rich in polysaccharides (hemicelluloses, cellulose, and pectin) and has a low lignin content. A lower lignin content in the feedstock facilitates pretreatment and decreases the bioethanol production costs. In the dried form, it is generally steady, and can be either utilized directly or stored for up to a year without any unfavorable effect on its quality [10,26,27]. Therefore, the aim of this research was to provide a simulation solution for a sustainable bioethanol production plant from spent sugar beet pulp with minimal waste generation. In this research, a process and cost model for a bioethanol production plant has been developed with the aim of applying it in the evaluation of new technologies and products based on lignocellulosic raw materials.

2. Materials and Methods

2.1. Process Overview

In this research, process design and economic analyses were performed using SuperPro Designer® v.11 (Intelligen Inc., Scotch Plains, NJ, USA). The spent sugar beet pulp is brought to the factory by trucks and is stored before being used as raw material in the bioethanol-production process. The spent sugar beet pulp is then transferred from the storage unit to the shredders, where it is ground to a size optimal for further processing. Ground sugar beet pulp is sent for pretreatment, where hemicellulose and a small part of cellulose are converted to soluble sugars, by exposing the pulp to high temperatures and dilute sulfuric acid. Under these conditions, a certain amount of lignin also dissolves, which improves the efficiency of the cellulose hydrolysis. The low lignin content of the spent sugar beet pulp makes this raw material suitable for bioethanol production [28]. After pretreatment, the mixture is cooled and the liquid part containing sulfuric acid is separated from the solid phase. Lime is added to the liquid fraction in order to neutralize the solution and obtain gypsum, which forms a precipitate. Filtration is used to separate the gypsum, and the filtrate is mixed again with the solid phase before enzymatic hydrolysis.

Fermentation and enzymatic hydrolysis are performed separately (separated hydrolysis and fermentation—SHF) using several vessels, which allows this process to be performed at a slightly elevated temperature, reducing the time and amount of enzymes required, and increasing enzymatic activity. The advantage of SHF is the ability to perform both hydrolysis and fermentation under optimal conditions, although the entire process time is longer [29]. The enzyme preparation used for cellulose hydrolysis consists of endoglucanase, exoglucanase, and β-glucosidase enzymes. The hydrolysate of spent sugar beet pulp and the production microorganism are introduced into the main bioreactor. For fermentation, a glucose- and xylose-fermenting yeast is used as a biocatalyst, and five cascade vessels are used to ferment the hydrolysate to ethanol. After fermentation, the broth containing bioethanol is sent to separate and purify the product.

The separation and purification of bioethanol from the fermentation broth are performed by distillation (in two columns) and molecular sieves. The first distillation column removes dissolved carbon dioxide and water, while the second (rectification) column concentrates the bioethanol solution to an almost azeotropic mixture. All the water from this mixture is removed by adsorption in the vapor phase in molecular sieves. Ultimately, the 99.6% bioethanol vapor is cooled in a heat exchanger, condensed, and stored until use or sale. The process water obtained from the distillation, rectification, and molecular sieves is recirculated and reused in the pretreatment reactors, which reduces the process costs.

2.2. Process Design

The process flow diagram of the bioethanol production process from spent sugar beet pulp is shown in Figures 1–3. The economic analysis was conducted based on the process design and on mass and energy balances by using SuperPro Designer software. Figure 1 represents the process flow diagram of the reception and preparation of spent sugar beet pulp.

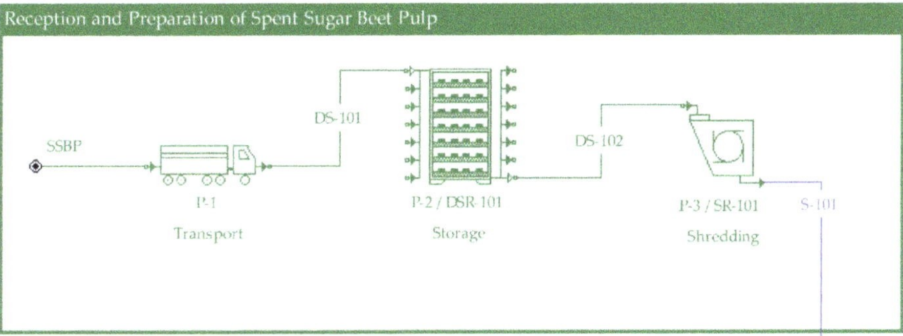

Figure 1. Process flow diagram of the process of bioethanol production from spent sugar beet pulp, consisting of reception (transportation and storage) and preparation (shredding) stages.

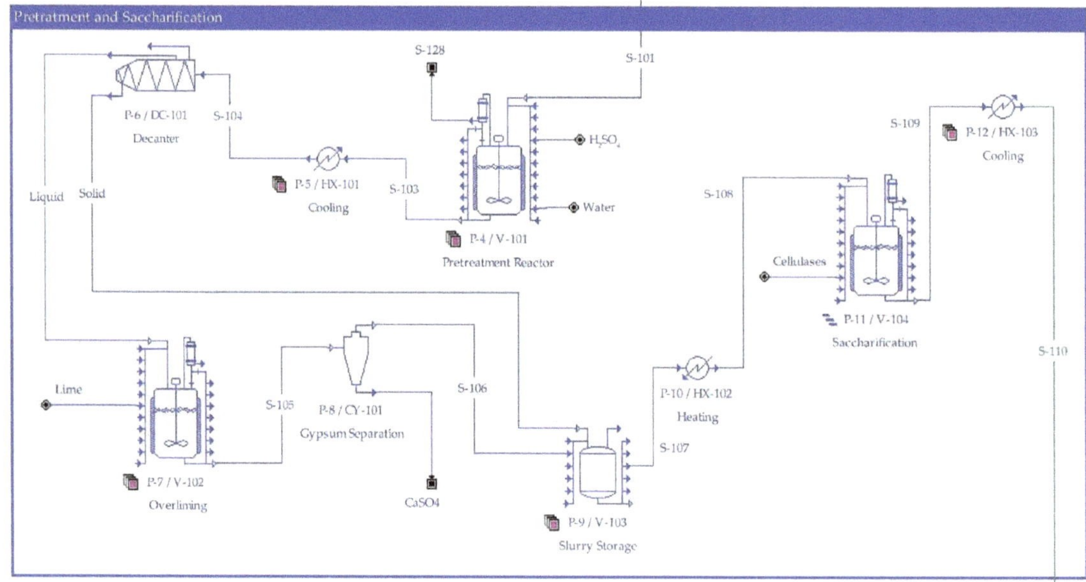

Figure 2. Process flow diagram of the bioethanol production process from spent sugar beet pulp, consisting of pretreatment (acid hydrolysis) and saccharification (enzyme hydrolysis) stages.

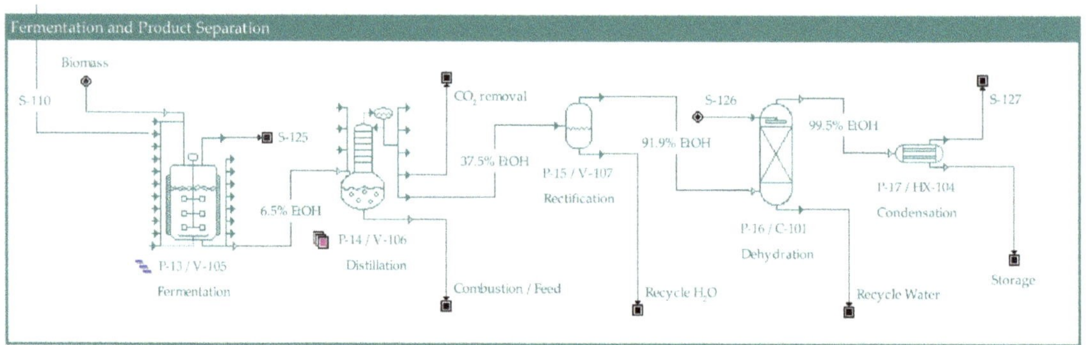

Figure 3. Process flow diagram of the bioethanol production process from spent sugar beet pulp, consisting of fermentation and product separation (distillation, rectification, and dehydration) stages.

The amount of spent sugar beet pulp required for one batch in this process is 20,000 kg. This quantity is calculated based on the amount of spent sugar beet pulp available from a local sugar factory processing 250,000 t of sugar beet, and on the possible number of batches per year. Spent sugar beet pulp (89.2% dry matter) contains: 21.7 (%dm) cellulose, 24.0 (%dm) hemicellulose, 7.6 (%dm) reducing sugars, 2.4 (%dm) lignin, and 9.3 (%dm) proteins [30]. Since the capacity of transport trucks (P-1) is 10 tons, less than 1700 deliveries are required annually. Sugar beet is harvested in a relatively short timeframe, and the obtained spent pulp requires storage in order to provide a constant source of raw material to the plant. Long-term storage can be in covered storages (P-2/DSR-101) located close to the plant itself. The spent sugar beet pulp from the storage is transferred into the shredder (P-3/SR-101), where it is reduced to an optimal size for pretreatment and hydrolysis. The shredder operating time is 15 min, with an energy consumption of 4000 kW and a throughput of 80,000 kg/h. Figure 2 shows the process flow diagram of the pretreatment and saccharification stages of the bioethanol production process from spent sugar beet pulp.

The ground beet pulp is transferred to two pretreatment reactors (P-4/V-101), after which around 45,000 kg of water and 718 kg concentrated sulfuric acid are added, in order to achieve the optimal acid concentration in the reactor of 1% H_2SO_4. At this point, the slurry contains 30% insoluble solids. The pretreatment reactors are operated at elevated temperatures (170 °C) and have a retention time of 2 min. The volume of each reactor is 35.2 m^3 (h = 6.545 m, d = 2.62 m). High-pressure steam is used as a medium for heat transfer, typically with a flow rate of around 200,000 kg/h.

Table 1 shows the reactions, with the corresponding reaction extents, in the pretreatment reactors.

Table 1. Pretreatment reactions with reaction extents.

Reaction	Referent Component	Reaction Extent (%)
Cellulose + nWater → nGlucose	Cellulose	7.7
Cellulose + 1/2nWater → 1/2nCellobiose	Cellulose	0.7
Hemicellulose + nWater → nXylose	Xylan	92.5
(Lignin)n → nSoluble Lignin	Hemicellulose	5

The slurry leaving the pretreatment reactors is cooled to 50 °C (neutralization of the slurry takes place at this temperature), by cooling water in two plate and frame heat exchangers (P-5/HX-101) with a surface area of 89.84 m^2 each. The slurry stays in the heat exchanger for 10 min, after which the treated slurry containing 22% insoluble solids is added to a decanter centrifuge (P-6/DC-101) to separate the solid from the liquid phase. This equipment unit operates at a volumetric throughput close to 378,000 L/h and a duration of 10 min. The reason for the separation of the liquid is the reduction in the acidity (sulfuric acid) of the liquid phase, which positively affects the fermentation process.

The separated liquid phase is neutralized in a vessel (P-7/V-102) by adding around 545 kg of lime (calcium hydroxide) and keeping for 1 h, which is a sufficient time for the required reaction to take place. Two 27,800 L vessels (h = 6.05 m, d = 2.42 m) are required, and the power consumption for mixing is 3.7 kW. The formed crystals are separated in a hydrocyclone (P-8/CY-101) with the following characteristics: inlet fluid velocity—5 m/s, pressure drop—1.2 bars, and body diameter—0.83 m. This procedure removes 99.5% of the formed gypsum crystals with a dry matter content of 83%, which means gypsum can be handled as a solid. After removing the gypsum, the neutralized liquid is mixed again with the solid fraction from the pretreatment in the slurry storage (P-9/V-103). The mixing power of these vessels is 24.6 kW. Two vessels with a volume of 34.7 m^3 (h = 7.36 m and d = 2.45 m) are required.

The neutralized and pretreated slurry, containing 22% solids, is introduced into a heat exchanger (P-10/HX-102; heat exchange surface 45.25 m^2) and heated to 65 °C or hydrolysis temperature (using low-pressure steam, whose throughput is 11,295 kg/h), before being transferred into hydrolysis vessels (P-11/V-104). The hydrolysis occurs in five 78.7 m^3 (h = 8.56 m, d = 3.42 m) vessels operating in a cascade for 36 h. For this model, cellulase was fed at the rate of 10 international filter paper units (IFPU) per gram of cellulose, assuming an enzyme concentration of 50 kU/m^3 [31]. Table 2 shows the reactions and their reaction extents for the hydrolysis process. After the hydrolysis process, the hydrolysate of spent sugar beet pulp contains 13.6% reducing sugars.

Table 2. Hydrolysis reactions and reaction extents.

Reaction	Referent Component	Reaction Extent (%)
Cellulose + 1/2nWater → 1/2n Cellobiose	Cellulose	1.2
Cellulose + nWater → nGlucose	Cellulose	90
Cellobiose + Water → 2Glucose	Cellobiose	100

The hydrolysate of spent sugar beet pulp is cooled to 30 °C in three heat exchangers, with a heat exchange surface of 93.5 m^2 each. Figure 3 shows the process flow diagram of the fermentation and product separation stages of the bioethanol production process.

Fermentation occurs in five 87.4 m^3 (h = 10 m, d = 3.3 m) fermenters (P-13/V-105) for 36 h. Table 3 shows the reactions and their reaction extents for the fermentation process. The concentration of bioethanol in the fermentation broth after the bioprocess is 6.5%, while the sugar concentration is 1.63%. After the bioprocess, the fermentation broth is introduced into two distillation columns (reboiler temperature of 85 °C, condenser temperature of 45 °C) of 37,900 L (P-14/V-106), with a reflux ratio of 3:1, and adjusted to emit CO_2 and as little bioethanol as possible at the top of the column, removing 86% of the water at the bottom of the column. A high percentage of bioethanol (>99%) from the feed is separated as a 37.55% mixture of water and bioethanol.

Table 3. Reactions and reaction extents for the fermentation process.

Reaction	Referent Component	Reaction Extent (%)
Glucose → 2Bioethanol + 2Carbon dioxide	Glucose	90
Glucose + 5.7Other compounds → 6Biomass + 2.87Oxygen + 2.4Water	Glucose	4
Glucose + 2Water → 2Glycerol + Oxygen	Glucose	0.4
Glucose + 2 Carbon dioxide → Oxygen + 2Succinic acid	Glucose	0.6
Glucose → 3Acetic acid	Glucose	1.5
Glucose → 2 Lactic acid	Glucose	0.2
3Xylose → 5Bioethanol + 5 Carbon dioxide	Xylose	80
Xylose + 4.67 Other compounds → 5Biomass + 2.35Oxygen + 2Water	Xylose	4
3Xylose + 5Water → 5Glycerol + 2.5Oxygen	Xylose	0.3
Xylose + Water → Xylitol + 0.5 Oxygen	Xylose	4.6
3Xylose + 5Carbon dioxide → 2.5Oxygen + 5Succinic acid	Xylose	0.9
2 Xylose → 5Acetic acid	Xylose	1.4
3 Xylose → 5Lactic acid	Xylose	0.2

Due to its composition, the contents from the bottom of the distillation column can be dried and burned or used as animal feed while reducing operating costs, which is examined in the economic analysis of the model.

The vapor phase from the column (a mixture of bioethanol and water) is introduced directly to the rectification column (P-15/V-107) with a working volume of 1800 L (heating steam throughput 5237 kg/h). The vapor phase at the top of the rectification column contains 91.9% bioethanol, while the content of bioethanol at the bottom of the column is 0.06%.

The vapor phase from the top of the rectification column is introduced into the adsorption unit of molecular sieves (P-16/C-101). The 7300 L column removes 95% of the water. Pure 99.56% bioethanol is cooled to 20 °C in a heat exchanger (P-17/HX-104) with a heat exchange surface of 71.9 m^2, and placed into storage. The heat transfer medium is chilled water, with a throughput of 345,444 L/h.

3. Results and Discussion

3.1. Economic Analysis

Figure 4 represents the results of the economic analysis, and provides a detailed breakdown of the capital investment costs for this process model. This form of presentation of these results was chosen in order to distinctly show how each cost item is generated by adding up the previous ones. For example, the capital investment cost is obtained by adding together the direct fixed capital (DFC), working capital, and start-up and validation costs; the DFC is the sum of the total plant cost (TPC) and contractor's fees and contingencies, etc.

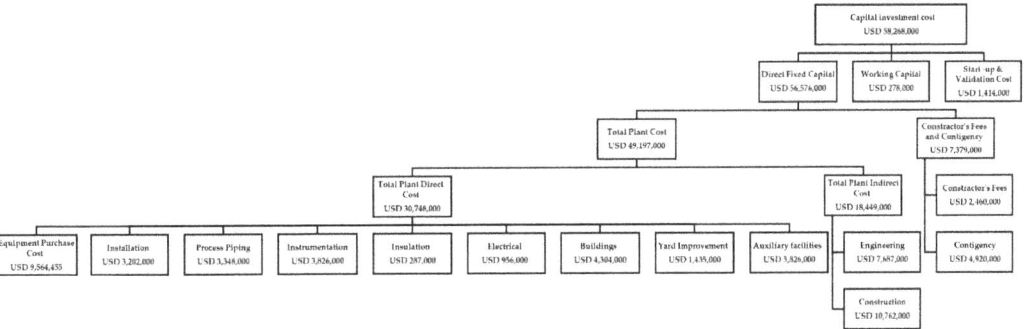

Figure 4. Detailed breakdown of the capital investment costs for the bioethanol-production plant from spent sugar beet pulp.

It should also be noted that the purchase price of major equipment in Figure 4 was obtained from local and foreign equipment suppliers. However, if the suppliers and modeled equipment capacities did not match, the following equation was used to obtain the appropriate price:

$$P_M = P_S \left(\frac{C_M}{C_S} \right)^{0.6} \tag{1}$$

where P_M is the estimated price (USD) of the modeled equipment item with the capacity C_M (L, kW, kg/h, or something else), while P_S is the supplier's price (USD) of the same equipment item with its available capacity C_S, which has the same units as the modeled capacity C_M.

As seen in Figure 4, the estimated capital investment that should be charged to this project is USD 58,268,000. However, the bioethanol industry practice is to multiply the total equipment purchase cost by three in order to obtain the total capital investment, which would lower the current capital expenses by almost 50%. Other studies also reported this type of difference in their capital cost estimations [22,32], demonstrating that this industry feedback is valid only in the initial phase of process modelling, when there is a lack of real data.

Unit production cost breakdown is shown in Figure 5, which shows that the two key parameters are the raw materials and the utilities, each with over 30% share in the bioethanol-production cost. Spent sugar beet pulp, as a primary feedstock, has a major impact on the cost of producing bioethanol, due to the high quantities used per batch. The price of spent sugar beet pulp changes with changes to the price of sugar beet, as a result of market and weather conditions. For this reason, a 10-year average price was used in the model. Likewise, the required quantities of biomass (yeast), cellulase enzymes, H_2SO_4, and lime were defined by the model, and their prices are 1.15, 0.08, 0.07, and 0.07 USD/kg, respectively.

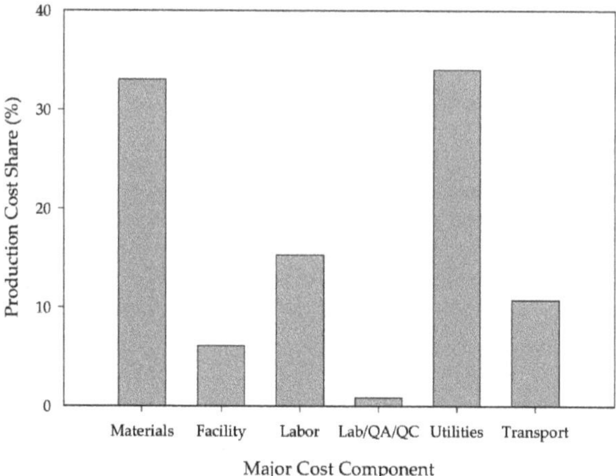

Figure 5. The share of the major cost components in the total operating costs for the modeled spent sugar beet pulp bioethanol production plant.

On the other hand, utilities also play a major role in unit production costs, due to the high energy demand in the pretreatment and separation phase of the process. Prices for steam, electricity and chilled water are 12.00 USD/MT, 0.1 USD/kWh, and 0.4 USD/MT, respectively.

The prices for all materials, raw material, utilities, and byproducts were obtained from official reports and personal consultations with suppliers [23,33].

Since the bottoms (water) of the rectifier column and molecular sieves are recycled into the pretreatment reactor, the need for industrial water is reduced to a minimum.

Since the variability of raw material prices has been taken into account, the energy efficiency (utilities exploitation) has been examined through different model scenarios.

3.2. *Scenario and Sensitivity Analysis*

There are several products that can be obtained in the process of spent sugar beet conversion to bioethanol. Bioethanol, as the main product, is intended to be sold as a renewable fuel, i.e., a substitute for fossil fuels (gasoline). Hence, its price was taken from the Global Petrol Prices website [34], used in the model as 1.07 USD/kg (~0.84 USD/L). Co-products of the examined model are carbon dioxide and animal feed. Carbon dioxide from fermentation can be sold to food and beverage producers for the price of 0.015 USD/kg. The bottoms from the distillation column, containing the nonfermented parts of the pulp and yeasts, can be dried, thus obtaining animal feed with a market price of 0.05 USD/kg. On the other hand, the dried distillation stillage can be used in combustion to generate heat for the process, thus lowering the need for buying steam and lowering operating costs. Table 4 shows the economic indices of the model for the two examined scenarios, i.e., when the stillage is used for feed or for combustion.

Table 4. Different model scenarios for process co-products.

Project Indices	Combustion Scenario	Animal Feed Scenario
Gross margin (%)	69.11	61.05
Return on investment (%)	16.08	11.27
Payback time (years)	6.22	8.88
Internal rate of return (%)	10.22	4.55
Net present value at 7.00% (USD)	11,887,591	−8,252,807

Gross margin helps a company assess the profitability of its manufacturing activities, i.e., the higher the value of this parameter, the more capital a company retains. It equates to revenue minus cost of goods sold divided by revenue. Return on investment (ROI) is used to calculate the investor's benefit compared to their investment cost, and it is determined as net income divided by the capital cost of the investment. In general, the higher the ROI, the greater the benefit earned. Payback time is the most important static method in investment calculations. It represents how long it will take to get back the money that has been invested, and it is often used because it is easy to apply and understand. If a project pays back its investment in five years, it is better than a project with a 10-year payback time. Net present value (NPV) is the present value of the cash flow at the required rate of return of a project compared to the initial investment. In other words, NPV considers the time value of money, translating future cash flows into today's dollars. A project is acceptable if it has a positive value of NPV. The internal rate of return (IRR) is defined as the discount rate which, when applied to the cash flow of a project, produces an NPV equal to zero. This discount rate can then be thought of as the forecast return for the project. If the IRR is greater than a preset percentage target (7% in this case), the project can be accepted. If the IRR is less than the target, the project is rejected.

Comparing the two scenarios from Table 4, it turns out that the one with the stillage combustion is more favorable. Moreover, the negative value of NPV, as well as the IRR value lower than 7% for the feed scenario, makes it arid for investment. Likewise, the payback time for acceptable projects in practice should be lower, or around 7 years, which is not the case for the scenario where the dried stillage is used as animal feed.

Since the only project index that became undesirable in the scenario analysis was NPV (IRR is tied to NPV), it was interesting to examine how using one part of the stillage for combustion and the other remaining part for animal feed would influence this economic parameter, i.e., at which the ratio of combustion/feed is NPV equal to zero. By varying the percentage of the amount of stillage going to combustion from 10 to 90% (by 20 increments), which meant that, on the other side, 90 to 10% of stillage was going to animal feed production, the effect of this split on NPV was obtained and is shown in Figure 6. The bars in Figure 6 represent the obtained data for NPV, while the line shows the linear connection between NPV and stillage to combustion percentage, which was obtained after fitting a linear equation into the data obtained. The equation is as follows:

$$\text{NPV} = 8,252,807.650 + 201,403.985\ \text{STC} \tag{2}$$

where STC is the percentage of stillage sent to combustion. From the intersection of the linear plot and X-axis in Figure 6, as well as from the above equation, the percentage of combusted stillage should be 40.97% for the NPV to be 0.

At this ratio of combustion/feed, the economic indices are as follows: gross margin—64.81%, return on investment—13.24%, payback time—7.55 years, and internal rate of return—6.99%. This means that the project is economically viable, with nearly 60% of the distillation stillage usable for animal feed.

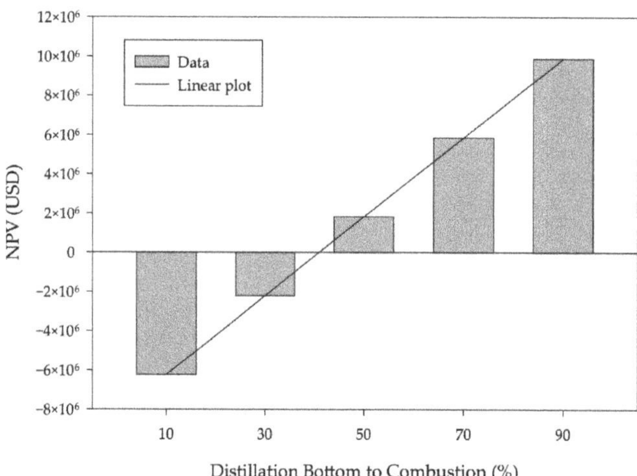

Figure 6. Spent sugar beet pulp bioethanol production process model net present value (NPV) as a function of stillage to combustion percentage.

4. Conclusions

The developed process model for bioethanol production presents the base case for processing about 17,000 tons of spent sugar beet pulp annually. A cost model has been developed for an economic analysis of this bioethanol production from spent sugar beet pulp. The obtained distillation stillage can be used as animal feed or for combustion to generate heat for the process, thus lowering the need for utilities and reducing operating costs. Therefore, two scenarios were assessed: when stillage is used for animal feed and when it is used for combustion, as well as at what split ratio of combustion/feed for the stillage is acceptable. Comparing the two scenarios, results showed that the scenario with the stillage combustion is more favorable. The results obtained for the stillage split factor showed that this project becomes economically viable when approximately 40% of the distillation stillage or more is used for generating power, with the remaining being exploited as animal feed.

Author Contributions: D.V., B.B. and S.D.; methodology, D.V. and B.B.; data collection and analysis, V.V. and R.J.-M.; writing—original draft preparation, B.B., V.V. and R.J.-M.; writing—review and editing, D.V. and S.D.; supervision, D.V. and S.D. All authors have read and agreed to the published version of the manuscript.

Funding: This work was supported by the Ministry of Education, Science and Technological Development of the Republic of Serbia (Grant no. 451-03-68/2022-14/200134).

Institutional Review Board Statement: Not applicable.

Informed Consent Statement: Not applicable.

Data Availability Statement: Not applicable.

Conflicts of Interest: The authors declare no conflict of interest. The funders had no role in the design of the study, nor in the collection, analyses, or interpretation of the data, nor in the writing of the manuscript, nor in the decision to publish the results.

References

1. Alexandri, M.; López-Gómez, J.P.; Olszewska-Widdrat, A.; Venus, V. Valorising Agro-industrial Wastes within the Circular Bioeconomy Concept: The Case of Defatted Rice Bran with Emphasis on Bioconversion Strategies. *Fermentation* **2020**, *6*, 42. [CrossRef]
2. Martins, F.; Felgueiras, C.; Smitkova, M.; Caetano, N. Analysis of Fossil Fuel Energy Consumption and Environmental Impacts in European Countries. *Energies* **2019**, *12*, 964. [CrossRef]
3. Dodić, S.; Vučurović, D.; Popov, S.; Dodić, J.; Zavrago, Z. Concept of cleaner production in Vojvodina. *Renew. Sust. Energ. Rev.* **2010**, *14*, 1629–1634. [CrossRef]
4. Tse, T.J.; Wiens, D.J.; Reaney, M.J.T. Production of Bioethanol—A Review of Factors Affecting Ethanol Yield. *Fermentation* **2021**, *7*, 268. [CrossRef]
5. Branco, R.H.R.; Serafim, L.S.; Xavier, A.M.R.B. Second Generation Bioethanol Production: On the Use of Pulp and Paper Industry Wastes as Feedstock. *Fermentation* **2019**, *5*, 4. [CrossRef]
6. Directive (Eu) 2018/2001 of the European Parliament and of the Council, Official Journal of the European Union. 2018. Available online: https://eur-lex.europa.eu/eli/dir/2018/2001/oj (accessed on 29 December 2021).
7. Energy Sector Development Strategy of the Republic of Serbia until 2025 with Projections until 2030, Official Gazette of RS, No. 101/2015, Belgrade. 2015. Available online: https://www.pravno-informacioni-sistem.rs/SlGlasnikPortal/eli/rep/sgrs/skupstina/ostalo/2015/101/1/r (accessed on 29 December 2021).
8. Anwar, Z.; Gulfraz, M.; Irshad, M. Agro-industrial lignocellulosic biomass a key to unlock the future bio-energy: A brief review. *J. Radiat. Res. Appl. Sci.* **2014**, *7*, 163–173. [CrossRef]
9. Ning, P.; Yang, G.; Hu, L.; Sun, J.; Shi, L.; Zhou, Y.; Wang, Z.; Yang, J. Recent advances in the valorization of plant biomass. *Biotechnol. Biofuels* **2021**, *14*, 102. [CrossRef]
10. Rezić, T.; Oros, D.; Marković, I.; Kracher, D.; Ludwig, R.; Šantek, B. Integrated Hydrolyzation and Fermentation of Sugar Beet Pulp to Bioethanol. *J. Microbiol. Biotechnol.* **2013**, *23*, 1244–1252. [CrossRef]
11. Gumienna, M.; Szambelan, K.; Jeleń, H.; Czarnecki, Z. Evaluation of ethanol fermentation parametersfor bioethanol production from sugar beet pulp and juice. *J. Inst. Brew.* **2014**, *120*, 543–549. [CrossRef]
12. Mood, S.H.; Golfeshan, A.H.; Tabatabaei, M.; Jouzani, G.S.; Najafi, G.H.; Gholami, M.; Ardjmand, M. Lignocellulosic biomass to bioethanol, a comprehensive review with a focus on pretreatment. *Renew. Sust. Energ. Rev.* **2013**, *17*, 77–93. [CrossRef]
13. Vasić, K.; Knez, Ž.; Leitgeb, M. Bioethanol Production by Enzymatic Hydrolysis from Different Lignocellulosic Sources. *Molecules* **2021**, *26*, 753. [CrossRef] [PubMed]
14. Awoyale, A.A.; Lokhat, D. Experimental determination of the effects of pretreatment on selected Nigerian lignocellulosic biomass in bioethanol production. *Sci. Rep.* **2021**, *11*, 557. [CrossRef] [PubMed]
15. Heinzle, E.; Biwer, A.P.; Cooney, C.L. *Development of Sustainable Bioprocesses: Modeling and Assessment*; John Wiley & Sons: Chichester, UK, 2006.
16. Sokolowski, J.A.; Banks, C.M. *Principles of Modeling and Simulation a Multidisciplinary Approach*; John Wiley & Sons: Hoboken, NJ, USA, 2009.
17. Kumar, D.; Murthy, G.S. Impact of pretreatment and downstream processing technologies on economics and energy in cellulosic ethanol production. *Biotechnol. Biofuels.* **2011**, *4*, 27. [CrossRef] [PubMed]
18. Mupondwa, E.; Li, X.; Tabil, L. Integrated bioethanol production from triticale grain and lignocellulosic straw in Western Canada. *Ind. Crops Prod.* **2018**, *117*, 75–87. [CrossRef]
19. Barrera, I.; Amezcua-Allieri, M.A.; Estupinan, L.; Martínez, T.; Aburto, J. Technical and economical evaluation of bioethanol-production from lignocellulosic residues in Mexico:Case of sugarcane and blue agave bagasses. *Chem. Eng. Res. Des.* **2016**, *107*, 91–101. [CrossRef]
20. Abdullah, S.S.S.; Shirai, Y.; Ali, A.A.M.; Mustapha, M.; Hassan, M.A. Case study: Preliminary assessment of integrated palm biomass biorefinery for bioethanol production utilizing non-food sugars from oil palm frond petiole. *Energy Convers. Manag.* **2016**, *108*, 233–242. [CrossRef]
21. Ferrari, M.D.; Guigou, M.; Lareo, C. Energy consumption evaluation of fuel bioethanol production from sweet potato. *Bioresour. Technol.* **2013**, *136*, 377–384. [CrossRef]
22. Vučurović, D.; Dodić, S.; Popov, S.; Dodić, J.; Grahovac, J. Process model and economic analysis of ethanol production from sugar beet raw juice as part of the cleaner production concept. *Bioresour. Technol.* **2012**, *104*, 367–372. [CrossRef]
23. *Statistical Yearbook of the Republic of Serbia*; Statistical Office of the Republic of Serbia: Belgrade, Serbia, 2021.
24. Hutnan, M.; Drtil, M.; Mrafkova, L. Anaerobic biodegradation of sugar beet pulp. *Biodegradation* **2000**, *11*, 203–211. [CrossRef]
25. Ivetić, D.; Šćiban, M.; Antov, M. Enzymatic hydrolysis of pretreated sugar beet shreds: Statistical modeling of the experimental results. *Biomass Bioenergy* **2012**, *47*, 387–394. [CrossRef]
26. Rana, A.K.; Gupta, V.K.; Newbold, J.; Roberts, D.; Rees, R.M.; Krishnamurthy, S.; Thakur, V.K. Sugar beet pulp: Resurgence and trailblazing journey towards a circular bioeconomy. *Fuel* **2022**, *312*, 122953. [CrossRef]
27. Balat, M.; Balat, H.; Oz, C. Progress in bioethanol processing. *Prog. Energy Combust. Sci.* **2008**, *34*, 551–573. [CrossRef]
28. Zheng, Y.; Yu, C.; Cheng, Y.S.; Zhang, R.; Jenkins, B.; VanderGheynst, J.S. Effects of ensilage on storage and enzymatic degradability of sugar beet pulp. *Bioresour. Technol.* **2011**, *102*, 1489–1495. [CrossRef]

29. Vohra, M.; Manwar, J.; Manmode, R.; Padgilwar, S.; Patil, S. Bioethanol production: Feedstock and current technologies. *J. Environ. Chem. Eng.* **2014**, *2*, 573–584. [CrossRef]
30. Asadi, M. *Beet-Sugar Handbook*; John Wiley & Sons: Hoboken, NJ, USA, 2007.
31. Marzo, C.; Diaz, A.B.; Caro, I.; Blandino, A. Conversion of Exhausted Sugar Beet Pulp into Fermentable Sugars from a Biorefinery Approach. *Foods* **2020**, *9*, 1351. [CrossRef] [PubMed]
32. Bajić, B.; Vučurović, D.; Dodić, S.; Grahovac, J.; Dodić, J. Process model economics of xanthan production from confectionery industry wastewaters. *J. Environ. Manag.* **2017**, *203*, 999–1004. [CrossRef] [PubMed]
33. Power Industry of Serbia, Annual Report for 2020. Available online: Epsdistribucija.rs/pdf/GI_2020.pdf (accessed on 29 December 2021).
34. Global Petrol Prices. Available online: https://www.globalpetrolprices.com/ethanol_prices/ (accessed on 29 December 2021).

fermentation

MDPI

Article

Bioethanol Production Optimization from KOH-Pretreated *Bombax ceiba* Using *Saccharomyces cerevisiae* through Response Surface Methodology

Misbah Ghazanfar [1], Muhammad Irfan [1,*], Muhammad Nadeem [2], Hafiz Abdullah Shakir [3], Muhammad Khan [3], Irfan Ahmad [4], Shagufta Saeed [5], Yue Chen [6] and Lijing Chen [7,*]

1. Department of Biotechnology, Faculty of Science, University of Sargodha, Sargodha 40100, Pakistan; misbahghazanfar@ymail.com
2. Food & Biotechnology Research Center, PCSIR Laboratories Complex Ferozepur Road, Lahore 54600, Pakistan; mnadeempk@yahoo.com
3. Institute of Zoology, Faculty of Life Science, University of the Punjab New Campus, Lahore 54590, Pakistan; shakir.zool@pu.edu.pk (H.A.S.); khan_zoologist@ymail.com (M.K.)
4. Department of Clinical Laboratory Sciences, College of Applied Medical Sciences, King Khalid University, Abha 62529, Saudi Arabia; irfancsmmu@gmail.com
5. Institute of Biochemistry & Biotechnology, University of Veterinary and Animal Sciences, Lahore 54000, Pakistan; shagufta.saeed@uvas.edu.pk
6. Shenyang No. 126 Middle School, Shenyang 110000, China; chenyue202203@126.com
7. College of Bioscience & Biotechnology, Shenyang Agricultural University, Shenyang 110065, China

* Correspondence: irfan.ashraf@uos.edu.pk (M.I.); chenlijing@syau.edu.cn (L.C.)

Citation: Ghazanfar, M.; Irfan, M.; Nadeem, M.; Shakir, H.A.; Khan, M.; Ahmad, I.; Saeed, S.; Chen, Y.; Chen, L. Bioethanol Production Optimization from KOH-Pretreated *Bombax ceiba* Using *Saccharomyces cerevisiae* through Response Surface Methodology. *Fermentation* **2022**, *8*, 148. https://doi.org/10.3390/fermentation8040148

Academic Editor: Konstantinos G. Kalogiannis

Received: 16 March 2022
Accepted: 23 March 2022
Published: 28 March 2022

Abstract: The present study was based on the production of bioethanol from alkali-pretreated seed pods of *Bombax ceiba.* Pretreatment is necessary to properly utilize seed pods for bioethanol production via fermentation. This process assures the accessibility of cellulase to the cellulose found in seedpods by removing lignin. Untreated, KOH-pretreated, and KOH-steam-pretreated substrates were characterized for morphological, thermal, and chemical changes by scanning electron microscopy (SEM), thermogravimetric analysis (TGA), X-ray diffraction (XRD), and Fourier transform infrared spectroscopy (FTIR). Hydrolysis of biomass was performed using both commercial and indigenous cellulase. Two different fermentation approaches were used, i.e., separate hydrolysis and fermentation (SHF) and simultaneous saccharification and fermentation (SSF). Findings of the study show that the maximum saccharification (58.6% after 24 h) and highest ethanol titer (57.34 g/L after 96 h) were observed in the KOH-steam-treated substrate in SSF. This SSF using the KOH-steam-treated substrate was further optimized for physical and nutritional parameters by one factor at a time (OFAT) and central composite design (CCD). The optimum fermentation parameters for maximum ethanol production (72.0 g/L) were 0.25 g/L yeast extract, 0.1 g/L K_2HPO_4, 0.25 g/L $(NH_4)_2SO_4$, 0.09 g/L $MgSO_4$, 8% substrate, 40 IU/g commercial cellulase, 1% *Saccharomyces cerevisiae* inoculum, and pH 5.

Keywords: ethanol; pretreatment; saccharification; *B. ceiba*; fermentation

1. Introduction

High dependence on conventional nonrenewable fuels and global warming have urged humankind to search for alternative renewable fuels. Ethanol from lignocellulosic biomass could be an encouraging substitute for gasoline, as it has a higher octane number which causes less emission of air pollutants. For this purpose, lignocellulosic biomass including grasses, agricultural wastes, and forest residues has gained much attention due to its ubiquitous availability and nature of being eco-friendly [1–4]. Bioethanol is a type of biofuel obtained after the hydrolysis and fermentation of lignocellulosic resources. Basically, three steps are involved in bioethanol production, which are pretreatment, saccharification, and fermentation. Pretreatment is the first step, which is achieved by various methods such as physical, chemical, biological, or a combination. The major aim of pretreatment is to alter the structure of biomass exposing maximum cellulose content. Therefore, pretreatment is essential for enzymatic saccharification of lignocellulosic biomass. The pretreated lignocellulosic matter becomes more digestible as compared to the untreated biomass, although it may have nearly the same amount of lignin as raw biomass. Pretreatment has both chemical and physical effects. Physically it damages the structure of lignin and increases the surface area, hence causing physical or chemical perforation of the plant cell wall. Chemically it alters the solubility and depolymerization of the biomass and decomposes cross-linking between macromolecules. The alkali causes swelling of the biomass, which leads to disruption or disintegration of the lignin. Therefore, lignocellulosic biomass requires pretreatment to make the substrate digestible for commercial cellulase, or cellulase-producing microorganisms, to release sugars for fermentation [5–7]. The second step is saccharification, which is conversion of cellulose into sugars by cellulase enzymes. The third step is fermentation of saccharified material to ethanol yeast such as *Saccharomyces cerevisiae* [8,9]. Different technical approaches such as X-ray diffraction (XRD), scanning electron microscopy (SEM), Fourier transform infrared spectroscopy (FTIR), and thermogravimetric analysis (TGA) have been extensively used to study the structural and chemical changes generated by pretreatments of lignocellulosic biomass [10].

The most common processes of fermentation used in ethanol production are simultaneous saccharification and fermentation (SSF) and separated hydrolysis and fermentation (SHF) [11]. In SSF, sugars produced by the action of cellulase are immediately converted by *S. cerevisiae* into ethanol [12]. Thus, the inhibition effect caused by the sugars over the cellulases is neutralized [13]. This process has various studied advantages such as cost-effectiveness, requirements for fewer enzymes, high saccharification efficacy, high yield of ethanol, reduced operational time, and low chances of contamination or inhibition, as well as it not requiring reactors with large volumes [14–16].

In the case of SHF, saccharification and fermentation proceed in separate units at their optimal conditions. However, it has some issues regarding inhibition and risk of contamination as it is a prolonged process [12,15]. Several fermenting microorganisms have been studied to convert sugars but *Saccharomyces* is the most commonly used microorganisms for this purpose because it can produce ethanol from glucose with almost 90% of theoretical yield [11,17]. In this study, KOH-pretreated seed pods of *Bombax ceiba* were hydrolyzed by commercial as well as indigenous cellulase, then hydrolysate was fermented into ethanol by *S. cerevisiae* in SSF and SHF. Ethanol production was further optimized by one factor at a time (OFAT) and central composite design (CCD). There are some reasons for the selection of this substrate as it is easily and abundantly available. It is an inexpensive source. It has a good polysaccharide content. Use of this tree waste is nature-friendly because it is a second-generation feedstock; it would not compete with food sources and would also lead to waste management. There is no research reported on this feedstock for bioethanol production.

2. Materials and Methods

2.1. Substrate

Seed pods of *B. ceiba* were picked from native areas of district Sargodha, Punjab, Pakistan. The substrate was processed and pretreated with KOH as described in our earlier report [1].

2.2. Substrate Characterization

Raw and two other samples with maximum cellulose contents, each from different treatments (chemical and thermochemical), were selected for further characterization through X-ray diffraction (XRD) D8 Advance model [18], thermogravimetric analysis (TGA) SDT Q600 V8.0 Build 95 [10], Fourier transform infrared spectroscopy (FTIR) Align technologies Cary 630 and scanning electron microscopy (SEM) S-3700 (Hitachi, Tokyo, Japan) [9].

2.3. *Saccharification and Fermentation*

Untreated substrate (raw) and pretreated substrates from each pretreatment with maximum cellulose contents, i.e., KOH-pretreated and KOH-steam-treated, were employed for ethanol production through SHF and SSF [9,19].

2.4. *Separate Hydrolysis and Fermentation (SHF)*

For separate hydrolysis, three parameters including time (2, 4, 6, 8, 24, 26, 28, 30 h), substrate loading (1, 2, 3, 4, 5, 6, 7, 8, 9, 10% (*w/v*)), and enzyme concentration (FPU range of 20, 40, 60, 80, 100, and 120 IU/mL) were optimized by following OFAT. For SHF, substrate loading (2%) was hydrolyzed with 100 IU/mL of indigenous cellulase (produced by *Bacillus aerius* MG597041) in a 250 mL Erlenmeyer flask. In parallel, substrate (2%) was also saccharified with 40 IU/mL of commercial cellulase (obtained from the microbiology lab, PCSIR) in citrate buffer of pH 5. Hydrolysis was conducted at 50 °C for the total time period of 24 h. Material was centrifuged at 10,000 rpm for 10 min at the end of saccharification. The supernatant was collected for the analysis of sugar. Saccharification (%) was calculated using the following formula [9]. One unit (U) of enzyme activity was described as the total extent of the enzyme, which released 1 micromole of glucose under the standard assay conditions.

$$\text{Saccharification } (\%) = \frac{\text{Reducing sugars released (mg/mL)}}{\text{Substrate used (mg/mL)}} \times 100 \quad (1)$$

2.5. *Inoculum Preparation of Saccharomyces Cerevisiae*

Locally isolated and identified strains of *S. cerevisiae* were revived on potato dextrose agar (PDA) slants. The inoculum medium used was composed of (%) 1 glucose, 0.25 $(NH_4)_2SO_4$, 0.1 KH_2PO_4, 0.05 $MgSO_4$, and 0.25 yeast extract. *S. cerevisiae* suspension (inoculum) was prepared by adding a loopful of *S. cerevisiae* culture from slant to *S. cerevisiae* growth media (inoculum media) at 30 °C for 24 h [9,20]. The vegetative cells obtained after 24 h were used as an inoculum source.

2.6. *Bioethanol Production*

The hydrolysates obtained from saccharification of untreated and pretreated substrates using indigenous and commercial cellulase were fermented in different flasks for bioethanol production. *S. cerevisiae* media components (%, 0.25 $(NH_4)_2SO_4$, 0.1 KH_2PO_4, 0.05 $MgSO_4$, and 0.25 yeast extract) were added to the hydrolysates and then autoclaved at 121 °C for 15 min. After sterilization, the media were allowed to cool at room temperature. Then, 1% (*v/v*) suspension of *S. cerevisiae* was inoculated in each hydrolysate media mixture and incubated anaerobically at 30 °C for a 96 h fermentation period. At the end of fermentation, ethanol produced was analyzed by HPLC [9].

2.7. *Simultaneous Saccharification and Fermentation*

SSF of untreated and pretreated substrate (with maximum cellulose content from both pretreatments) was performed in a 1 L fermenter (Eyla, Japan). About 8% of substrate was mixed in 1 L citrate buffer and then sterilized [21]. After autoclaving, indigenously produced cellulase (FPU 100 IU/mL) from *Bacillus aerius*, accession number MG597041, was added to the substrate to make a mixture of enzyme and substrate at 40 °C at 200 rpm. After 24 h, the mixture was aseptically incorporated with 1% culture of *S. cerevisiae* containing various nutrients, i.e., 10 g glucose, 2.5 g $(NH_4)_2SO_4$, 1g KH_2PO_4, 0.5 g $MgSO_4$, and 2.5 g yeast extract, and incubated at 30 °C at 200 rpm. In parallel, the same experiment was performed with commercial cellulase (FPU 40 IU/mL). The samples were withdrawn periodically at intervals of 24 h for estimation of sugars and ethanol.

2.8. Optimization of Physical Parameters for Ethanol Production in SSF

KOH-steam-pretreated seedpods offered maximum ethanol production in SSF when hydrolyzed with commercial cellulase. Ethanol production from this KOH-steam treated-substrate in SSF was further optimized through OFAT [22] by varying different physical parameters, i.e., concentration of substrate (2%, 4%, 6%, 8%, 10%), pH (4, 5, 6, 7, 8), cellulase concentration FPU (20, 40, 60, 80, 100, 120) and *S. cerevisiae* inoculum size (1%, 2%, 3%, 4%, 5% *v/v*) in media containing nutrients (g/L), i.e., glucose 10, $(NH_4)_2SO_4$ 2.5, KH_2PO_4 1, $MgSO_4$ 0.5, and yeast extract 2.5.

2.9. Optimization of Nutritional Parameters for Ethanol Production in SSF

Central composite design was used to optimize the different components of the medium for ethanol production in SSF [23]. Each variable was designated and used with a high (+) and a low (−) concentration. The nutrient factors tested included concentrations of yeast extract, K_2HPO_4, $(NH_4)_2SO_4$, and $MgSO_4$ (Table 1). CCD was conducted by the experiment of 31 runs and ethanol was measured by HPLC.

Table 1. Range of parameters used for central composite design.

Sr. No.	Parameters (g/L)	Label	Codes	
			+1	−1
1	Yeast extract	A	0.2	0.3
2	K_2HPO_4	B	0.05	0.15
3	$(NH_4)_2SO_4$	C	0.2	0.3
4	$MgSO_4$	D	0.03	0.07

2.10. Ethanol Estimation

Samples were taken aseptically after every 24 h during the fermentation process. Consumed glucose and ethanol produced were determined by HPLC (PerkinElmer, Waltham, MA, USA) using a BioRad Aminex HPX 87H (250 mm × 4.6 mm) column with a mobile phase of 5 mM H_2SO_4, flow rate of 0.7 mL/min, and column temperature of 60 °C. All samples were passed through a 0.2 µm sterile membrane filter and an injection volume of 20 µL was used for estimation. Concentrations of ethanol and glucose were determined by a calibration curve [9].

2.11. Ethanol Fermentation Kinetics

Kinetic parameters for biomass and bioethanol were measured as described by Pirt [24] and Okpokwasili and Nweke [25]. Different kinetic parameters such as μ (h^{-1}), $Y_{p/x}$, $Y_{p/s}$, $Y_{x/s}$, q_s, and q_p were examined in the fermentation process.

3. Results and Discussion

3.1. SEM of KOH-Pretreated B. ceiba

Scanning electron microscopy was used to observe the structural modifications in *B. ceiba* biomass after KOH and KOH steam pretreatment. SEM micrographs revealed that the surface texture and morphology after both pretreatments were significantly different from those of untreated *B. ceiba* (Figure 1). The SEM micrograph of the untreated specimen exhibits a non-porous, smooth, and more compact surface, while a greater degree of porosity is seen on both the pretreated samples. The size and number of pores are greater in the thermochemically treated substrate, showing more lignin breakdown. This indicates that a large portion of lignin and hemicellulose can be eliminated by pretreatment. Relative to untreated substrate, remarkable changes were observed in the morphology of treated samples. A possible reason may be the breaking of the lignin xylan bond caused by acid/base pretreatment [26,27]. Tsegaye et al. [10] examined NaOH-pretreated rice straw by FE-SEM, showing that a significant amount of lignin was removed, which helps in releasing cellulose tangled in lignin. Jabasingh and Nachiyar [28] also observed such changes in bagasse. Irfan et al. [9] and Kusmiyati et al. [29] noticed a rough surface with holes in pretreated wheat straw and palm tree trunk waste, respectively, through SEM.

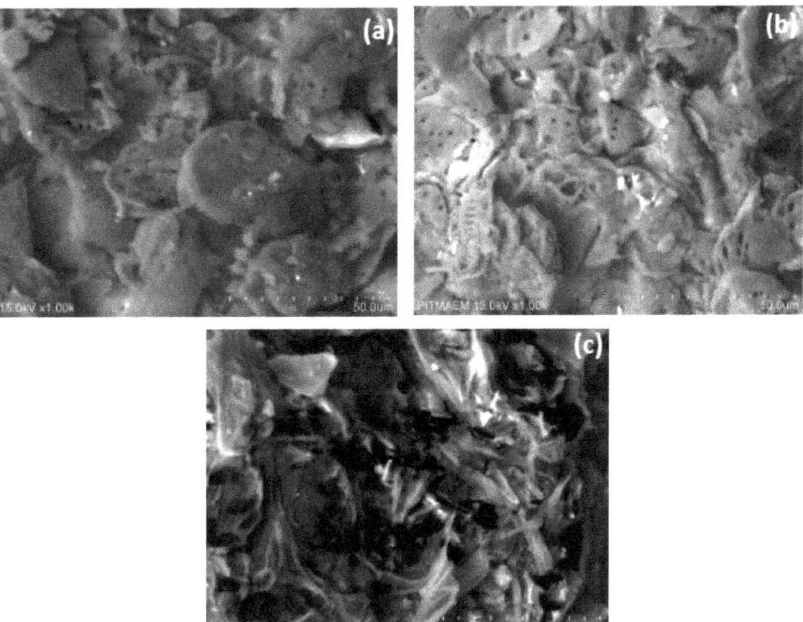

Figure 1. SEM images of *B. ceiba* biomass. (**a**) KOH-treated biomass, (**b**) KOH-steam-treated biomass, (**c**) untreated biomass.

3.2. FTIR of KOH-Pretreated B. ceiba

FTIR of *B. ceiba* substrate (seed pods) pretreated with KOH and KOH steam was carried out to observe the alterations in the structural composition. The FTIR spectra of untreated and pretreated substrate were in the range 4000–400 cm^{-1}. FTIR analysis revealed differences in untreated and pretreated *B. ceiba* (Figure 2). Many high- and low-intensity peaks were examined for all sample spectra. The highest peak seen in the untreated specimen was 1023.2 cm^{-1}, which increased up to 1028.7 cm^{-1} and 1026.9 cm^{-1} in KOH and KOH steam treated substrates, respectively. This peak shift represents changes in C-O stretching in cellulose. The peak at 3352.7 cm^{-1} in untreated *B. ceiba* was seen at 3341.6 cm^{-1} after chemical treatment, whereas in thermochemical treatment this band was stretched to 3334.1 cm^{-1}. The peak at 1593.4 cm^{-1} in untreated *B. ceiba* shifted to 1591.6 cm^{-1} in both treated samples, representing breakdown of lignin due to pretreatment. In the result of the alkalization process, OH bond distortion in the absorption region around 1518 cm^{-1} occurred, which illustrated the water immersion by cellulose and may also represent the occurrence of bands of the lignin and guasil ring.

The ester and acetyl groups in the hemicellulose, the COOH in the ferulic, and *p* coumeric bands in lignin shown in the spectra should be around 1740 cm^{-1}, specified by C=O groups, where the 1236 cm^{-1} peak may specify the existence of lignin siringil groups [30]. Carbon hydrogen bond vibrations in cellulose and carbon oxygen bond vibrations in syringyl derivatives were illuminated by the peak observed at 1317.6 cm^{-1}, where syringyl derivatives are salient constituents of lignin. When poplar substrate was pretreated with acid followed by steam, the C–O–C vibrations in cellulose and hemicellulose were demonstrated by the peak at 1157.3 cm^{-1}. C–O vibrations in cellulose and hemicellulose were denoted by the band at 1028 cm^{-1} [20].

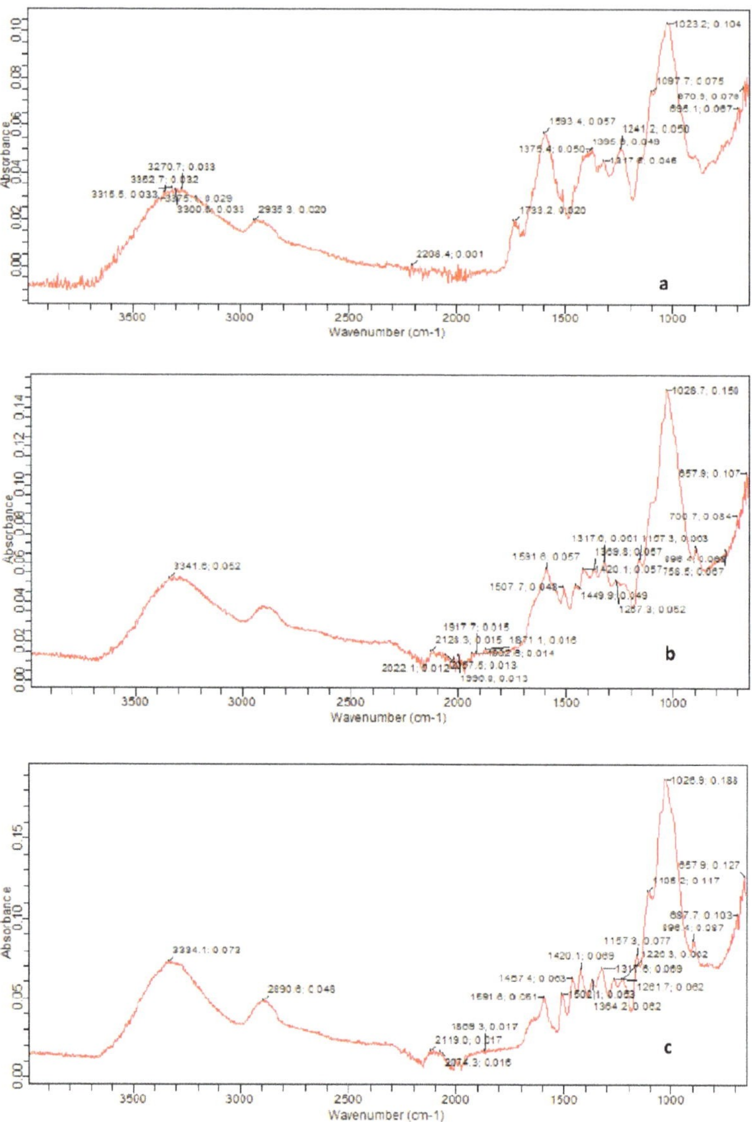

Figure 2. FTIR spectroscopy images of (**a**) untreated, (**b**) KOH-pretreated, and (**c**) KOH-steam-pretreated *B. ceiba*.

3.3. TGA of KOH-Pretreated B. ceiba

Thermal degradation behavior of raw and treated (KOH and KOH steam pretreatment) *B. ceiba* was studied by performing thermogravimetric analysis. Figure 3a reveals decomposition of raw substrate with time and temperature; 9.194% degradation was observed at 100–200 °C (first stage), 50.02% at 300–400 °C (second stage), and 33.39% at 500–600 °C (third stage). During the first stage the KOH-treated substrate showed 10.61% conversion, and it showed 63.06% during the third stage, as shown in Figure 3b, whereas the KOH-steam-treated substrate exhibited degradation of 10.99% during the first stage, 63.38% during the second stage, and 32.22% at 500–600 °C (Figure 3c). The KOH-steam-treated substrate exhibited a maximum degradation of 63.38% during the second stage. In an earlier investigation, TGA revealed the highest (74.48%) decomposition of *Pinus ponderosa* (sawdust) followed by *Shorea robusta* (sawdust) (70.03%) and *Areca catechu* (nut husk) (69.09%) in the temperature range 200–500 °C (second stage). Hemicellulose degraded at temperatures in the range 180–340 °C, cellulose conversion occurred at 230–450 °C, and lignin decomposed at temperatures greater than 500 °C [31]. A recent study by Tsegaye et al. [10] reported that the rate of loss of weight was very high (nearly 80%) in the temperature range 200–500 °C for all treatments considered.

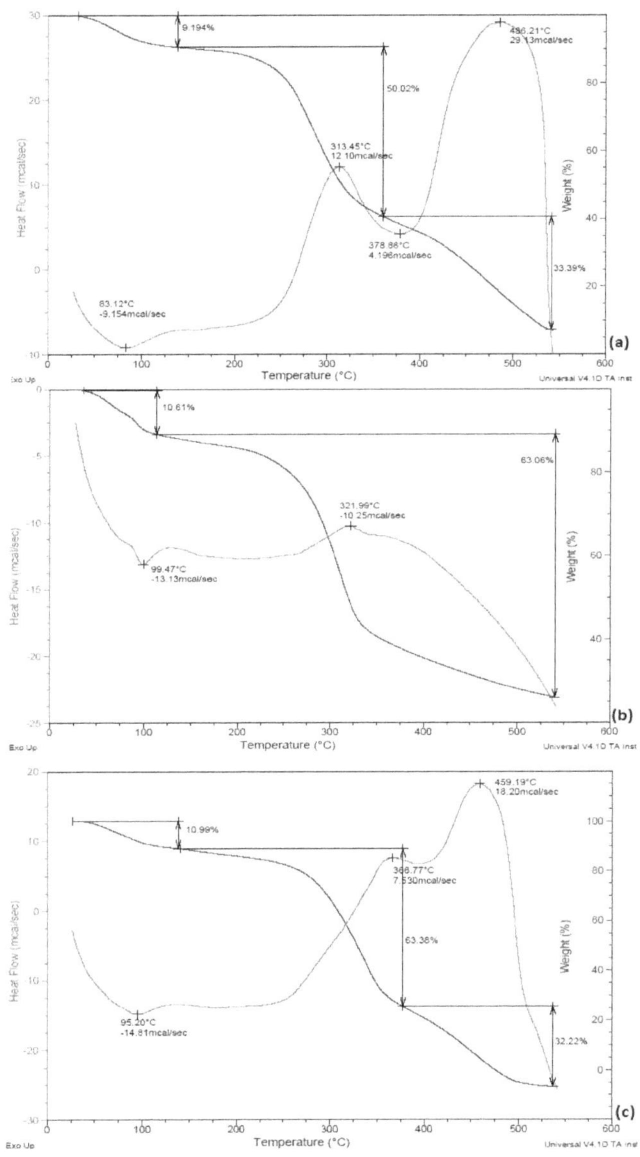

Figure 3. TGA of *B. ceiba*: (a) untreated, (b) KOH-treated, (c) KOH-steam-treated.

3.4. XRD of KOH-Pretreated *B. ceiba*

Figure 4 reveals the XRD spectra of controlled and treated (both KOH and KOH steam) samples. The crystallinity index presents the crystalline features of cellulose. Two peaks obtained at 2θ = 22° and 2θ = 18° represent the crystalline part (cellulose only) and amorphous part (cellulose, hemicellulose, and lignin) of *B. ceiba* biomass, respectively. The crystallinity index of raw *B. ceiba* substrate was 34.5%, which increased in KOH-treated (44.6%) and KOH steam pretreatment (50.5%). Removal of the amorphous portion, such as hemicellulose and lignin, from biomass caused an increase in the crystallinity index. Irfan et al. [9] performed XRD of biomass and concluded that the crystallinity index of pretreated biomass was increased relative to the control, which specified the elimination of lignin and hemicellulose. Our findings were in accordance with a previous study by Barman et al. [32]. They reported that the crystallinity index (53.3%) of raw wheat straw increased up to 60.3% after 1.5% NaOH pretreatment. A recent study performed XRD of the NaOH-pretreated pith of coconut husk and determined that the removal of the aromatic layer increased the crystallinity index from 65% to 81.7% [33].

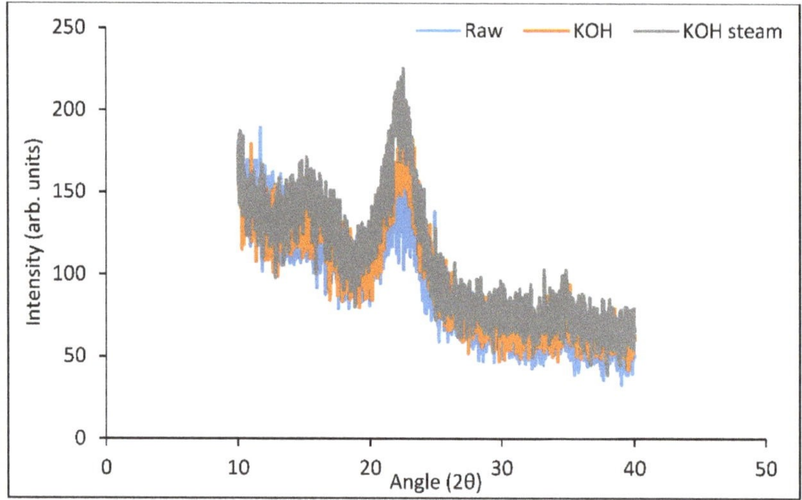

Figure 4. XRD analysis of *B. ceiba* substrate pretreated with KOH and KOH followed by steam.

3.5. Optimization of Saccharification

The process of saccharification was optimized for both commercial and indigenous cellulase. Optimization was investigated with three parameters, i.e., time, substrate concentration, and cellulase concentration. The study found maximum saccharification (25.5%) with commercial cellulase and maximum saccharification (16.8%) with indigenous cellulase after 24 h. A gradual increase in hydrolysis (%) was observed until 24 h and after this optimum time, a decline in hydrolysis was observed in both cases (Figure 5).

For optimization of substrate concentration, maximum saccharification (28% with commercial cellulase and 14.4% with indigenous cellulase) was found at 2% substrate concentration in both hydrolysis with commercial cellulase as well as with indigenous enzymes. A decline in saccharification percentage was observed by increasing substrate concentration from 2–4%. Hydrolysis (%) remained constant on further increases in substrate concentration (Figure 6).

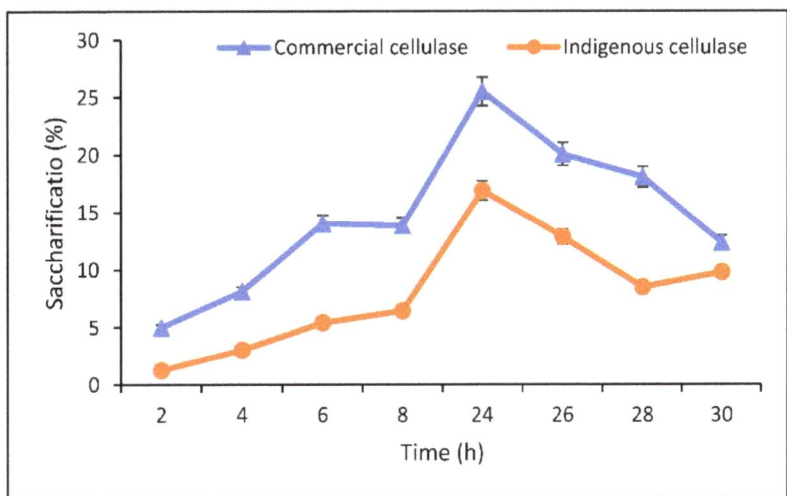

Figure 5. Optimization of time (h) for saccharification.

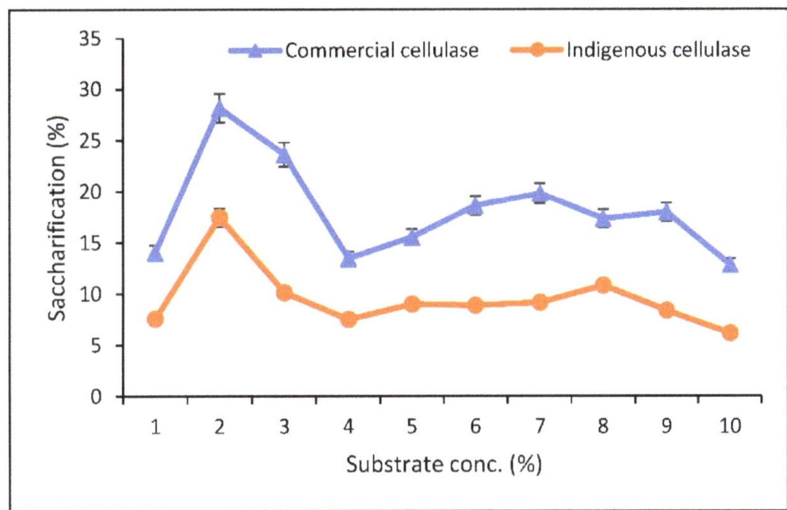

Figure 6. Optimization of substrate concentration (%) for saccharification.

In the case of optimization of enzyme concentration, maximum saccharifications of 43% and 25.8% were found at FPU 40 IU/mL of commercial cellulase and 100 IU/mL of indigenous cellulase, respectively. Beyond these optimal conditions a distinct decline in saccharification (%) was noticed (Figure 7). In other research, the optimized conditions observed for maximum saccharification (40.15%) of wheat straw were 2% wheat straw, 0.5% cellulase loading, and a time period of 6 h [9]. Sindhu et al. [34] used BBD for optimizing hydrolysis and obtained maximum RS (0.651 g/g) at 11.25% (*w/w*) of substrate concentration, 50 FPU of commercial cellulase, and an incubation period of 42 h. Asghar et al. [35] obtained maximum hydrolysis (52.93%) with 2.5% biomass loading, 0.5% enzyme loading, and an incubation period of 8 h at 50 °C.

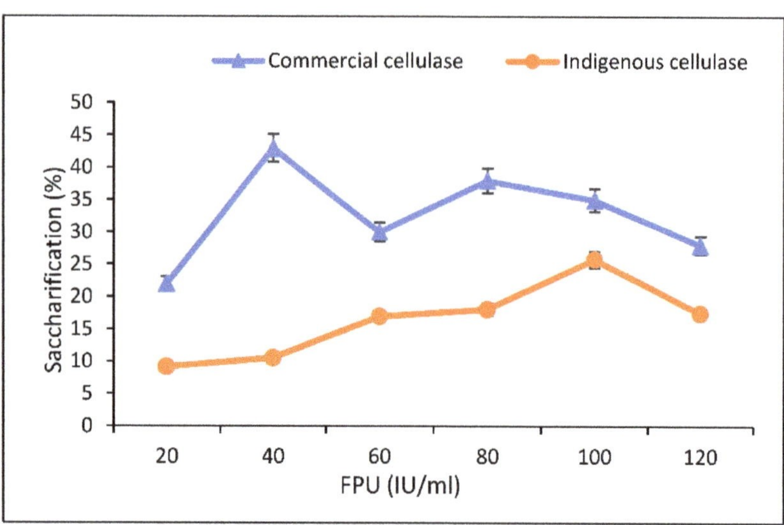

Figure 7. Optimization of enzyme concentration (IU/mL) for saccharification.

3.6. Separate Hydrolysis and Fermentation (SHF)

Saccharification was carried out in the optimized conditions by using both indigenous cellulase and commercial cellulase. Using indigenous cellulase, maximum saccharification (38%) was obtained in substrate B (KOH-pretreated followed by steam). Maximum saccharifications with indigenous cellulase in raw and KOH-treated samples were 10% and 28.4%, respectively (Figure 8a). Hydrolysates of this saccharification were fermented using *S. cerevisiae*. Fermentation resulted in production of ethanol; maximum ethanol production of 29.8 g/L was seen on the fermentation of hydrolysate of KOH-treated substrate after 4 days of incubation. Ethanol yields in the hydrolysate of untreated and KOH-treated substrates were 8.73 g/L and 18.04 g/L, respectively (Figure 8b). As compared to indigenous enzymes, maximum fermentable sugars were obtained with saccharification performed by commercial enzymes.

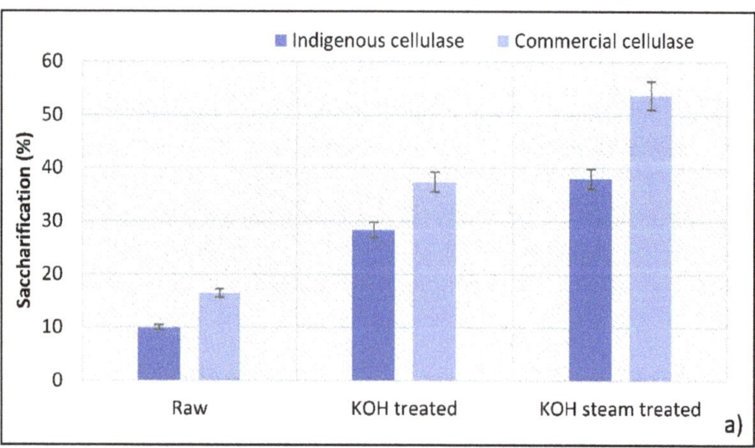

Figure 8. *Cont.*

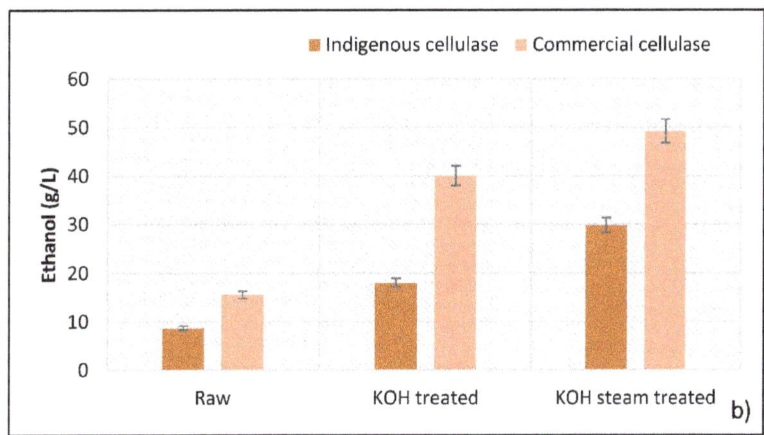

Figure 8. Separate hydrolysis and fermentation. (**a**) Saccharification after 24 h, (**b**) ethanol produced after 96 h.

Hydrolysis with commercial cellulase offered maximum saccharification in the KOH-steam-treated substrate (53.7%) followed by the KOH-pretreated (37.3%) and raw (16.4%) substrates, as shown in Figure 8a. Fermentation (with *S. cerevisiae*) of sugars obtained from this saccharification gave a significant ethanol yield. Maximum ethanol production (49.2 g/L) was seen in the KOH-pretreated substrate, followed by the KOH-treated (40.06 g/L) and raw (15.6 g/L) substrates, as illustrated in Figure 8b. Our results corroborated the findings of Irfan et al. [36] that the commercial cellulase offered better saccharification as compared to the indigenous cellulase. They noticed 63.3% and 33.6% saccharifications in pretreated sugarcane bagasse and wheat straw, respectively, with commercial enzymes. The saccharification recorded with indigenously produced cellulase was in the range 6–14%.

3.7. Simultaneous Saccharification and Fermentation (SSF)

SSF of untreated and treated *B. ceiba* biomass was conducted in a 1 L fermenter with both indigenous cellulase and commercial cellulase separately. Samples were taken every 24 h aseptically for estimation of glucose and ethanol. Estimation of glucose and ethanol was performed by HPLC. With indigenous cellulase, untreated biomass offered maximum saccharification (17.3%) after 48 h of hydrolysis. Among pretreated substrates, maximum saccharification of 42.9% was seen in KOH-steam-pretreated substrate followed by KOH treated (32.7%) after 24 h of hydrolysis. After 24 h, sugar contents started to decline due to the consumption of sugars by *S. cerevisiae*, as the fermentation process proceeded. KOH-steam-treated *B. ceiba* offered the highest ethanol production of 41.5 g/L after 96 h of fermentation. Maximum ethanol production (g/L) in raw (11.2) and KOH-treated (23.1) biomass was also observed after 96 h of fermentation at 40 °C (Figure 9).

In SSF with commercial cellulase enzymes, results showed maximum saccharification in KOH-steam-treated (58.6%), KOH-treated (37.9%), and untreated seedpods (20.2%) after 24 h of hydrolysis. Maximum ethanol yield (57.34 g/L) was observed in KOH-steam-pretreated substrate followed by KOH-treated (29.67 g/L) and raw (19.87 g/L) after 96 h of fermentation at 40 °C (Figure 9). The findings of Sukhang et al. [37] and Vintila et al. [38] corroborate our results that the SSF offered a higher yield of ethanol from lignocellulosic material than that of SHF. The findings show that SSF is more effective than SHF in terms of energy consumption, time, cost, and greater bioethanol yield. Kusmiyati and coworkers [29] reported 2.648% bioethanol production from pretreated palm tree trunk waste through SSF using *S. cerevisiae* and cellulase enzymes at 37 °C temperature, 4.8 pH, 10% substrate, and 100 rpm for 120 h. Another study noticed a remarkable ethanol titer from the SHF of an oil palm empty fruit bunch. Hydrolysis at 50 °C, pH 4.8, and 150 rpm of agitation for 96 h yielded 75.48% glucose, which subsequently produced 78.95% ethanol [39].

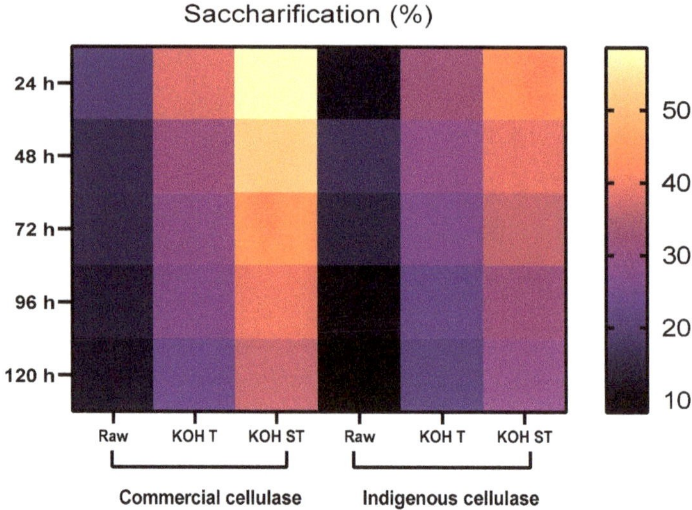

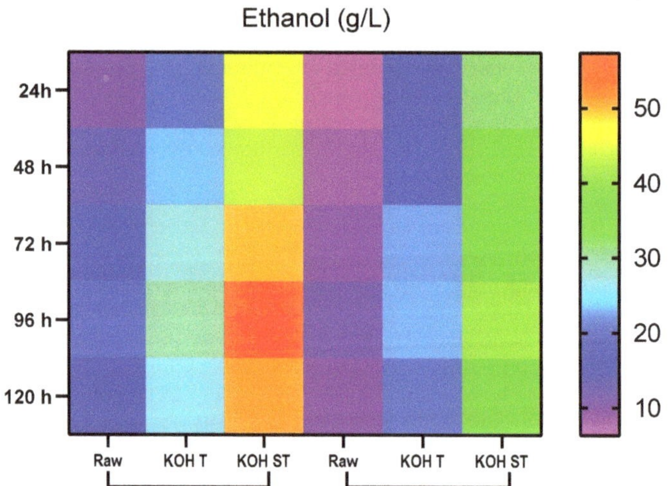

Figure 9. Saccharification (%) and ethanol production (g/L) in SSF (KOH T = KOH-treated substrate, KOH ST = KOH-steam-treated substrate).

3.8. Optimization of Physical and Nutritional Parameters for Ethanol Production in SSF

KOH-steam-pretreated seedpods offered maximum ethanol production in SSF when hydrolyzed with commercial cellulase. Physical parameters of SSF for this substrate were further optimized by OFAT for improved yield of ethanol. HPLC was used to check ethanol production. The ethanol titer increased gradually with an increase in the concentration of the substrate. Maximum yield (57.53 g/L) was observed with 8% substrate. A further increase in the substrate caused a sudden drop in activity (Figure 10). Error bars in the graphs indicate variation among triplicates.

For cellulase optimization, the best ethanol titer of 59.07 g/L was attained when 40 FPU of commercial cellulase was used in SSF. A gradual decline in ethanol production was recorded with an increase in the FPU of enzymes until 120 (Figure 10). Maximum ethanol (59.96 g/L) in the case of pH optimization was observed at pH 5; a decline in activity was seen as the pH increased towards neutrality. A sharp decline in ethanol production was observed at pH 7 and 8 (Figure 10). Optimization of the inoculum size of *S. cerevisiae* resulted in the maximum ethanol yield of 61.74 g/L at 1% inoculum. Ethanol production decreased from inoculum size 2% to 3% and then it remained almost constant from 3% to 4% and 5% as inoculum size increased as shown in Figure 10.

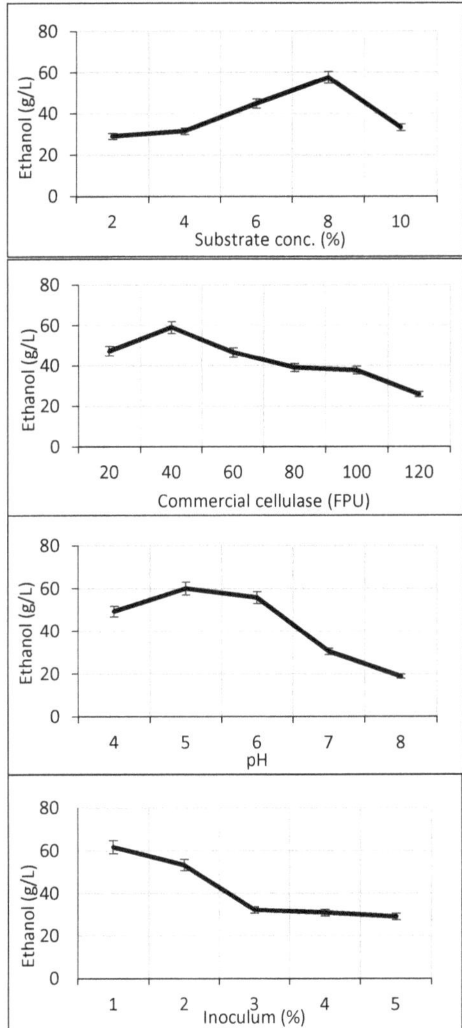

Figure 10. Optimization of physical parameters for ethanol production in SSF after 96 h.

To obtain the maximum ethanol titer, four nutritional parameters (yeast extract, K_2HPO_4, $(NH_4)_2SO_4$, and $MgSO_4$) were optimized by CCD. The optimum medium composition for maximum ethanol production (72.0 g/L) was 0.25 g/L yeast extract, 0.1 g/L K_2HPO_4, 0.25 g/L $(NH_4)_2SO_4$, and 0.09 g/L $MgSO_4$ (Table 2). ANOVA was performed, which depicts the F-value 26.26 and p-value 0.00 (Table 3). The regression equation indicates the significance of the results (Equation 2). Contour plots for interactions of yeast extract, K_2HPO_4, $(NH_4)_2SO_4$, and $MgSO_4$ for ethanol production are displayed in Figure 11. Tan and Lee [40] reported a higher bioethanol yield in SSF (90.9%) than in SHF (55.9%). They suggested that the SSF of seaweed biomass using *S. cerevisiae* had various merits over SHF, as the former technique is a simple single-step process that can save energy, time, and cost while attaining a high production of bioethanol. A study obtained the maximum ethanol titer of 85.71% at 30 °C with 2% wheat straw and 30 FPU of enzyme loading in SSF [41]. The results of a study on the production of bioethanol from rice husk also supports our findings that SSF was better than SHF in yielding ethanol titer [42]. Berłowska and coworkers [43] employed *S. cerevisiae* in SSF and achieved the highest ethanol concentration reaching 26.9 ± 1.2 g/L and 86.5 ± 2.1% fermentation efficiency relative to the theoretical yield. Ballesteros et al. [44] reported maximum production of ethanol at 72 h of fermentation period. They also described that the reason for a good yield of enzymes in SSF may be due to the immediate conversion of formed sugars into ethanol, thus avoiding any feedback inhibition. Wang et al. [45] obtained 5.16 g/L of ethanol using commercial cellulase after 24 h in SSF from biologically delignified poplar chips and the yield was 75%. Kusmiyati and coworkers [29] reported 2.648% bioethanol production from HNO_3-pretreated palm tree trunk waste through SSF using *S. cerevisiae* and cellulase enzymes at 37 °C temperature, 4.8 pH, 10% substrate, and 100 rpm for 120 h.

$$\text{Ethanol (g/L)} = 109.6 - 346.9A - 171.2B + 48.5C - 292D + 768.0A{*}A + 464.0B{*}B - 15.0C{*}C + 5853D{*}D + 172.5A{*}B - 87.5A{*}C - 334A{*}D + 137.0B{*}C + 298B{*}D - 675C{*}D \quad (2)$$

Table 2. CCD for optimizing nutritional parameters (g/L) for production of ethanol in SSF.

Run No.	A	B	C	D	Ethanol (g/L)		
					Observed	Predicted	Residual
1	0.25	0.1	0.25	0.05	61.01	60.75	0.252
2	0.3	0.05	0.2	0.07	68.72	67.87	0.845
3	0.25	0	0.25	0.05	63.58	64.01	−0.432
4	0.35	0.1	0.25	0.05	69.66	70.01	−0.357
5	0.25	0.1	0.25	0.05	60.71	60.75	−0.047
6	0.2	0.05	0.3	0.03	64.18	64.14	0.037
7	0.25	0.1	0.25	0.05	60	60.75	−0.757
8	0.25	0.1	0.25	0.05	60	60.75	−0.757
9	0.3	0.05	0.3	0.07	65.48	65.31	0.165
10	0.25	0.1	0.25	0.09	72	72.95	−0.959
11	0.25	0.1	0.25	0.05	60	60.75	−0.757
12	0.2	0.15	0.2	0.03	62.33	62.37	−0.04
13	0.3	0.15	0.2	0.03	67.17	65.91	1.251
14	0.2	0.15	0.2	0.07	68	67.82	0.18
15	0.2	0.15	0.3	0.03	65	64.75	0.245
16	0.3	0.15	0.3	0.03	67.01	67.42	−0.417
17	0.25	0.1	0.25	0.05	61.23	60.75	0.472
18	0.2	0.05	0.3	0.07	65.54	65.7	−0.161
19	0.25	0.1	0.25	0.05	62.35	60.75	1.592
20	0.25	0.1	0.25	0.01	67.03	67.28	−0.255
21	0.25	0.1	0.35	0.05	60	60.51	−0.519
22	0.3	0.05	0.2	0.03	63.82	64.95	−1.13
23	0.3	0.05	0.3	0.03	66	65.09	0.91
24	0.3	0.15	0.2	0.07	70.12	70.03	0.087
25	0.3	0.15	0.3	0.07	69.06	68.84	0.218
26	0.25	0.2	0.25	0.05	66	66.78	−0.782
27	0.2	0.05	0.2	0.03	64	63.12	0.871
28	0.2	0.05	0.2	0.07	67.93	67.38	0.542
29	0.25	0.1	0.15	0.05	60	60.69	−0.695
30	0.15	0.1	0.25	0.05	66	66.85	−0.857
31	0.2	0.15	0.3	0.07	68.76	67.5	1.255

A = yeast extract, B = K_2HPO_4, C = $(NH_4)_2SO_4$, D = $MgSO_4$.

Table 3. Analysis of variance for ethanol production by *S. cerevisiae*.

Source	DF	Adj SS	Adj MS	F-Value	*p*-Value
Model	14	352.618	25.187	26.26	0.000
Linear	4	74.815	18.704	19.50	0.000
A	1	14.978	14.978	15.62	0.001
B	1	11.509	11.509	12.00	0.003
C	1	0.047	0.047	0.05	0.828
D	1	48.280	48.280	50.34	0.000
Square	4	261.697	65.424	68.21	0.000
A×A	1	105.425	105.425	109.92	0.000
B×B	1	38.484	38.484	40.12	0.000
C×C	1	0.040	0.040	0.04	0.841
D×D	1	156.758	156.758	163.44	0.000
2-Way Interaction	6	16.106	2.684	2.80	0.047
A×B	1	2.976	2.976	3.10	0.097
A×C	1	0.766	0.766	0.80	0.385
A×D	1	1.782	1.782	1.86	0.192
B×C	1	1.877	1.877	1.96	0.181
B×D	1	1.416	1.416	1.48	0.242
C×D	1	7.290	7.290	7.60	0.014
Error	16	15.346	0.959		
Lack-of-Fit	10	10.799	1.080	1.43	0.345
Pure Error	6	4.547	0.758		
Total	30	367.964			

A = yeast extract, B = K_2HPO_4, C = $(NH_4)_2SO_4$, D = $MgSO_4$.

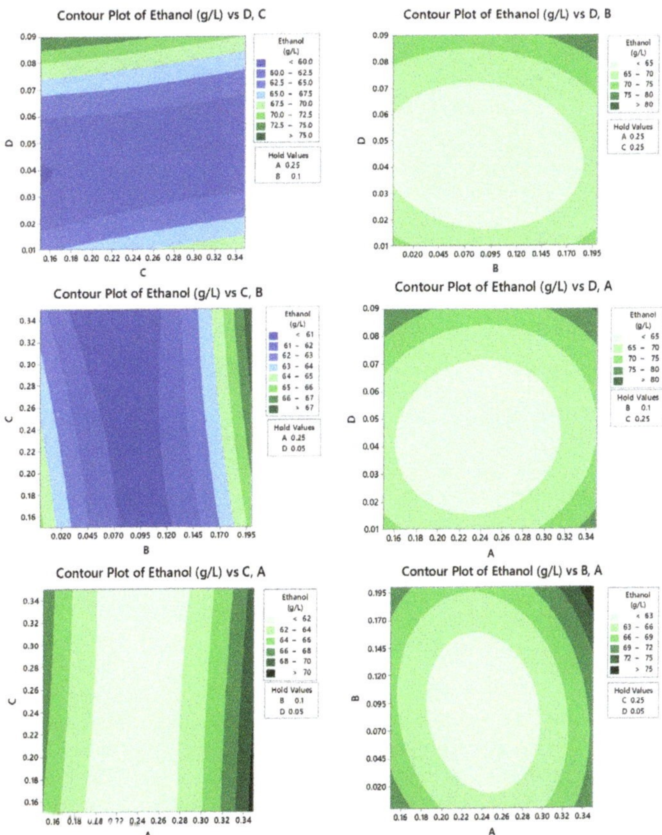

Figure 11. Contour plots of interactions of yeast extract, K_2HPO_4, $(NH_4)_2SO_4$, and $MgSO_4$ for ethanol production.

3.9. Fermentation Kinetics

The results presented in Tables 4 and 5 show that ethanol yield was increased and the substrate (glucose) concentration decreased with the increase in fermentation time. It was observed that the specific growth rate also increased with the passage of fermentation time. However, after 96 h of fermentation, a noticeable decline in specific growth rate was observed. Maximum ethanol yield (0.451) per substrate utilization was observed after 96 h of simultaneous saccharification and fermentation using commercial cellulase, while using indigenous cellulase the maximum ethanol yield (0.434) was also observed at 96 h of fermentation in SSF (Table 4). Maximum ethanol yields of 0.443 and 0.413 were observed after 96 h in SHF using commercial and indigenous enzymes, respectively (Table 5).

Table 4. Kinetic parameter estimation for ethanol fermentation in simultaneous saccharification and fermentation (SSF).

	Fermentation Time (h)	**Kinetic Parameters**					
		μ	$Y_{x/s}$	q_s	$Y_{p/s}$	$Y_{p/x}$	q_p
KOH + Steam (commercial cellulase)	24	0.0079	0.178	0.031	0.403	0.279	0.014
	48	0.0131	0.199	0.152	0.424	3.41	0.071
	72	0.0177	0.207	0.123	0.436	4.22	0.063
	96	0.0186	0.219	0.117	0.451	5.37	0.059
	120	0.0100	0.119	0.128	0.431	6.86	0.061
KOH + Steam (indigenous cellulase)	24	0.0067	0.173	0.024	0.397	0.272	0.012
	48	0.0099	0.191	0.141	0.402	3.01	0.062
	72	0.0111	0.198	0.112	0.418	3.93	0.053
	96	0.0120	0.206	0.107	0.434	4.71	0.052
	120	0.0064	0.201	0.119	0.417	5.78	0.053

μ (h − 1), specific growth rate; $Y_{x/s}$, g of cell biomass/g of glucose consumed; q_s, g of glucose consumed/g of cell biomass per h; $Y_{p/x}$, ethanol produced/g of cells formed; Y_{p}/s, ethanol produced/g of glucose consumed; q_p, ethanol produced/g of cells per h.

Table 5. Kinetic parameter estimation for ethanol fermentation in separate hydrolysis and fermentation (SHF).

	Fermentation Time (h)	**Kinetic Parameters**					
		μ	$Y_{x/s}$	q_s	$Y_{p/s}$	$Y_{p/x}$	q_p
KOH + Steam (commercial cellulase)	24	0.0075	0.173	0.025	0.395	0.261	0.009
	48	0.0112	0.194	0.147	0.417	3.21	0.054
	72	0.0160	0.203	0.115	0.434	4.01	0.049
	96	0.0182	0.219	0.109	0.443	4.92	0.045
	120	0.096	0.118	0.123	0.415	5.76	0.050
KOH + Steam (indigenous cellulase)	24	0.0065	0.169	0.021	0.387	0.268	0.010
	48	0.0097	0.187	0.139	0.396	2.98	0.058
	72	0.0108	0.193	0.108	0.406	3.91	0.049
	96	0.0115	0.209	0.105	0.413	4.65	0.054
	120	0.0061	0.195	0.117	0.389	5.81	0.051

Our findings are in accordance with a study by Irfan et al. [9], which reported that production of ethanol increased with an increase in fermentation time, whereas glucose concentration declined with time. Specific growth rate also improved with the passage of fermentation time. Maximum ethanol titer (0.497) per substrate consumption was recorded after 96 h of fermentation. Sathendra et al. [46] achieved a maximum biomass and ethanol yield at pH 4.5 and 40 °C with biomass yield ($Y_{x/s}$) 14.7 g/L^{-1}, specific growth rate (μ) 0.021 h, and bioethanol yield ($Y_{p/s}$) 21.89 g/L^{-1}. Hadiyanto et al. [47] found a maximum specific growth rate of 0.186 h^{-1}, $Y_{x/s}$ 0.32 g g^{-1}, and $Y_{p/s}$ 0.21 g g^{-1} at 30 °C.

4. Conclusions

Alkali-pretreated substrate *B. ceiba* (64% cellulose) was explored in the present research for the production of bioethanol. A set of optimum parameters offered the highest ethanol yield of 72.0 g/L using commercial cellulase and *S. cerevisiae* during SSF. The findings of the research highly recommend this cheap and novel biomass as a promising feedstock for pilot-scale production of second-generation bioethanol.

Author Contributions: M.G. performed experiments and wrote the first draft; M.I. conceived the study design and critically edited the manuscript; M.N. helped in the analysis; H.A.S. and M.K. helped in the analysis and critical revision; I.A. and L.C. acquired funding; S.S. helped with data interpretation; Y.C. helped with the literature survey. All authors have read and agreed to the published version of the manuscript.

Funding: This work was supported by the Scientific Research Deanship at King Khalid University, Abha, Saudi Arabia, with financial support through the Large Research Group Project under grant number RGP.02/205/42, the National Key R & D Program of China (2019YFD1001002), and the China Agriculture Research system of MOF and MARA (CARS-23).

Institutional Review Board Statement: Not applicable.

Informed Consent Statement: Not applicable.

Data Availability Statement: Data is available within the article.

Acknowledgments: The authors are also grateful to the Department of Biotechnology, University of Sargodha, Pakistan, for technical help with this study.

Conflicts of Interest: The authors declare no conflict of interest.

References

1. Ghazanfar, M.; Irfan, M.; Nadeem, M. Statistical modeling and optimization of pretreatment of *Bombax ceiba* with KOH through Box– Behnken design of response surface methodology. *Energy Sour. Part A Recov. Utiliz. Environ. Eff.* **2018**, *40*, 1114–1124. [CrossRef]
2. Mejica, G.F.C.; Unpaprom, Y.; Whangchai, K.; Ramaraj, R. Cellulosic-derived bioethanol from Limnocharis flava utilizing alkaline pretreatment. *Biomass Conver. Biorefin.* **2021**, 1–7. [CrossRef]
3. Wannapokin, A.; Ramaraj, R.; Unpaprom, Y. An investigation of biogas production potential from fallen teak leaves (Tectona grandis). *Emergent Life Sci. Res.* **2017**, *3*, 1–10.
4. Sophanodorn, K.; Unpaprom, Y.; Whangchai, K.; Duangsuphasin, A.; Manmai, N.; Ramaraj, R. A biorefinery approach for the production of bioethanol from alkaline-pretreated, enzymatically hydrolyzed Nicotiana tabacum stalks as feedstock for the bio-based industry. *Biomass Conver. Biorefin.* **2020**, *12*, 1–9. [CrossRef]
5. Desvaux, M. *Clostridium cellulolyticum*: Model organism of mesophilic cellulolytic Clostridia. *FEMS Microbial. Rev.* **2005**, *29*, 741–764. [CrossRef]
6. Bajpai, P. Structure of lignocellulosic biomass. In *Pretreatment of Lignocellulosic Biomass for Biofuel Production*; Springer: Singapore, 2016; pp. 7–12.
7. Rajendran, K.; Drielak, E.; Varma, V.S.; Muthusamy, S.; Kumar, G. Updates on the pretreatment of lignocellulosic feedstocks for bioenergy production–A review. *Biomass Conver. Bioref.* **2018**, *8*, 471–483. [CrossRef]
8. Maceiras, R.; Alfonsín, V.; Seguí, L.; González, J.F. Microwave Assisted Alkaline Pretreatment of Algae Waste in the Production of Cellulosic Bioethanol. *Energies* **2021**, *14*, 5891. [CrossRef]
9. Irfan, M.; Asghar, U.; Nadeem, M.; Nelofer, R.; Syed, Q.; Shakir, H.A.; Qazi, J.I. Statistical optimization of saccharification of alkali pretreated wheat straw for bioethanol production. *Waste Biomass Valor.* **2016**, *7*, 1389–1396. [CrossRef]
10. Tsegaye, B.; Balomajumder, C.; Roy, P. Microbial delignification and hydrolysis of lignocellulosic biomass to enhance biofuel production: An overview and future prospect. *Bull N Res. Cent.* **2019**, *43*, 51. [CrossRef]
11. Gupta, A.; Verma, J.P. Sustainable bio-ethanol production from agro-residues: A review. *Renew. Sustain. Energy Rev.* **2015**, *41*, 550–567. [CrossRef]
12. Kamzon, M.A.; Abderafi, S.; Bounahmidi, T. Promising bioethanol processes for developing a biorefinery in the Moroccan sugar industry. *Internat. J. Hydro. Energy* **2016**, *41*, 20880–20896. [CrossRef]
13. Stenberg, K.; Galbe, M.; Zacchi, G. The influence of lactic acid formation on the simultaneous saccharification and fermentation (SSF) of softwood to ethanol. *Enz. Microb. Technol.* **2000**, *26*, 71–79. [CrossRef]
14. Sun, Y.; Cheng, J. Hydrolysis of lignocellulosic materials for ethanol production: A review. *Biores. Technol.* **2002**, *83*, 1–11. [CrossRef]

15. Chen, H.; Fu, X. Industrial technologies for bioethanol production from lignocellulosic biomass. *Renew. Sustain. Energy Rev.* **2016**, *57*, 468–478. [CrossRef]
16. Hakim, A.; Chasanah, E.; Uju, U.; Santoso, J. Bioethanol production from seaweed processing waste by simultaneous saccharification and fermentation (SSF). *Squ. Bull. Mar. Fish. Postharv. Biotechnol.* **2017**, *12*, 41–47. [CrossRef]
17. Tong, Z.; Pullammanappallil, P.; Teixeira, A.A. How ethanol is made from cellulosic biomass constituents of cellulosic biomass. *Agricul. Biol. Engin.* **2012**, *2012*, 12.
18. Segal, L.G.J.M.A.; Creely, J.J.; Martin Jr, A.E.; Conrad, C.M. An empirical method for estimating the degree of crystallinity of native cellulose using the X-ray diffractometer. *Textile Res. J.* **1959**, *29*, 786–794. [CrossRef]
19. Tabassum, F.; Irfan, M.; Shakir, H.A.; Qazi, J.I. Statistical optimization for deconstruction of poplar substrate by dilute sulfuric acid for bioethanol production. *Green Chem. Lett. Rev.* **2017**, *10*, 69–79. [CrossRef]
20. Iram, M.; Asghar, U.; Irfan, M.; Huma, Z.; Jamil, S.; Nadeem, M.; Syed, Q. Production of bioethanol from sugarcane bagasse using yeast strains: A kinetic study. *Energy Sour. Part A Recov. Util. Environ. Eff.* **2018**, *40*, 364–372. [CrossRef]
21. Gul, A.; Irfan, M.; Nadeem, M.; Syed, Q.; Haq, I.U. Kallar grass (*Leptochloa fusca L. Kunth*) as a feedstock for ethanol fermentation with the aid of response surface methodology. *Environ. Prog. Sustain. Energy* **2018**, *37*, 569–576.
22. Sharma, N.; Kalra, K.L.; Oberoi, H.S.; Bansal, S. Optimization of fermentation parameters for production of ethanol from kinnow waste and banana peels by simultaneous saccharification and fermentation. *Indian J. Microbiol.* **2007**, *47*, 310–316. [CrossRef] [PubMed]
23. Thontowi, A.; Perwitasari, U.; Kholida, L.N.; Prasetya, B. Optimization of simultaneous saccharification and fermentation in bioethanol production from sugarcane bagasse hydrolyse by *Saccharomyces cerevisiae* BTCC 3 using response surface methodology. *IOP Conf. Ser. Earth Environ. Sci.* **2018**, *183*, 012010. [CrossRef]
24. Pirt, S.J. *Principles of Microbe and Cell Cultivation*; Blackwell Scientific Publications: Hoboken, NJ, USA, 1975.
25. Okpokwasili, G.C.; Nweke, C.O. Microbial growth and substrate utilization kinetics. *Afr. J. Biotechnol.* **2006**, *5*, 305–317.
26. Ravindran, R.; Jaiswal, A.K. A comprehensive review on pre-treatment strategy for lignocellulosic food industry waste: Challenges and opportunities. *Biores. Technol.* **2016**, *199*, 92–102. [CrossRef]
27. Gao, C.; Xiao, W.; Ji, G.; Zhang, Y.; Cao, Y.; Han, L. Regularity and mechanism of wheat straw properties change in ball milling process at cellular scale. *Biores. Technol.* **2017**, *241*, 214–219. [CrossRef]
28. Jabasingh, S.A.; Nachiyar, C.V. Utilization of pretreated bagasse for the sustainable bioproduction of cellulase by *Aspergillus nidulans* MTCC344 using response surface methodology. *Indust. Crops Prod.* **2011**, *34*, 1564–1571. [CrossRef]
29. Kusmiyati, K.; Anarki, S.T.; Nugroho, S.W.; Widiastutik, R.; Hadiyanto, H. Effect of dilute acid and alkaline pretreatments on enzymatic saccharfication of palm tree trunk waste for bioethanol production. *Bull. Chem. Reac. Engin. Catal.* **2019**, *14*, 705–714. [CrossRef]
30. Gunam, I.B.W.; Setiyo, I.Y.; Antara, I.N.S.; Wijaya, I.M.M.; ST, I.; Wijaya, M.M.; Arnata, I.W.; Arnata, I.W. Enhanced delignification of corn straw with alkaline pretreatment at mild temperature. *Rasayan J. Chem.* **2020**, *13*, 1022–1029. [CrossRef]
31. Mishra, R.K.; Mohanty, K. Pyrolysis kinetics and thermal behavior of waste sawdust biomass using thermogravimetric analysis. *Biores. Technol.* **2018**, *251*, 63–74. [CrossRef]
32. Barman, D.N.; Haque, M.A.; Kang, T.H.; Kim, M.K.; Kim, J.; Kim, H.; Yun, H.D. Alkali pretreatment of wheat straw (*Triticum aestivum*) at boiling temperature for producing a bioethanol precursor. *Biosci, Biotechnol. Biochem.* **2012**, *76*, 120480–120486. [CrossRef]
33. Gundupalli, M.P.; Kajiura, H.; Ishimizu, T.; Bhattacharyya, D. Alkaline hydrolysis of coconut pith: Process optimization, enzymatic saccharification, and nitrobenzene oxidation of Kraft lignin. *Biomass Conver. Bioref.* **2020**, 1–19. [CrossRef]
34. Sindhu, R.; Kuttiraja, M.; Binod, P.; Sukumaran, R.K.; Pandey, A. Bioethanol production from dilute acid pretreated Indian bamboo variety (*Dendrocalamus* sp.) by separate hydrolysis and fermentation. *Indus. Crops Prod.* **2014**, *52*, 169–176. [CrossRef]
35. Asghar, U.; Irfan, M.; Iram, M.; Huma, Z.; Nelofer, R.; Nadeem, M.; Syed, Q. Effect of alkaline pretreatment on delignification of wheat straw. *Nat. Prod. Res.* **2015**, *29*, 125–131. [CrossRef] [PubMed]
36. Irfan, M.; Gulsher, M.; Abbas, S.; Syed, Q.; Nadeem, M.; Baig, S. Effect of various pretreatment conditions on enzymatic saccharification. *Songklan. J. Sci. Technol.* **2011**, *33*, 397–404.
37. Sukhang, S.; Choojit, S.; Reungpeerakul, T.; Sangwichien, C. Bioethanol production from oil palm empty fruit bunch with SSF and SHF processes using *Kluyveromyces marxianus* yeast. *Cellulose* **2020**, *27*, 301–314. [CrossRef]
38. Vintila, T.; Vintila, D.; Neo, S.; Tulcan, C.; Hadaruga, N. Simultaneous hydrolysis and fermentation of lignocellulose versus separated hydrolysis and fermentation for ethanol production. *Rom. Biotechnol. Lett.* **2011**, *16*, 106–112.
39. Triwahyuni, E. Valorization of oil palm empty fruit bunch for bioethanol production through separate hydrolysis and fermentation (SHF) using immobilized cellulolytic enzymes. *IOP Conf. Ser. Earth Environ. Sci.* **2020**, *439*, 12–18. [CrossRef]
40. Tan, I.S.; Lee, K.T. Enzymatic hydrolysis and fermentation of seaweed solid wastes for bioethanol production: An optimization study. *Energy* **2014**, *78*, 53–62. [CrossRef]
41. Ruiz, H.A.; Silva, D.P.; Ruzene, D.S.; Lima, L.F.; Vicente, A.A.; Teixeira, J.A. Bioethanol production from hydrothermal pretreated wheat straw by a flocculating Saccharomyces cerevisiae strain–Effect of process conditions. *Fuel* **2012**, *95*, 528–536. [CrossRef]
42. Pabón, A.M.; Felissia, F.E.; Mendieta, C.M.; Chamorro, E.R.; Area, M.C. Improvement of bioethanol production from rice husks. *Cellulose Chem. Technol.* **2020**, *54*, 689–698.

43. Berłowska, J.; Kręgiel, D.; Klimek, L.; Orzeszyna, B.; Ambroziak, W. Novel yeast cell dehydrogenase activity assay in situ. *Pol. J. Microbiol.* **2006**, *300*, 127–131.
44. Ballesteros, I.; Negro, M.; Dominguez, J.; Cabañas, A.; Manzanares, P.; Ballesteros, M. Ethanol production from steam-explosion pretreated wheat straw. *Appl. Biochem. Biotechnol.* **2006**, *129–132*, 496–508. [CrossRef]
45. Wang, K.; Yang, H.; Wang, W.; Sun, R.C. Structural evaluation and bioethanol production by simultaneous saccharification and fermentation with biodegraded triploid poplar. *Biotechnol. Biofuels* **2013**, *6*, 42–50. [CrossRef] [PubMed]
46. Sathendra, E.R.; Baskar, G.; Praveenkumar, R. Production of bioethanol from lignocellulosic banana peduncle waste using *Kluveromyces marxianus*. *J. Environ. Biol.* **2019**, *40*, 769–774. [CrossRef]
47. Hadiyanto, H.; Ariyanti, D.; Aini, A.; Pinundi, D. Optimization of ethanol production from whey through fed-batch fermentation using *Kluyveromyces marxianus*. *Energy Proc.* **2014**, *47*, 108–112. [CrossRef]

 fermentation

Article

Techno-Economic Analysis of Integrating Soybean Biorefinery Products into Corn-Based Ethanol Fermentation Operations

Kurt A. Rosentrater * and Weitao Zhang

Department of Agricultural and Biosystems Engineering, Iowa State University, Ames, IA 50011, USA
* Correspondence: karosent@iastate.edu

Abstract: With the development of agricultural biorefineries and bioprocessing operations, understanding the economic efficiencies and environmental impacts for these have gradually become popular for the deployment of these industrial processes. The corn-based ethanol and soybean oil refining industries have been examined extensively over the years, especially details of processing technologies, including materials, reaction controls, equipment, and industrial applications. The study focused on examining the production efficiency changes and economic impacts of integrating products from the enzyme-assisted aqueous extraction processing (EAEP) of soybeans into corn-based ethanol fermentation processing. Using SuperPro Designer to simulate production of corn-based ethanol at either 40 million gallons per year (MGY) or 120 MGY, with either oil separation or no oil removal, we found that indeed integrating soy products into corn ethanol fermentation may be slightly more expensive in terms of production costs, but economic returns justify this integration due to substantially greater quantities of ethanol, distillers corn oil, and distillers dried grains with solubles being produced.

Keywords: biofuels; corn; extraction; enzyme-assisted; protein; soybean

Citation: Rosentrater, K.A.; Zhang, W. Techno-Economic Analysis of Integrating Soybean Biorefinery Products into Corn-Based Ethanol Fermentation Operations. *Fermentation* **2021**, 7, 82. https://doi.org/10.3390/fermentation7020082

Academic Editor: Alessia Tropea

Received: 22 April 2021
Accepted: 20 May 2021
Published: 25 May 2021

1. Introduction

As a renewable energy resource, bio-based ethanol has been successful as a partial replacement to gasoline fuel over the last few decades. It has been shown to be relatively benign to the environment, provided many jobs to rural economies in the USA, made substantial contributions to the global feed industry, and provided improved energy security for U.S. agriculture [1]. In fact, the corn-based ethanol industry and related upstream and downstream industries have been very dynamic in recent years and have been developing various new technologies to increase economic returns. To obtain better efficiency and lower production costs, it may be prudent to combine the corn-based ethanol process with other processes. For example, if a biorefinery operation can make several types of products simultaneously, it may have lower costs for the combined products rather than producing the same products at different facilities. Soybean processing may be one type of bioprocessing system that might make sense to incorporate into corn processing operations. In the soybean oil extraction process, enzyme-assisted aqueous extraction processing (EAEP) is a new method to obtain soybean oil, which uses water as an extraction media to remove oil from ground soybeans. However, to date, it is not yet commercialized extensively.

Due to cell walls and membranes around oil and protein bodies, which create barriers to freeing oils and proteins in soybeans, EAEP utilizes the insolubility of oil in water and uses water as a media to fractionate oil, protein, and fiber. Enzymes can be added to assist in breaking down cell walls [2]. EAEP denatures some proteins and destabilizes the cream (oil), which can increase the final oil extraction yield to nearly 90% [3]. The enzymes in the skim can be recycled in the extraction process. Soy skim has the potential to be used as a partial water replacement in the corn-based ethanol process. Actually, in addition to being used as a water source, soy skim has been shown to be an effective nutrient source and has increased ethanol yield and the protein content in final coproducts in laboratory-scale fermentations [4,5]. The high fiber content of soybean fiber from the extraction process is another advantage of EAEP, which can potentially be pretreated and then used directly in saccharification prior to ethanol fermentation. During soybean EAEP, fiber fractions contained 60–70% moisture, and the solids were mainly cellulose, hemicellulose, and insoluble proteins. There may also be some advantages to using soy fiber in terms of resulting DDGS value.

Laboratory-scale research [6] was conducted using coproducts from EAEP on ethanol production in laboratory-scale fermentations. This study indicated that adding soy skim and untreated insoluble fiber from EAEP significantly increased ethanol production rates and ethanol yields. Thus, integrating EAEP products (fiber and skim) into corn-based ethanol production might improve the efficiency for producing ethanol as well as the value of the DDGS.

The composition of skim has been found to be 9% solids (91% moisture content), 6.3% (db) lipids, 57.6% (db—dry basis) protein, 10.1% (db) ash, and 26% (db) carbohydrates; the composition of insoluble fiber has been found to be 15.1% solids (85% moisture), 4.7% (db) lipids, 7.7% (db) protein, 4.0% (db) ash, and 83.6% (db) carbohydrates [6].

Even though integrating corn–soybean fermentation has effectively used for corn-based ethanol production on a lab scale, there is a lack of economic information on efficiency and profit at larger scales. In order to determine if this type of biorefining is economically viable, a techno-economic analysis for combining corn and soybean biorefinery processes is necessary. The objective of this study was to use techno-economic analysis (TEA) for developing complete estimates of all costs associated with the construction and operation of this type of integrated system. In addition, this study compared an integrated corn and soybean biorefinery with an original corn-based ethanol process in economic performance, to explore the effect of new applications on the corn-based ethanol production under 40- and 120-million-gallon ethanol production scales.

2. Materials and Methods

2.1. Computer Model

SuperPro Designer v8.5 (Intelligen, Inc., Scotch Plains, NJ, USA) was utilized to model biofuels processing for an integrated corn and soybean biorefinery. This industrial design software facilitated modeling, evaluation, and optimization of integrated processes for a wide range of industries [7]. Based on the structure of a modified model [8,9], this model was updated by adding operations for integrating soy insoluble fiber (UIF) and soy skim, thus developing an integrated corn and soybean biorefinery. Due to the UIF and skim added as raw material, the energy and mass balances for the individual unit operations were reset, and the recycling index was also updated so the system obtained the mass and economic balances for the entire manufacturing process.

This study utilized 330 working days per year to mirror real industrial processing operations, which generally operate 24 h per day year-round. All annual calculations were based on these factors and were included in the range of reports available by the program. After setting basic data into the model, SuperPro produced a variety of reports based on simulation data changes for each scenario and facilitated judging the economic feasibility of the various scenarios. These reports were produced and compared for each year for each processing scenario, as well as sensitivities for each affected factor. The model framework and structure of dry grind ethanol from corn processing is shown in Figure 1.

2.2. Simulation Scenarios

Scenarios were updated and modified based on the basic corn-based ethanol process plant model, which was developed by [8] and expanded by [9]. Two scenarios were set below:

- Integrated EAEP with corn-based ethanol process producing 40 million gallons ethanol per year with distillers corn oil removal vs. no oil separation.
- Integrated EAEP with corn-based ethanol process producing 120 million gallons ethanol per year with distillers corn oil removal vs. no oil separation.

Based on the pilot scale research of [5], 75 kg of raw soybeans in the EAEP process could yield 14.28 kg oil, 50.64 kg UIF, and 363.81 kg soy skim. According to previous studies, [6] indicated maximum ethanol production could be achieved when UIF and skim were mixed together with at a rate of corn-to-UIF ratio 1:0.16 and skim-to-UIF ratio 6.5:1. To be appropriate for the scales of 40 and 120 million gallons ethanol production, the 75 kg per hour soybean pilot scale with EAEP process was scaled up to commercial scales, which were 17 million and 51 million kg annual soybean oil production, respectively [10]. This equated to UIF of 7596 kg per hour and soy skim of 54,572 kg per hour for the 40 MGY; while large (120 MGY) scale equated to UIF of 22,788 kg per hour and soy skim of 16,3716 kg per hour. After adding UIF and soy skim in the process at these rates, the ratio of water-to-solids in the fermenter increased from 2.0:1 to 2.5:1, which obtained optimal ethanol yield and maximum ethanol production (based upon information from [5,8]).

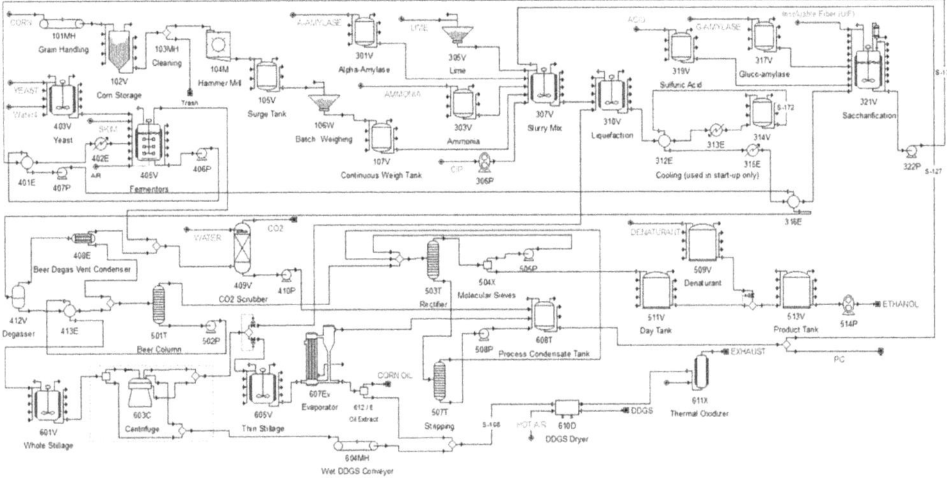

Figure 1. Process flow diagram used to model integrated corn-EAEP ethanol fermentation processing. Note that soy skim is added directly to the fermentor, but soy insoluble fiber is added to the saccharification step immediately prior to fermentation. This SuperPro model was adapted from [8,9].

In this model, operating cost included raw material cost, labor cost, facility cost and utility cost. Corn, UIF and skim were the main raw materials used, which made a significant contribution on materials costs. The market price of corn was USD 145.67 per metric ton, and untreated insoluble fiber was USD 30.66 per metric ton in 2015 [11,12]. The market price of soy skim was unavailable due to lack of data and was generally disposed of as waste trash—since commercial production of EAEP has not really begun yet. In this model, the price of soy skim was treated as water with a price of USD 0.04 per metric ton in 2015 and was mainly used to reduce the water requirement in the fermentation process [13]. For the utility cost, steam was mainly utilized as a heat transfer agent, while natural gas and electricity were used as the energy resource. In this model, steam was set as USD 12.86 per metric ton, while the industrial price of natural gas and electricity in 2015 was set as USD 4.4533 per MBtu (million British Thermal Units) and USD 0.0691 per kW·h [13,14]. Labor cost and inflation rate were set according to data form the U.S. Department of Labor. Installation cost depended on various types of equipment. Loan interest was set at 7.0% per year as a common assumption.

Similar to previous studies, the facility costs in this study were composed of maintenance costs, equipment depreciation, interest on debt, insurance, taxes, and other industrial expenses. Based on basic parameters set by [8], maintenance expenses were determined as 3% of total capital costs, while insurance and other industrial expenses were set to 0.8% and 0.75% of the capital cost. Depreciation was set as an initial index, and taxes were set as 24%, because corn-based ethanol plants belong to green and renewable energy industrial sectors, which have a lower tax rate than basic chemical industrial plants [15].

The cost of labor was determined based upon a lump estimate of number of working hours per year (330 day per year). The hourly wage was based on U.S. Department of Labor minimum wage data [16]. This model multiplied the minimum wage by available workers and automatically created labor costs for all scenarios.

Ethanol was the main product of the entire process, with DDGS and DCO extracted from DWG, both of which were treated as coproducts for revenue estimation. The market prices of ethanol (USD 594.91 per metric ton) and DDGS (USD 157.64 per metric ton) were collected from the USDA database in 2015. Corn distillers oil (USD 611.32 per metric ton) from the oil extraction process was collected from The Jacobsen Company [17]. In addition, physical properties, material combinations, and other basic indices were maintained at a similar level as the original corn-based ethanol model [8,9]. Among profitability analyses parameters, unit production cost, unit production revenue, net profit, and payback time were the most important results to explore between the various scenarios. Specific information about how costs, profits, payback, etc. were calculated in this model have been provided elsewhere [10].

3. Results

3.1. Capital Costs

In this study, the total capital cost was composed of the following individual process operations: grain handling and milling, starch to sugar conversion, fermentation, ethanol processing, coproduct processing, and common support systems. For each individual process's capital cost, the final result was determined based on the equipment purchase price, setting a material factor by the model, and an installation factor [8,9]. For simplifying some indirect support equipment, steam generation and cooling water equipment were not included in capital cost and were treated as purchased utilities. All settings were based on previous data provided by [8,9], which reflected commonly used technologies at US corn-based ethanol plants.

The effect that each of these scenarios had on total capital costs are presented in Figure 2. The simulation data indicated that the cost of starch to sugar conversion decreased in the integrated EAEP models due to soy skim partially replacing the water requirement, which caused a decrease in reactor size requirements. In contrast, the capital cost in fermentation, ethanol processing and coproduct processing increased, which was caused by more products being produced by the additional UIF and skim from EAEP. Overall, the total capital cost required increased at both scales (40 and 120 million gallon) ethanol production which integrated EAEP products and increased to 95.27 million and 162.78 million USD. Coproduct processing represented the largest portion of fixed capital costs in both the 40- and 120-million-gallon ethanol plant models. For the 40-million gallon plant, the coproduct processing costs were 56.5 million USD with the integrated EAEP scenario, were 26.2 million USD for the traditional corn-only processing, and were 37.8 million USD for corn ethanol processing using oil separation systems. For the 120-million gallon plant, the coproduct processing costs were, respectively, 84.4, 68.9, and 68.9 million USD. It was apparent that factory scale clearly impacted the proportion of processing costs that were due to coproduct processing.

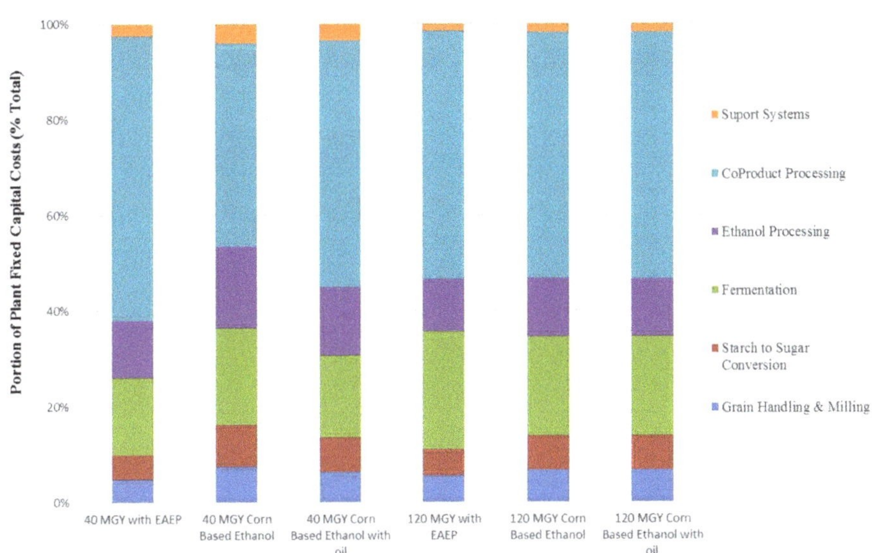

Figure 2. Capital costs for integrated corn-soy ethanol fermentation processing. Scenarios include a 40 million gallon per year (MGY) ethanol plant with or without corn oil separation, a 120 MGY ethanol with or without corn oil separation, and a 40 or 120 MGY ethanol plant, without corn oil separation but with EAEP products used for fermentation.

3.2. Annual Operating Costs

Similar to previous studies, annual operating costs in this study consisted of labor, facility, utility costs, and raw material costs in all three scenarios. In this model, consumables, advertising, royalties, and failed product disposal were not estimated in this techno-economic analysis. Due to added skim and UIF from the EAEP process, integrated EAEP with a corn-based ethanol process required greater operating costs. The 40- and 120-million-gallon integrated EAEP with corn-based ethanol production required 86.71 million and 233.80 million dollars per year, respectively which was around 8% more than the models which did not integrate EAEP products (Figure 3). Differently from original corn-based ethanol models and corn-based ethanol with oil extraction processes, the corn-based ethanol process with integrated EAEP had lower portions of operating costs in raw materials, but higher rates in facility costs and utility costs. The reason for this was that integrating EAEP into corn-based ethanol production required higher liquid-to-solid conditions in fermentation, which meant more DDGS was produced in the coproduct processing operation.

3.2.1. Facility and Labor Costs

A portion of the facility in the EAEP 40 MGY scenario was 15.98%, and the portion of the facility in EAEP 120 MGY scenario was 10.13% (Figure 3). Both scenarios were around 3% higher than other scenarios.

Compared to other indexes of operating costs, labor cost was relatively stable in all scenarios, and was around 2% of total operating cost (Figure 3).

3.2.2. Material Costs

Differently from previous studies, the raw materials for integrated EAEP into a corn-based ethanol plant included corn, water, yeast, caustic, lime, octane, ammonia, sulfuric acid, gluco-amylase, alpha-amylase, as well as untreated insoluble fiber and soy skim from the EAEP process. All material prices were set using marketing prices of 2015. The simulation results are shown in Figure 4. Differently from other scenarios, untreated insoluble fiber made a significant contribution to the annual material cost in the 40 and 120 MGY with integrated EAEP products. The corn portion cost was decreased in those two scenarios, as it was replaced by the coproducts from EAEP.

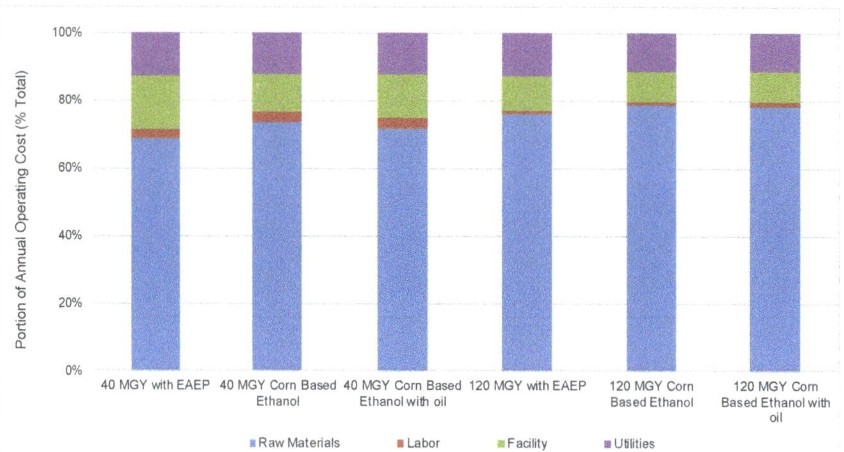

Figure 3. Operating costs for integrated corn-soy ethanol fermentation processing. Scenarios include a 40 million gallon per year (MGY) ethanol plant with or without corn oil separation, a 120 MGY ethanol with or without corn oil separation, and a 40 or 120 MGY ethanol plant, without corn oil separation but with EAEP products used for fermentation.

For the 40-million gallon plant, the raw material costs were 59.6 million USD/year with the integrated EAEP scenario, were 57.6 million USD/year for the traditional corn-only processing, and were 57.6 million USD/year for corn ethanol processing using oil separation systems. For the 120-million gallon plant, the raw material costs were, respectively, 178.1, 172.1, and 172.1 million USD/year. It was apparent that factory scale clearly impacted the proportion of annual operating costs that were due to raw materials.

3.2.3. Utility Costs

Similar to previous studies, utility costs were mainly from electricity, natural gas, steam, and chilled water. The price of electricity was set at USD 0.0691 per kW h, and natural gas was set at USD 4.4533 per MBtu (million British thermal unit). According to Figure 3, utility costs increased slightly in the two EAEP scenarios, which was required to treat more DDGS coproducts which were being produced. Due to the relatively stable rate of composition for utilities, integrating EAEP within a corn-based ethanol process had only a slight effect on the portion of utility in annual operation costs.

3.3. *Annual Revenues*

In this study, annual revenues were defined as the total income from all of the final products and coproducts, which included ethanol, distillers corn oil (DCO) and DDGS. The average corn price was USD 594.91 per metric ton in 2015, and the average DDGS price was USD 157.64 per metric ton in 2015 [11]. According to data from [17], the marketing price of DCO was USD 611.32 per metric ton in 2015.

3.3.1. Ethanol

For these simulations, ethanol was approximately 30% of the total mass produced annually by the ethanol process but contributed more than 70% of the total annual revenues in all scenarios. Compared with corn-based ethanol models and corn-based ethanol with oil extraction processes, 40 and 120 MGY with integrated EAEP produced more ethanol, which was a 7.5% increase for ethanol production (Figure 5). The main reason for this increase was that UIF provided more carbon for fermentation, thus integrating EAEP with a corn-based ethanol process increased the ethanol yield and production.

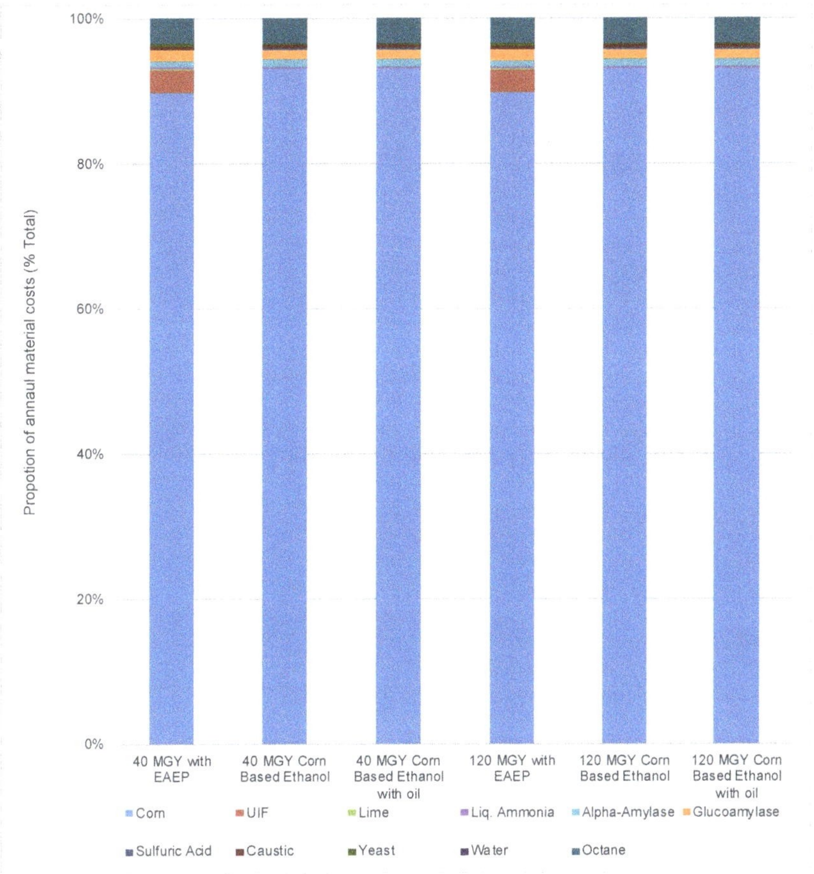

Figure 4. Material costs for integrated corn-soy ethanol fermentation processing. Scenarios include a 40 million gallon per year (MGY) ethanol plant with or without corn oil separation, a 120 MGY ethanol with or without corn oil separation, and a 40 or 120 MGY ethanol plant, without corn oil separation but with EAEP products used for fermentation.

3.3.2. Distillers Dried Grains with Solubles (DDGS)

DDGS made up about 55% of the total mass produced by the ethanol plant. Sales price was determined using [11]. Comparing corn-based ethanol models and corn-based ethanol with oil extraction processes, the revenues for DDGS in the 40 and 120 MGY with integrated EAEP products significantly increased, which was a 20% increase for DDGS annual revenue (Figure 5). The main reason for this increase was that skim and UIF supplied more DDGS mass and thus resulted in more DDGS sales.

3.3.3. Distillers Corn Oil (DCO)

Similar to previous studies, oil extraction rates for this model were set at 80%, which was a reasonable rate for current industrial production. Compared to corn-based ethanol with an oil extraction process (both production scales), the revenue for oil in the 40 and 120 MGY with integrated EAEP products scenarios significantly increased, which was around a 23% increase for the oil annual revenue (Figure 5). The main reason was skim and UIF from the EAEP process supplied extra oil content, which facilitated oil extraction from the thin stillage.

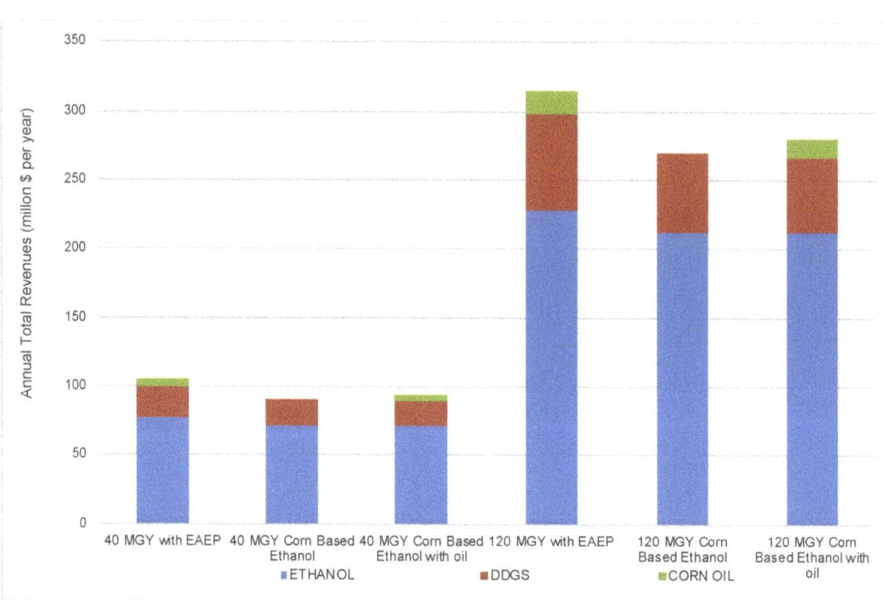

Figure 5. Annual total revenues for integrated corn-soy ethanol fermentation processing. Scenarios include a 40 million gallon per year (MGY) ethanol plant with or without corn oil separation, a 120 MGY ethanol with or without corn oil separation, and a 40 or 120 MGY ethanol plant, without corn oil separation but with EAEP products used for fermentation.

3.4. Gross Operating Margins and Payback Time

Similar to previous studies, gross profit was defined as the annual revenues minus the annual operating costs. The payback period was the length of time required to recover the cost of an investment. Gross operating margins are seen in Figure 6, which contains capital costs, operating costs, revenues, and profits in millions of dollars per year. Figure 6 clearly indicates that the 40 and 120 MGY with integrated EAEP products scenarios had higher amounts in capital investment and operating cost, which was directly affected by the addition of soy skim and UIF from the EAEP process. However, due to more resources for fermentation and coproduct processes, the corn-based ethanol process with integrated EAEP obtained more revenues from the higher amounts of ethanol, DDGS and distillers corn oil production. The 40 MGY with EAEP scenario obtained USD 23.33 million per year at the scale of a 40 million gallons ethanol per year, while the 120 MGY with integrated EAEP scenario obtained USD 77.17 million per year for the 120 million gallons per year ethanol process. The net profit results indicated that the 40 and 120 MGY with integrated EAEP products scenarios had better performance, while corn-based ethanol models with oil extraction also obtained more profit than the original corn-based ethanol process (i.e., without oil separation).

The annual operating costs and annual revenues were then divided into dollar per kg ethanol basis, which directly reflected the efficiency of how costs are related to each kilogram of ethanol produced by the plant. Unit production cost, unit production revenue, and payback time are shown in Figure 7. Not surprisingly, due to economies of scale, the unit production cost decreased when increasing the production scale. The corn-based ethanol process with integrated EAEP required more equipment and utility capacity, causing small increases in unit production. Due to the addition of UIF and skim from EAEP, unit production revenues increased with more ethanol and other coproducts being produced. Payback time also indicated that integrating EAEP with a corn-based ethanol process had economic feasibility in industrial applications. In addition, larger scales owned a higher efficiency for unit production.

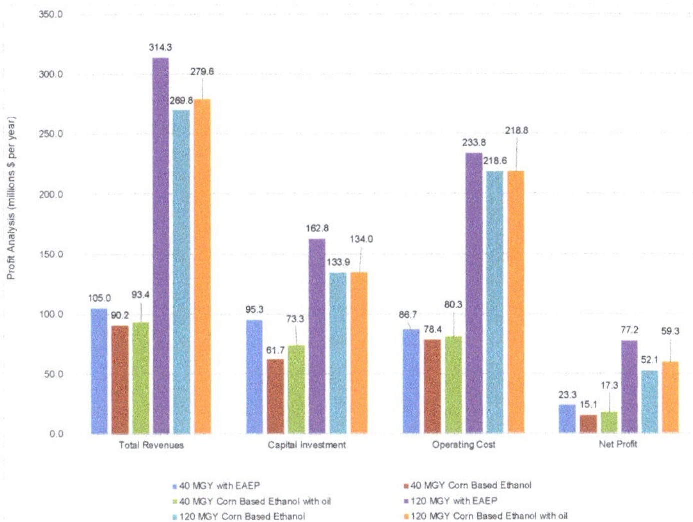

Figure 6. Profitability analysis for integrated corn-soy ethanol fermentation processing. Scenarios include a 40 million gallon per year (MGY) ethanol plant with or without corn oil separation, a 120 MGY ethanol with or without corn oil separation, and a 40 or 120 MGY ethanol plant, without corn oil separation but with EAEP products used for fermentation.

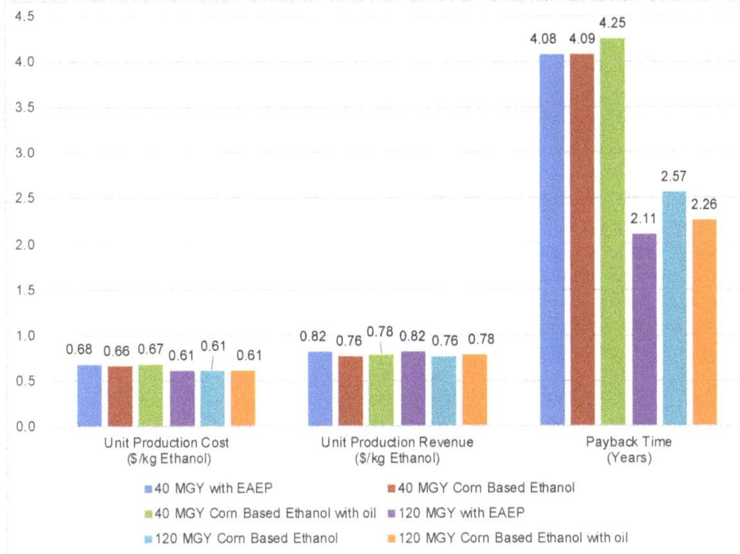

Figure 7. Unit analysis and payback time for integrated corn-soy ethanol fermentation processing. Scenarios include a 40 million gallon per year (MGY) ethanol plant with or without corn oil separation, a 120 MGY ethanol with or without corn oil separation, and a 40 or 120 MGY ethanol plant, without corn oil separation but with EAEP products used for fermentation.

4. Conclusions

To perform economic calculations for integrating EAEP products into a corn-based ethanol processes, SuperPro Designer was used for techno-economic analysis on two industrial scales—40 and 120 MGY ethanol production—both without and with corn oil separation operations. According to the simulation results, integrating EAEP with corn-based ethanol processing required more capacity for equipment and utilities, causing small increases in unit production costs. Due to the addition of UIF and skim from EAEP, unit production revenues increased by generating more ethanol and other coproducts. Payback time also indicated that integrating EAEP with corn-based ethanol processing has economic feasibility for industrial applications. Even though economic analyses look promising, to date, this integration has not yet been commercially realized. In fact, EAEP soy processing has not been widely commercially deployed either. Future research should aim to optimize process efficiencies and lower production costs in order to make this type of processing more economically attractive.

Author Contributions: W.Z. built and tested the model, and drafted the paper; K.A.R. supervised the project, verified the model results, and edited the paper. Both authors have read and agreed to the published version of the manuscript.

Funding: United States Department of Agriculture, National Institute of Food and Agriculture.

Institutional Review Board Statement: Not applicable.

Informed Consent Statement: Not applicable.

Data Availability Statement: Not applicable.

Acknowledgments: The authors would like to thank Andy McAloon and Winnie Yee at the United States Department of Agriculture-Agricultural Research Service (USDA-ARS) for developing the original SuperPro model of the corn ethanol plant, which they generously shared, and which we built and expanded upon during this research project.

Conflicts of Interest: The authors declare no conflict of interest.

References

1. Alinia, R.; Zabihi, S.; Esmaeilzadeh, F.; Kalajahi, J.F. Pretreatment of wheat straw by supercritical CO_2 and its enzymatic hydrolysis for sugar production. *Biosyst. Eng.* **2010**, *107*, 61–66. [CrossRef]
2. Lamsal, B.P.; Murphy, P.A.; Johnson, L.A. Flaking and extrusion as mechanical treatments for enzyme-assisted aqueous extraction of oil from soybeans. *J. Am. Oil Chem. Soc.* **2006**, *83*, 973–979. [CrossRef]
3. Chabrand, R.M.; Glatz, C.E. Destabilization of the emulsion formed during the enzyme-assisted aqueous extraction of oil from soybean flour. *Enzyme Microb. Technol.* **2009**, *45*, 28–35. [CrossRef]
4. Yao, L.; Wang, T.; Wang, H. Effect of soy skim from soybean aqueous processing on the performance of corn ethanol fermentation. *Bioresour. Technol.* **2011**, *120*, 9199–9205. [CrossRef] [PubMed]
5. Yao, L.; Lee, S.L.; Wang, T.; de Moura, J.M.; Johnson, L.A. Effects of fermentation substrate conditions on corn-soy co-fermentation for fuel ethanol production. *Bioresour. Technol.* **2012**, *120*, 140–148. [CrossRef] [PubMed]
6. Sekhon, J.K.; Jung, S.; Wang, T.; Rosentrater, K.A.; Johnson, L.A. Effect of co-products of enzyme-assisted aqueous extraction of soybeans on ethanol production in dry-grind corn fermentation. *Bioresour. Technol.* **2015**, *192*, 451–460. [CrossRef] [PubMed]
7. Ngo, H.L.; Yee, W.C.; McAloon, A.J.; Haas, M.J. Techno-economic analysis of an improved process for producing saturated branched-chain fatty acids. *J. Agric. Sci.* **2014**, *6*, 158. [CrossRef]
8. McAloon, A.; Yee, W. *Ethanol Plant Model*; USDA, ARS: Wyndmoor, PA, USA, 2011.
9. Wood, C.; Rosentrater, K.A.; Muthukumarappan, K. Techno-economic modeling of a corn based ethanol plant in 2011/2012. *Ind. Crops Prod.* **2014**, *56*, 145–155. [CrossRef]
10. Cheng, M.H.; Rosentrater, K.A.; Sekhon, J.; Wang, T.; Jung, S.; Johnson, L.A. Economic feasibility of soybean oil production by enzyme-assisted aqueous extraction processing. *Food Bioprocess Technol.* **2019**, *12*, 539–550. [CrossRef]
11. USDA (United States Department of Agriculture). Fuel Ethanol, Corn and Gasoline Prices, Marketing Year. Available online: http://marketnews.usda.gov/portal/lg (accessed on 21 May 2021).
12. Alibaba. Alibaba.com. Retrieved from Untreated Insoluble Fiber. 2016. Available online: http://www.alibaba.com/showroom/untreatedinsolublefiber.html (accessed on 21 May 2021).
13. EIA (U.S. Energy Information Administration). U.S. Natural Gas Wellhead Price. 2016. Available online: http://www.eia.gov/dnav/ng/hist/n9190us3a.htm (accessed on 21 May 2021).

14. EIA (U.S. Energy Information Administration). Average Retail Price of Electricity to Ultimate Customers. 2016. Available online: http://www.eia.gov/electricity/data.cfm#traderel (accessed on 21 May 2021).
15. Damodaran, A. *Investment Valuation: Tools and Techniques for Determining the Value of Any Asset*; John Wiley & Sons: Hoboken, NJ, USA, 2012; Volume 666.
16. Index, C.P. Bureau of Labor Statistics, U.S. Department of Labor. Washington, DC. 2012. Available online: http://data.bls.gov/cgi-bin/surveymost/ (accessed on 21 May 2021).
17. Jacobsen. (2016, July). Distillers Corn Oil. Retrieved from The Jacobsen Research Database. Available online: https://thejacobsen.com (accessed on 21 May 2021).

 *fermentation*

Article

Simultaneous Saccharification and Fermentation of Empty Fruit Bunches of Palm for Bioethanol Production Using a Microbial Consortium of *S. cerevisiae* and *T. harzianum*

Eryati Derman [1], Rahmath Abdulla [1,*], Hartinie Marbawi [1], Mohd Khalizan Sabullah [1], Jualang Azlan Gansau [1] and Pogaku Ravindra [2]

[1] Faculty of Science and Natural Resources, University Malaysia Sabah, Kota Kinabalu 88400, Malaysia; eryatiderman@gmail.com (E.D.); hartinie@ums.edu.my (H.M.); khalizan@ums.edu.my (M.K.S.); azlanajg@ums.edu.my (J.A.G.)

[2] Chemical and Bioprocess Engineering, Rowan University, Glassboro, NJ 08322, USA; dr_ravindra@hotmail.com

* Correspondence: rahmath@ums.edu.my; Tel.: +60-8-832-0000 (ext. 5592)

Abstract: A simultaneous saccharification and fermentation (SSF) optimization process was carried out on pretreated empty fruit bunches (EFBs) by employing the Response Surface Methodology (RSM). EFBs were treated using sequential acid-alkali pretreatment and analyzed physically by a scanning electron microscope (SEM). The findings revealed that the pretreatment had changed the morphology and the EFBs' structure. Then, the optimum combination of enzymes and microbes for bioethanol production was screened. Results showed that the combination of *S. cerevisiae* and *T. harzianum* and enzymes (cellulase and β-glucosidase) produced the highest bioethanol concentration with 11.76 g/L and a bioethanol yield of 0.29 g/g EFB using 4% (*w/v*) treated EFBs at 30 °C for 72 h. Next, the central composite design (CCD) of RSM was employed to optimize the SSF parameters of fermentation time, temperature, pH, and inoculum concentration for higher yield. The analysis of optimization by CCD predicted that 9.72 g/L of bioethanol (0.46 g/g ethanol yield, 90.63% conversion efficiency) could be obtained at 72 h, 30 °C, pH 4.8, and 6.79% (*v/v*) of inoculum concentration using 2% (*w/v*) treated EFBs. Results showed that the fermentation process conducted using the optimized conditions produced 9.65 g/L of bioethanol, 0.46 g/g ethanol yield, and 89.56% conversion efficiency, which was in close proximity to the predicted CCD model.

Keywords: empty fruit bunches; response surface methodology; central composite design; simultaneous saccharification and fermentation; bioethanol

Citation: Derman, E.; Abdulla, R.; Marbawi, H.; Sabullah, M.K.; Gansau, J.A.; Ravindra, P. Simultaneous Saccharification and Fermentation of Empty Fruit Bunches of Palm for Bioethanol Production Using a Microbial Consortium of *S. cerevisiae* and *T. harzianum*. *Fermentation* **2022**, *8*, 295. https://doi.org/10.3390/fermentation8070295

Academic Editor: Alessia Tropea

Received: 12 May 2022
Accepted: 9 June 2022
Published: 23 June 2022

1. Introduction

Biofuel has attracted lots of attention among renewable energy resources due to its potential to replace existing fossil fuels in order to alleviate the global energy crisis and its demand [1]. This awareness has led to a dramatic increase in biofuel production and research [2]. Sustainable and renewable liquid biofuels such as bioethanol are seen as an alternative to fossil gasoline substitution and replacement [3]. Bioethanol is considered a natural and ecological fuel, can be produced from renewable energy sources, and is widely used in automobile engines [4,5]. This can be done mainly by reducing the operational cost as well as using cheaper and sustainable feedstocks [6]. Thus, research on bioethanol production using renewable, sustainable, and non-food feedstock is important to overcome the issue of fossil fuel demand.

Empty fruit bunches (EFBs) are cheap, readily available, and accessible biomass wastes in the oil palm industries in Malaysia [7–9]. Recently, they emerged as a potential biomass feedstock in producing bioethanol because of their great abundance and favorable physiochemical characteristics [10,11]. Three important components in EFBs, such as lignin, hemicellulose, and cellulose, make it possible for the EFBs to be converted into bioethanol [12,13]. However, an in-depth study into the bioconversion process is needed to fully utilize EFBs for bioethanol production. An efficient bioconversion process of EFBs into bioethanol is crucial as it affects the ethanol yield and also the overall cost of bioethanol production [8,14]. One of the strategies to reduce the production cost is by operating the fermentation process at a high loading substrate and low enzyme requirement [15].

Bioethanol production from EFBs can be carried out in two ways, which are the separate hydrolysis and fermentation (SHF) and simultaneous saccharification and fermentation (SSF) processes. However, SSF is preferred over SHF as the whole process of SSF is performed in a single vessel combining both processes of hydrolysis and fermentation to produce bioethanol [16,17]. This helps in reducing the chances of contamination in the fermentation medium that occur during SHF [18]. Moreover, this process is a fitting technique for the production of bioethanol, as sugars formed from biomass are rapidly converted into bioethanol at higher concentrations and yields [19,20]. Thus, diminishing the accumulation of inhibitory sugars, end-product inhibition and bioethanol presence in the medium will also make it less vulnerable to contamination [21,22]. In the SSF process, both enzymes and microorganisms are used at the same time. Hence, the optimization of process parameters should be investigated to obtain the maximum amount of sugars that can be converted to bioethanol during the process of saccharification [23]. For example, the optimal conditions for hydrolysis using cellulolytic enzymes is between 40 °C and 50 °C, but microorganisms for fermentation work best around 30 °C and 40 °C [24–26]. Therefore, it is important to strike a balance between the optimal conditions for the enzymes and microorganisms used in the SSF process. Choosing an ideal EFB bioconversion process into bioethanol is also very important to establish optimal fermentation conditions for both enzymes and microorganisms in order to develop a cost-efficient bioethanol production.

In this study, a microbial consortium of *S. cerevisiae* and *T. harzianum* were used in the simultaneous saccharification and fermentation (SSF) process of EFBs. A microbial consortium was used in the SSF process instead of using a single microbe, as it not only utilizes substrate more efficiently but also increases the product yield [23]. In a study by Polprasert et al. [27], palm EFBs were used as a substrate to produce ethanol using a microbial consortium of *Saccharomyces cerevisiae* and *Pichia stipitis* at a 1:1 ratio for bioethanol production. In another study conducted by Ali et al. [28], it is highlighted that higher bioethanol production from date palm fronds was achieved by using the same microbial consortium of *S. cerevisiae* and *P. stipitis*. Mishra and Ghosh [29] reported that the maximum theoretical ethanol production from Kans grass biomass was achieved at 78.6% with 0.45 g/g ethanol yield by using a microbial consortium of *Zymomonas mobilis* and *Scheffersomyces shehatae*. Similarly, Izmirlioglu and Demirci [30] produced 35.19 g/L ethanol from 92.37 g/L industrial waste potato mash, which corresponds to 0.38 g ethanol/g starch when *Aspergillus niger* and *S. cerevisiae* co-cultured in the fermentation process. Kabbashi et al. [31] compared the compatibility of several fungi and yeast to develop direct solid-state bioconversion using the potential mixed culture to produce bioethanol. From the study, the mixed culture of a fungus (*T. harzianum*) and a yeast (*S.cerevisiae*) showed the best ethanol production with 14.1% (v/v) bioethanol concentration compared to other mixed culture combinations, which produced bioethanol concentrations in the range of 6.4 to 7.5% (v/v). At present, finding ideal optimization parameters for the simultaneous saccharification and fermentation process for all the concerned microbial strains and enzymes are important to enhance the utilization of substrate and increase the ethanol production yield.

To the best of the authors' knowledge, there has been no published optimization study for the fermentation process using RSM for bioethanol production from EFBs employing a mixed microbial consortium. Meanwhile, the SSF process using microbial strains had been well studied using a wide range of lignocellulosic biomass, but reports on using a microbial consortium for EFB fermentation are limited. Therefore, the aim of this study is to optimize the production of bioethanol using a microbial consortium of *S. cerevisiae* and *T. harzianum* during the simultaneous saccharification and fermentation (SSF) process of

EFBs by employing the central composite design (CCD) of Response Surface Methodology (RSM). The employment of CCD for optimization would benefit researchers, as by using this design, the expensive cost of the analysis could be reduced as it provides a large amount of information from a few experimental runs. RSM is also able to overcome the limitation of one-at-a-time parameter optimization.

2. Materials and Methods

2.1. Raw Materials

Empty fruit bunches (EFBs) were provided by a local palm oil processing mill in Beaufort, Sabah (Lumadan Palm oil Mill). The collected samples in the form of whole bunches were initially shredded before washing with tap water to remove salts, dirt, oil, and debris. Then, the EFBs were dried at 70 °C for 72 h to remove residual moisture until a constant weight was obtained. They were then blended using a laboratory blender (Waring Commercial), sieved, and separated into fractions using a test sieve [32]. The particle size of EFBs used for this study is 0.1–0.5 mm to maximize the contact area of the substrate and to facilitate the pretreatment and enzymatic hydrolysis process [19,27]. The samples were stored in sealed plastic bags and in a dry place until further use.

2.2. Chemicals and Microorganisms

The enzymes cellulase (cellulase from Trichoderma reesei ATCC 26921, aqueous solution, 50 mL) and β-glucosidase (EC 3.2.1.21 from almonds, 0.88 g solid, crude, lyophilized powder) were purchased from Sigma Aldrich, St. Louis, MO, USA. The cellulase had an activity of 700 units/g while the β-glucosidase had an activity of 2.85 units/mg solid. The enzymes, cellulase and β-glucosidase, were used in the saccharification process. The yeast strain *Saccharomyces cerevisiae* Type II (YSCII) and fungi strain *Trichoderma harzianum* W2(4)-1(2) were employed for this research. Yeast from Saccharomyces cerevisiae Type II (YSII) was purchased from Sigma Aldrich, USA. *Trichoderma harzianum* was supplied by Dr. Syafiquezzaman from the Biotechnology Research Institute, UMS.

Microorganisms Cultivation

Saccharomyces cerevisiae Type II (YSII) and *Trichoderma harzianum* W2(4)-1(2) were used as an ethanol fermentation strain. Both the yeast and fungi strains were cultured in potato dextrose agar (PDA) at 30 °C, which was then maintained and stored at 4 °C until further use [33]. In this study, the growth rates of the *S. cerevisiae*, *T. harzianum*, and co-culture of *S. cerevisiae* and *T. harzianum* were evaluated by measuring the optical densities (OD) at a wavelength of 600 nm [34]. The approximate number of cells in the culture can be determined with a spectrophotometer by measuring the optical density (OD) at 600 nm [35] every three hours for 48 h using a microplate reader (Multiskan go, Thermo Scientific) to identify the growth phases of both microorganisms. A growth curve was drawn based on the OD_{600} measured.

Then, the fermentation inoculums were prepared by inoculating a loopful of the microbial consortium of *S. cerevisiae* and *T. harzianum* cells into a 50 mL sterile potato dextrose broth (PDB) medium and harvested at the exponential phase [36]. At the exponential growth of co-cultured *S. cerevisiae* and *T. harzianum*, the active cells were centrifuged in a refrigerated centrifuge (10,000 rpm at 4 °C for 10 min), washed with sterile distilled water three times, and then the precipitated cells were collected under aseptic conditions and added to the fermentation stage as inoculums [37,38].

2.3. Pretreatments of EFBs

Dried EFBs were soaked in 2% (v/v) sulphuric acid (H_2SO_4) and incubated in an autoclave at 121 °C, 15 psi for 20 min. The dilute acid-treated EFBs fibers were then soaked in water and occasionally mixed for 1 h [25]. The washed EFBs were then dried at 70 °C overnight. The dried acid-treated EFBs were soaked in 10% (w/v) sodium hydroxide (NaOH) solution [39,40], stirred at ambient temperature for 4 h, and then recovered from the alkali solution. The EFBs in the wet alkali solid-state were heated again at 121 °C, 15 psi, for 20 min. The thermal-treated biomass was soaked in the water and stirred occasionally to remove NaOH from the surface. The samples were washed several times with distilled water to neutralize the pH, after which they were dried in an oven at 70 °C overnight [25]. The pretreated EFBs were then stored in a sealed plastic bag until further use.

2.4. EFBs Analysis

2.4.1. Scanning Electron Microscope (SEM) Analysis of EFBs

The untreated and pretreated EFBs were subjected to microscopic observation. The samples were washed with distilled water before drying at 70 °C for 24 h [32]. The dried samples were subjected to SEM using a Carl Zeiss MA10 model brand which has elemental analysis and chemical characterization with element surface mapping via EDX (Energy Dispersive X-ray Spectroscopy). The EFB samples were mounted on conductive tape and coated with gold particles prior to analysis.

2.4.2. Fourier Transform Infrared (FTIR) Analysis of EFBs

FTIR analysis was performed to evaluate the infrared spectrum that shows the chemical composition of the samples. The difference between the untreated and pretreated EFBs was studied using FTIR analysis (Perkin Elmer). The spectra for the samples were recorded in the wavelength range of 400 to 4000 cm^{-1} with the direct transmittance at the rate of 4 scans/min [41]. The FTIR spectra were smoothened and corrected to the baseline correction. The formation, breaking, and shifting of bands were observed. Functional groups associated with major vibration bands were also determined.

2.5. Enzymatic Saccharification of EFBs

The enzymatic saccharification of the acid/alkali pretreated EFBs was performed using cellulase derived from *Trichoderma reesei* (*Trichoderma reesei* ATCC 26921) and β-glucosidase (EC 3.2.1.21 from almonds). The amount of enzyme used was 50 U/g of cellulase and 10 U/g of β-glucosidase. The pretreated EFBs of 4% (*w*/*v*) were hydrolyzed in a 50 mM citrate buffer (pH 4.8). The samples were then incubated at 50 °C, 150 rpm, for 72 h. Sample aliquots were withdrawn at 24 h intervals and analyzed for reducing sugar glucose [32].

2.6. Selection of Microorganisms and Enzyme Combinations

In this study, the combination of microorganisms of *S. cerevisiae* and *T. harzianum* and also the combination of cellulase from T. reesei and β-glucosidase from almonds were employed for the simultaneous saccharification and fermentation (SSF) process. The selection was done to determine which combination can enhance the conversion of EFBs into ethanol during the fermentation process. Different microorganism and enzyme combinations (Table 1) at constant inoculum loadings of the microorganisms at 10% (*v*/*v*)—50 U/g for cellulase and 10 U/g for β-glucosidase—were added under baseline parameters of 4% (*w*/*v*) of pretreated empty fruit bunches at a fixed volume (50 mL) of sodium citrate buffer (pH 4.8) at a temperature of 30 °C for 72 h in a 250 mL Erlenmeyer flask and placed in an orbital shaker (Heidolph Incubator 1000) operated at an agitation speed of 150 rpm (triplicates for each run). After 72 h of fermentation time, the fermentation product was immediately heated for 5 min in a boiling water bath to end the enzymatic reaction. The fermentation product was then centrifuged (Thermo Scientific, Heraeus Megafuge 16R) at 10,000 rpm for 10 min. The supernatant was taken and used in the distillation process to obtain the ethanol. Then, the ethanol produced was subjected to ethanol analysis using gas chromatography–mass spectrometry (GC–MS). The combination which produced the higher ethanol concentration was selected for the fermentation process.

Table 1. Different combinations of microorganisms (*S. cerevisiae* and *T. harzianum*) and enzymes (Cellulase and β-glucosidase).

Combination	Microorganisms and Enzymes			
	S. cerevisiae	*T. harzianum*	Cellulase	β-glucosidase
M1	✓	✓	✓	✓
M2	✓	✓	✓	
M3	✓	✓		✓
M4		✓	✓	✓
M5		✓	✓	
M6	✓	✓		✓
M7	✓		✓	✓
M8	✓		✓	
M9	✓			✓

Operating conditions: 4% (*w*/*v*) pretreated EFBs, at constant inoculums loading of 10% (*v*/*v*) for each combination, pH 4.8, temperature 30 °C, for 72 h at 150 rpm.

2.7. Optimization of Simultaneous Saccharification and Fermentation

The statistical analysis of Response Surface Methodology (RSM) was utilized to optimize the simultaneous saccharification and fermentation process by employing the central composite design (CCD). Various parameters or factors affecting the SSF process of EFBs for bioethanol production were optimized.

Central Composite Design (CCD) for Optimization

The CCD of RSM was applied to determine the optimum conditions of the significant parameters for the SSF process. The effect of fermentation time (24–72 h), temperature (30–50 °C), pH (4.8–6.0), and inoculum concentration (5–10% *v*/*v*) on the production of bioethanol were studied at five experimental levels (−2 (α), −1, 0, +1, +2 (α)). The design matrix of 30 sets of experimental runs was generated from the CCD of RSM software. All the 30 experiments with three replicates were carried out according to the design matrix to screen the best optimum value of each parameter for bioethanol production [42]. The response surface graphs were obtained using the software to understand the effect of variables individually and in combination, in order to determine their optimum levels.

The experimental runs were carried out according to a 2^4 full factorial design for the four identified design independent variables with low (−1) and high (+) levels. The total number of experiments (runs) was given by the simple formula [$30 = 2^k + 2k + 6$], where k is the number of independent variables (k = 4); this includes the following: 16 factorial points from 24 full factorial CCDs were augmented with 6 replicates at the center point to assess the pure error. The response was selected based on preliminary study results. The design factors (variables) with low −1 and high +1 levels are, namely, A (24 and 72), B (30 and 50), C (4.8 and 6), and D (5 and 10). The central values (zero levels) chosen for experimental design were as follows: 48 h, 40 °C, pH 5.4, and 7.5 % (*v*/*v*) for A, B, C, and D, respectively (Table 2) [43].

Table 2. Experimental range and levels of variables used in the Central Composite Design for the optimization of fermentation.

	Parameters	Levels				
		−2 (α)	−1	0	+1	+2 (α)
A	Fermentation Time, (h)	0.0	24.0	48.0	72.0	96.0
B	Temperature, (°C)	20.0	30.0	40.0	50.0	60.0
C	pH	4.2	4.8	5.4	6.0	6.6
D	Inoculum concentration, % (*v*/*v*)	2.5	5.0	7.5	10.0	12.5

2.8. Simultaneous Saccharification and Fermentation

The simultaneous saccharification and fermentation of the acid-alkali-pretreated EFBs were performed in a fixed volume of 100 mL of citrate buffer broth (1% (w/v) yeast extract, 2% (w/v) peptone, and 4% (w/v) pretreated EFBs) in a 250 mL Erlenmeyer flask in an orbital incubator shaker (Heidolph, Uimax 1010 and Incubator 1000) at an agitation speed of 150 rpm. Combinations of different microorganisms (*S. cerevisiae* and *T. harzianum*) and enzymes (Cellulase and β-glucosidase) were used in the bioethanol fermentation. The sample obtained at the end of the fermentation process was centrifuged at 10,000 rpm for 10 min. The pellet was discarded and only the supernatant was transferred to the new Falcon tube. The fermenting products were then quantified for their bioethanol concentration after undergoing the distillation process to obtain the bioethanol.

2.9. Analytical Methods

2.9.1. Bioethanol Determination

The ethanol contents of the samples after the distillation process were analyzed using gas chromatography–mass spectrometry (GC–MS) (Model 6890N, Agilent Technologies, CA, USA) equipped with a thermal conductivity detector and an HP−5MS column, 0.25 mm × 30 m, 0.25 µm ID. Samples were filtered through a Durapore (PVDF) syringe-driven filter unit (0.2 µm) into 1.5 mL glass vials, sealed with a cap, and kept at 5–8 °C before being analyzed using GC–MS. The sample (1.0 µL) was injected into the GC–MS in split mode with a split ratio of 100:1. Helium gas with 99.995% purity was used as the carrier gas and its flow rate was set to 10.0 mL/min. The initial temperature of the oven was 40 °C and was increased at a rate of 10 °C/min up to 100 °C [44]. Hexane was used as the solvent for the standard and sample dilution.

2.9.2. Statistical Analysis of the Experiment

The bioethanol yield (g/g) was calculated based on the experiment and expressed as g of bioethanol per total g of glucose utilizing Equation (1) and g of bioethanol per total g of EFBs utilizing Equation (2). The bioethanol conversion efficiency or theoretical ethanol yield (%) was calculated based on the ratio of ethanol yield obtained against the theoretical maximum ethanol yield using Equation (3) [32,45].

$$\text{Bioethanol yield (g/g) of glucose} = \frac{\text{Bioethanol concentration (g/L)}}{\text{Initial glucose concentration (g/L)}} \quad (1)$$

$$\text{Bioethanol yield/g of EFBs} = \frac{\text{Bioethanol concentration (g/L)}}{\text{Substrate (EFBs) used (g)}} \quad (2)$$

$$\text{Conversion efficiency (\%)} = \frac{[\text{EtOH}]}{0.51} \times 100\% \quad (3)$$

3. Results and Discussion

3.1. Pretreatment of EFBs

The chemical composition of EFBs includes cellulose, hemicellulose, and lignin fractions. The approximate percentage compositions of EFBs depend on the source of the EFBs. Table 3 shows the chemical composition of EFBs from a previous study before and after the pretreatment process. It can be seen that cellulose has the highest content (%), followed by hemicelluloses, lignin, and ash. The amount of cellulose in EFB increases while the hemicellulose and lignin content decrease after the pretreatment process. The previous study by Burhani et al. [46] obtained 90.5% cellulose, no trace of hemicellulose, and 9.13% lignin after the pretreatment process. In a study by Campioni et al. [47], it was reported that there was an increase in EFB cellulose content after acid-alkali pretreatment from 42.2 to 62.6%. Different authors observed different EFB compositions obtained after the acid-alkali pretreatment process. Akhtar et al. [48] reported that in the first step of the pretreatment of

EFB using dilute acid, 90% of hemicellulose and 10% of lignin were removed and further treatment using dilute alkali with a microwave achieved 71.9% delignification.

The development of pretreatment is one of the crucial steps in bioethanol production to minimize the sugar loss, limit the inhibitor formation, and maximize the lignin removal [49]. Most of the hemicellulose contents of EFBs are usually lost after the acid-alkali pretreatment. A study by Kim and Kim [25] demonstrated that sequential acid-alkaline pretreatment efficiently reduced the hemicellulose and lignin content in EFBs. EFB biomass normally has 50 to 80% complex carbohydrates containing C6 and C5 sugar units. According to Abdul et al. [50], oil palm EFB fibers have about 60% (w/w) sugar components. However, no sugar loss was observed in the EFBs when they were pretreated using the ammonia fiber expansion (AFEX) method. In addition, Taherzadeh and Karimi [51], reported that the chemical pretreatment of lignocellulosic material should remove maximum lignin contents with no more than 5% sugar loss. In this work, the authors used a chemical acid agent, 2% (v/v) H_2SO_4, an alkaline agent, and 10% (w/v) NaOH solution.

Table 3. Chemical composition of EFBs before and after the pretreatment process.

	EFB Components			**Content (%)**		
Untreated	Cellulose	25.71	42.2	41.8	32.26	36.59
	Hemicellulose	17.37	29.4	35.6	17.62	24.97
	Lignin	34.02	13.8	18.8	33.02	26.53
	Ash	-	-	-	1.82	1.79
Treated	Cellulose	90.5	62.6	85.4	65.91	75.05
	Hemicellulose	0.00	5.6	3,5	15.55	10.19
	Lignin	9.13	24.3	5.3	11.70	8.11
	Ash	-	-		0.62	2.22
References		[46] [a]	[47] [a]	[48] [a]	[52] [b]	[53] [b]

The chemical composition of treated EFBs is based on the best result of the pretreatment process taken from the respective journal. [a] Sequential acid-alkaline pretreatment using H_2SO_4 and NaOH. [b] Alkaline pretreatment using NaOH.

The effects of sequential acid-alkali pretreatment on EFBs were measured by comparing the physical characteristics of the EFBs before and after pretreatment, shown in Figure 1. Moreover, changes in the EFBs' structure were also analyzed by using the scanning electron microscope (SEM) and Fourier Transform Infrared Spectroscopy (FTIR).

3.1.1. Physical Analysis of EFBs

The physical characteristics of the pretreated EFBs and the non-treated EFBs were observed and are presented in Figure 1. In general, the visual observation, which can be seen between the non-treated and treated EFBs, is the color and structure of the EFBs. From the figure, it can be seen that the surfaces of the non-treated EFB fibers (Figure 1a) have clear, well-ordered, and rigid fibrils, while the pretreated EFB fibers (Figure 1b) showed porous, rough, and irregularly ordered fibrils after the pretreatment process.

Morphological differences between the EFBs occurred due to the removal of the fibril components during the pretreatment process. Physical changes occurred on the surface of treated EFBs that enable easier enzyme access to hydrolyze the cellulose components into glucose and further facilitate the performance of enzymatic hydrolysis [54]. The treated EFBs also changed color to dark brown. This is due to an increase in steam temperature, which caused the degradation of carbohydrates when the EFBs were autoclaved at 121 °C [55]. Furthermore, the treated EFBs were more fragile compared to the non-treated EFBs.

(**a**) Non-treated EFBs

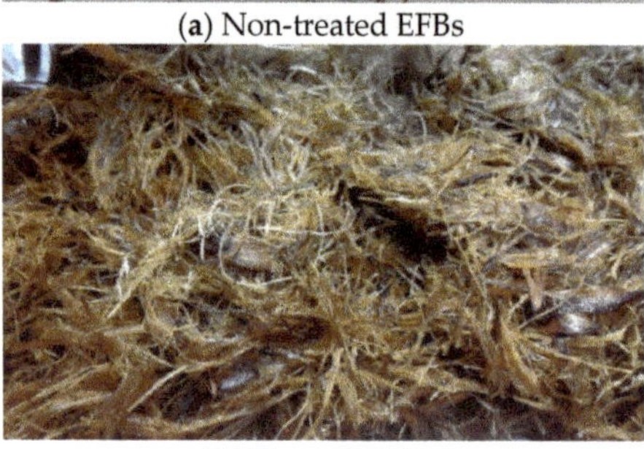

(**b**) Treated EFBs

Figure 1. Physical characteristics of the EFBs: (**a**) non-treated EFB fiber; (**b**) sequential acid/alkali-pretreated EFBs fiber.

The composition of biomass plays an important role in the pretreatment methods selection [56]. Musatto et al. [57] reported that the sequential acid-alkali pretreatment technique was used in order to eliminate the protective lignin-hemicellulose wrapper of the EFBs. The sequence of pretreatment in combined form gave a high impact on reducing sugar production by increasing the cellulose and reducing the hemicellulose and lignin content [58]. In the work performed by Campioni et al. [47], EFBs' cellulose content was increased from 42.2 to 62.6% after acid and alkali treatment, while their hemicellulose component had a mass loss of about 90% and a lignin loss of about 25%.

In the pretreatment process, chemical pretreatment using acid (low pH) and alkali (high pH) techniques can be used to boost the hydrolytic reactivity [59]. The acid pretreatment technique helps in the hydrolysis of hemicellulose fractions and lignin content reduction in biomass [60,61]. On the other hand, the alkaline pretreatment of lignocelluloses with NaOH can modify or remove lignin content in the feedstocks by fracturing the ester bonds, which are cross-links between lignin and xylan, so that the porosity of the biomass can be increased [40]. Furthermore, the alkali (NaOH) pretreatment technique is effective in exposing the cellulose to cellulose digestion by breaking the hemicelluloses–lignin linkage in the amorphous-crystalline structure of cellulose, thus enabling easier conversion of EFBs into glucose [62]. During pretreatment, NaOH penetrates and swells the substrate and solubilizes the hemicellulose, lignin, and the other non-cellulose components [63].

EFBs are usually incubated in an autoclave at 121 °C for 20 min to maximize the effect of NaOH and H_2SO_4 on lignin extraction [64]. Autoclaving at 121 °C and 15 psi is the best way to alter the chemical composition and physical structure of the EFBs, as well as increasing the reducing sugar production. High temperature promotes the removal of both hemicelluloses and lignin (delignification). Akhtar et al. [48] found that 90% of EFBs' hemicellulose was removed after the EFBs were soaked in dilute H_2SO_4 with additional autoclave heating. The combination of NaOH treatment at 10 MPa pressure and 121 °C during pretreatment disintegrated EFB fibers into pliable fibers. It also cleans up the fiber surface and thus exposes more cellulose components in the EFB fibers. Moreover, mass losses of EFBs occur due to the heating of the EFBs at a high temperature when autoclaved at 121 °C, as this causes degradation in the EFBs' hemicelluloses and lignin contents [65].

3.1.2. Scanning Electron Microscope (SEM) Analysis of EFBs

The SEM analysis of EFBs was conducted in order to study the effects of the pretreatment process based on its microscopic morphology differences. A distinct change in the EFBs' physical appearance can be seen in the structure of the untreated EFBs in Figure 2a,b and treated EFBs in Figure 2c,d. In Figure 2a, the untreated sample structures are complete, compact, rigid, and have a smooth surface. This is because no pretreatment process was used to destruct the lignocellulose component of the EFBs [64,66]. According to Tye et al. [67], untreated biomass usually shows low enzymatic hydrolyzability because the enzyme accessibility is restricted by the recalcitrance polymer lignin and hemicellulose.

For the treated sample in Figure 2c, there is a formation of pores on the EFB surface. The presence of pores occurs due to the removal of hemicelluloses [58]. It was reported that the pretreated lignocellulose, which has fractions of pores, was more accessible for enzymatic attack [32]. This is because pretreatment effectively degraded and exposed more surface area of fermentable sugars for the enzymatic hydrolysis process [56]. Pores present in the EFBs are also thought to be effective in the swelling of the EFBs' structure, thus attracting the enzymatic and microbe reactions for the bioconversion process [55,68]. It is revealed that the sequential acid-alkali pretreatment process changed the morphology of the EFBs and gave the biggest impact on the alteration of the EFB structure by removing the silica, which is the chemical composition barrier, causing pore formation.

The SEM micrographs for non-treated EFB surfaces (Figure 2a,b) showed a silica body embedded on the surface. From the figure, it can be seen that the silica bodies were attached to the circular craters, which were spread relatively uniformly over the EFB strands, as in a study by Isroi et al. [69]. This was also similar to the SEM micrograph shown in the study by Nurul Hazirah et al. [70]. The silica present in the cell wall acts as a barrier in the enzymatic digestibility and fermentation process [48]. However, after the pretreatment process was performed, the silica bodies were mostly removed from the EFBs' structure (Figure 2c,d). The remaining holes had homogenous dimensions of around 10 μm in diameter on the EFBs' outer surface [55]. The EFBs' structure became cleaner and smother where almost all the impurities on the EFBs surface were removed, as in the study by Norul Izani et al. [65]. The silica bodies also can be dislodged by an extensive treatment of the EFBs, such as hammering, washing, and crushing [69].

3.1.3. Fourier Transform Infrared Spectroscopy (FTIR) Analysis of EFBs

The structure of EFBs before and after the pretreatment was analyzed using the FTIR spectroscopy method. Based on Figure 3, the pattern of the graph and the existence peaks were different before and after the pretreatment. The basic elements and functional groups present in EFBs were obtained by FTIR analysis [70]. From the FTIR analysis performed by Eliza et al. [71], the presence of a new group was proven after the EFB pretreatment.

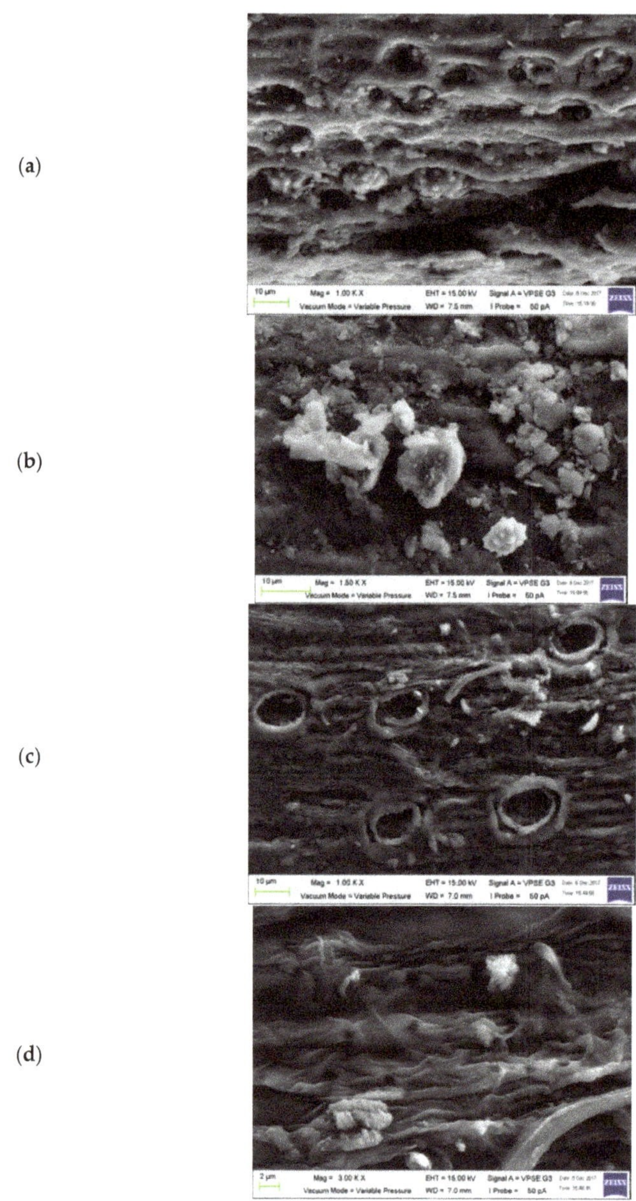

Figure 2. The EFB samples' structure from SEM analysis before and after the pretreatment process: (**a**) untreated EFBs at 1.0K× magnification; (**b**) untreated EFBs at 1.50K× magnification); (**c**) treated EFBs at 1.0K× magnification and (**d**) treated EFBs at 3.0K× magnification).

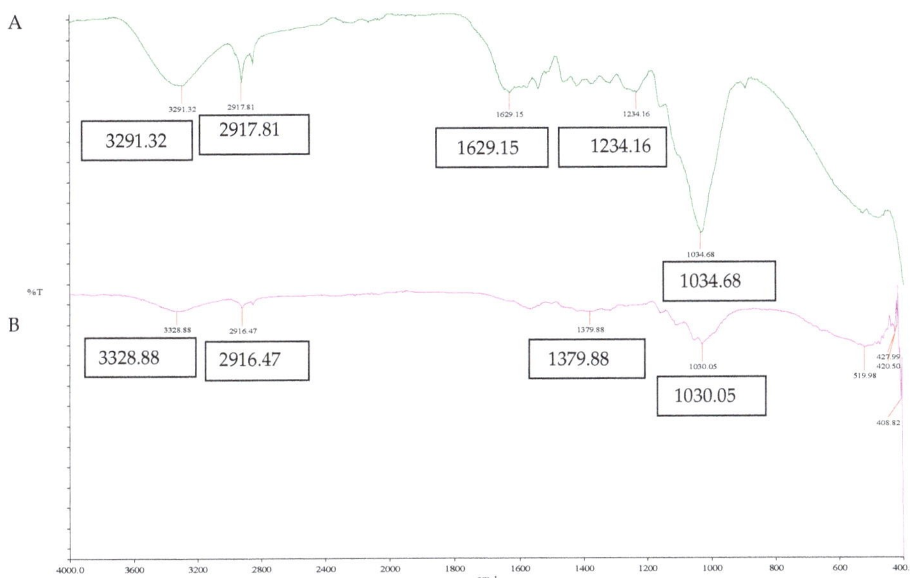

Figure 3. The EFB samples' FTIR analysis before and after the pretreatment process: (**A**) untreated EFBs, (**B**) treated EFBs.

From the figure, absorption bands at 1629.15, 1234.16, and 1034.68 cm^{-1} are shown to have disappeared or diminished, while other bands at 1379.88 and 1030.05 cm^{-1} notably decreased. According to Baharuddin et al. [55], the disappearance of the absorption occurs due to the decomposition of the hemicellulose component in the EFBs. The reduction in the peak intensity shows an indication that the functional group was disturbed or altered [72]. The difference in spectra also can be seen between the untreated and treated EFBs. Changes in the absorption bands were also visible, as some of the peaks became broader after the pretreatment process. The absorption of bands at 3291.32 and 2917.81 cm^{-1} of untreated EFBs was sharp but became broader in the treated EFBs at absorption bands of 3328.88 and 2916.47 cm^{-1}. These changes suggested a decrease in the silica component after the pretreatment process [55].

From the FTIR result, the EFB spectrum shows a strong similarity in the first peak before and after the pretreatment process at absorption bands of 3291.32 and 3328.88 cm^{-1} indicating the presence of hydroxyl (OH) groups in the aromatic and aliphatic compounds [64]. The absorption peak at 2917.81–2916.47 cm^{-1} (second peak) was also identified, which is attributed to the stretchiness of the C-H bonds of the methyl group. The peaks at 1629.15 and 1379.88 cm^{-1} represent the stretching of (C=C) and (C-C), respectively, in aromatics derived from EFBs. Peaks at 1234.16 cm^{-1} could be assigned to the (C-O) bonds of alcohol groups in ethers. The peaks at 1034.68 and 1030.05 cm^{-1} are attributed to glycosidic bonds, indicating the characteristic of cellulose [70]. In another study [55], the most intensive broad absorption band appeared in the carbohydrate region at 1034.68 cm^{-1}, assigned to the vibrations of $C_6H_20_6H$ and C_3HO_3H of the cellulose and pyranosyl ring.

3.2. Enzymatic Saccharification of Pretreated EFBs

The glucose production was determined using high-performance liquid chromatography (HPLC) every 24 h, up to 72 h of the saccharification process. The enzymatic saccharification was performed using cellulase and β-glucosidase, as reported by Hamzah et al. [73]. The highest initial glucose concentration from the pretreated EFBs was achieved at 72 h with 21.14 ± 1.49 g/L. Meanwhile, the initial glucose concentration at 24 and 48 h were

13.827 ± 2.813 g/L and 20.295 ± 1.308 g/L respectively. During the saccharification process, the cellulose in the EFBs was converted to glucose [39].

The enzymatic saccharification of pretreated EFBs was performed to determine the maximum glucose concentration which can be produced during the saccharification. The maximum glucose production was observed at 72 h of incubation with 21.14 ± 1.49 g/L. A similar result has also been reported by Abu Bakar et al. [74], in which the maximum reducing sugars reported was 6.86 g/L at 72 h. According to Adela et al. [32] the longer the enzymatic saccharification time, the higher the glucose yield obtained from the saccharification process. In another study by Hossain et al. [75] the result showed that the glucose content for the oil palm waste residue continuously increased with the increase in the hydrolysis time. The high concentration of the reducing sugars was not only due to the cellulase activity, which produces glucose, but it also can be attributed to the hemicellulases in the biomass [73]. The characteristics of lignocellulosic biomass feedstocks and their pretreatment method in the research influence the performance of cellulase during the enzymatic saccharification process [76].

3.3. Microbial Consortium of S. cerevisiae and T. harzianum

3.3.1. Morphology of the Microbial Consortium of *S. cerevisiae* and *T. harzianum*

Microbes in a consortium are able to use a broad range of carbon sources. Therefore, the microbes can perform complex functions that are impossible for a single type of microorganism [77]. A microbial consortium of *S. cerevisiae* and *T. harzianum* was used as the fermenting microorganisms during the fermentation process. Each microbial strain was cultured independently and then co-cultured together in the same plate, as in Figure 4a–c. The morphologies of yeast and fungi strains were also studied based on their microscopic morphology, as in Figure 4d–f. The morphology of the microbes cultured was observed under the microscope before being used as inoculums in the fermentation process to ensure healthy and pure cells were used in this research. This is to avoid unrelated microbes being inoculated and isolated into the fermentation broth during the fermentation.

(a)

(b)

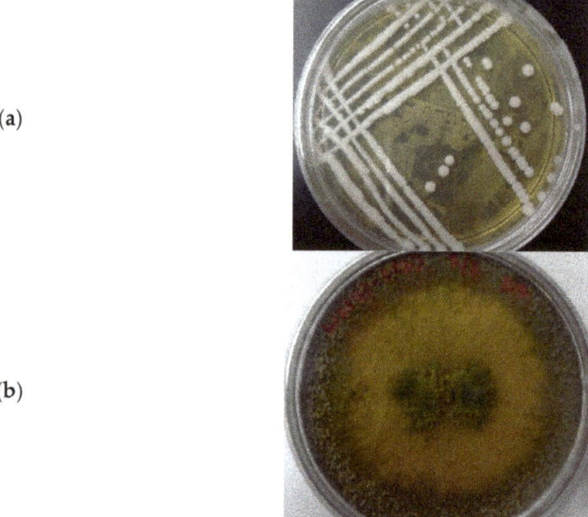

Figure 4. *Cont.*

(c)

(d)

(e)

(f)

Figure 4. Pure culture: (**a**) yeast *S. cerevisiae*; (**b**) fungi strain *T. harzianum*; (**c**) microbial consortium of *S. cerevisiae* and *T. harzianum*. Morphology: (**d**) *S. cerevisiae*; (**e**) *T. harzianum*; (**f**) microbial consortium of *S. cerevisiae* and *T. harzianum*.

Figure 4a,c shows the pure culture of *S. cerevisiae* and the cells' microscopic view on day 3 of culturing. The *S. cerevisiae* cells that were observed under the microscope were generally round, globular, and ellipsoid in shape, having a diameter of approximately 2–8 µm in length, and most of the cells were attached and elongated to each other. Kusfanto et al.'s [78] result showed that the *S. cerevisiae* cells were usually round or oval-shaped with various sizes. Cells reproduce through a process called budding, and a typical yeast cell is around 5–10 µm in diameter [79]. From the figure, some of the cells observed formed budding. Budding formation indicates the cell division process, in which the "mother" cells produce an ellipsoidal daughter cell. *S. cerevisiae* is one of the most common microbes used in producing bioethanol while *T. harzianum* is reported to produce the cellulase enzyme, which helps in the fermentation process [80].

The microscopic morphology of *Trichoderma* isolates was observed with 100X magnification under the light microscope in Figure 4d. The shapes, colors, and sizes of conidia were also observed. The conidia cells have ovoidal shapes and were mostly single-celled. The colors of the conidia of *Trichoderma* were found to be green. Conidiophores were many-branched, hyaline, and bearing a single or group of phialides. Phialides were usually flask-shaped, had a slightly narrowed base, and were also swollen in the middle with a pointed tip. Conidia were single-celled, green, and ovoid with rough or smooth walls generally borne in small terminal clusters. A few conidia cells were found to be slightly ovoidal shaped [81]. The *T. harzianum* colonies, which were grown in the PDA plates, should be white at the early stage but turn to a dark green color after 7 days of culturing [82]. The production of *T. harzianum* green conidia on the PDA plate was denser in the center [83]. Different intensities of green colors of mature conidia which were light green, dark green, yellowish-green, and grayish-green can be observed on the PDA plate, as in Figure 4b. PDA was the best medium in terms of biomass yield and growth spore production [84].

The morphological characteristics of the microbial consortium of *S. cerevisiae* and *T. harzianum* were also observed under the light microscope at 100× (Figure 4f). From the figure, both the fungal hyphae of the *T. harzianum* and yeast *S. cerevisiae* cells were observed. In the co-culture of *S. cerevisiae* and *T. reesei* on PDA and LM mixed with cassava, the fungal hyphae also grew with yeast cells when observed under a compound microscope at 100X magnification [85]. From the figure, it is shown that the co-culture has conidiophores with paired primary branches where their phialides were flask or cylindrical in shape. In a study conducted by Prajankate and Sriwasak [85], the white colonies of the *S. cerevisiae* were covered by the green *T. reesei* mycelium after culturing on the PDA plates at 37 °C for 5 days, as in Figure 4c.

In recent years, research has been more focused on bioprocesses using the *S. cerevisiae* as a co-culture with *Trichoderma* spp., due to better fermentation attributes in the conversion of a complex form of carbohydrates into glucose and then the conversion of glucose to ethanol and CO_2 [86]. A microbial consortium is considered a prospective bioprocess if each microorganism metabolizing its substrate is not disturbed by the presence of another microorganism [19]. According to Kumar et al. [87], *S. cerevisiae* and *Actinomyces* co-culture fermentation resulted in higher bioethanol production from apple pomace with 49.64 g/L, while employing a culture of *S. cerevisiae* alone produced only 37.6 g/L ethanol. Swain et al. [33] mentioned that the ability of bioethanol production from un-saccharified sweet potato flour using *S. cerevisiae* and *Trichoderma* spp. co-culture was 65% higher than employing a single culture of *S. cerevisiae*.

3.3.2. Growth Curve of *S. cerevisiae*, *T. harzianum*, and the Co-Culture of *S. cerevisiae* and *T. harzianum*

Figure 5 shows the growth curve of S. cerevisiae, *T. harzianum*, and the co-culture of *S. cerevisiae* and *T. harzianum* by measuring the optical densities of the suspension cultures every 3 h for 48 h at a wavelength of 600 nm [34]. OD is generally used to determine the inhibitory activity of antifungal compounds [88]. Microbes should be harvested at the exponential phase before being inoculated into the fermentation medium for bioethanol production. It is difficult to obtain a higher yield of bioethanol due to the slow growth of microbes from the depletion of nutrients [20]. Hence, the growth of the yeast and fungi was studied.

The growth curve of yeast *S. cerevisiae* showed a short lag phase while the log phase had the sharpest slope and lasted nine hours. From the figure, the logarithmic phase of the yeast *S. cerevisiae* was between the 3rd to 12th hours after the onset of the inoculation. In the first three hours, there was a slight increase in the growth of the yeast culture. Subsequently, the growth increased gradually from the 3rd h, (0.395 ± 0.013) to the 12th h (1.029 ± 0.005).

From the 12th to the 18th h, a slow increase in the growth of the yeast was observed with absorbances of 1.029 ± 0.005, 1.035 ± 0.006, and 1.037 ± 0.007, respectively. The growth curve reached a maximum point at the 21st h, at which the absorbance was recorded at 1.129 ± 0.003. After the maximum growth was achieved, the absorbance of yeast culture was in a stationary pattern until the 48th h (1.089 ± 0.009). There were no major observable changes shown in the growth curve of yeast *S. cerevisiae* from hours 21 to 48.

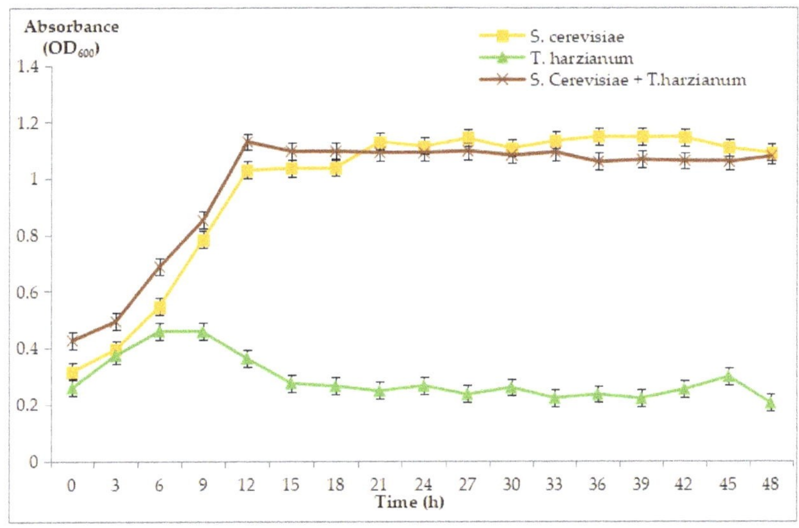

Figure 5. Growth curve of *S. cerevisiae, T. harzianum*, and a microbial consortium of *S. cerevisiae* and *T. harzianum*.

The growth curve of fungi *T. harzianum* demonstrated an increasing trend from the 3rd h until the 6th h. These can be seen in the absorbance reading, increasing from 0.374 ± 0.009 to 0.460 ± 0.007. However, the absorbance reading started to decrease from the 9th (0.457 ± 0.004) to the 15th (0.275 ± 0.003) hours. Then, the growth pattern of the fungi was in a stationary state until the 48th h (0.204 ± 0.002). The absorbance reading of the fungi *T. harzianum* showed a much lower reading compared to the yeast *S. cerevisiae* and the co-culture of *S. cerevisiae* and T. harzianum. Absorbance reading or using OD for the filamentous fungi was not so accurate because the hyphae that were growing were not distributed evenly in the microplate well. Thus, there are uncertainties in the estimation of the fungal growth in the medium. Moreover, a higher OD reading might occur due to the sporulation occurring on the surface of the wells, which gives an overestimation of growth. OD reading is, therefore, more suitable for growth vs. no growth studies or for the initial detection of mold growth [88].

For the growth curve of the co-culture *S. cerevisiae* and *T. harzianum*, the growth was increased from the 3rd h to the 12th h and started to enter the stationary phase from the 15th to the 48th h. There was a gradual increase in the growth of the co-culture for the first three observations (3rd, 6th, and 9th hour) with an absorbance reading of 0.494 ± 0.048, 0.688 ± 0.038, and 0.851 ± 0.002 respectively. At the 12th h, the absorbance reading was the highest growth of the co-culture with an absorbance reading of 1.129 ± 0.051. From the 15th (1.095 ± 0.005) to 48th (1.077 ± 0.015) hours, the yeast growth was slowed down, which eventually became a stationary phase.

The growth curve of yeast *S. cerevisiae* and the co-culture *S. cerevisiae* and *T. harzianum* was similar compared to the growth curve of fungi *T. harzianum*. The yeast and co-culture cells had a predictable pattern of growth which can be divided into lag, log, deceleration, and stationary phases [89]. In the lag phase, no growth occurs as the cell culture is adapting to its environment. Microorganisms are biochemically active in the lag phase but they are not dividing [44]. During the log phase, the cells are growing and dividing rapidly [89]. The cells then reach a stationary phase, where no growth occurs. This is because the cell numbers reach a maximum point at which the cell numbers stop increasing [44]. For the inoculation into the EFBs during the fermentation process, the microorganism cells were harvested at the early exponential phase, which was after 12 h of incubation.

3.4. Selection of Microorganisms and Enzyme Combinations

The selection of microbes (*S. cerevisiae* and *T. harzianum*) and enzymes (cellulase and β-glucosidase) was carried out by comparing the bioethanol concentration after the fermentation process, as in Figure 6 From the figure, the combination of *S. cerevisiae* and *T. harzianum* and enzymes (Cellulase and β-glucosidase) had better results in the conversion of the EFBs into bioethanol production.

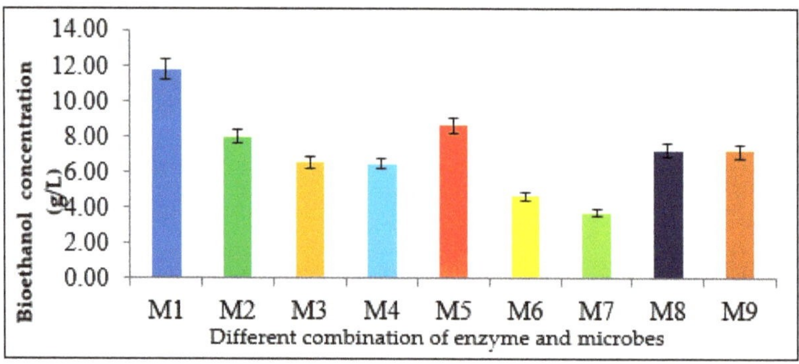

Figure 6. Comparison of Bioethanol concentration from the selection experiment.

From the results obtained, it can be seen that there is a significant difference in the bioethanol production of each run using the empty fruit bunches. According to the figure, M1 had the highest bioethanol concentration with a mean of 11.76 ± 0.79 g/L. Based on previous studies, a combination of the enzymes cellulase and β-glucosidase was successfully employed as the main enzymes for bioethanol production, according to the studies reported by Cui et al. [24], Jung et al. [90], Raman and Gnansounou [91], and Sudiyani et al. [40]. Enzyme cellulase possesses a different catalytic potential for cellulose breakdown and saccharification into fermentable sugar glucose [92]. The Addition of the β-glucosidase enzyme will help in attaining good cellulose hydrolysis performance by breaking down the cellobiose and cellotriose into glucose monomers [93]. Shokrkar et al. [94] described that β-glucosidase promoted the enzymatic hydrolysis process of algal cellulose by increasing the production rate of glucose and decreasing the cellobiose inhibition. A previous study by Poornejad et al. [95] reported that the glucose yield of untreated straw was increased significantly from 25.7% to over 75% for the treated straw during the saccharification process using cellulase and β-glucosidase enzymes. The results of these studies proved that the combinations of cellulase and β-glucosidase were better in enhancing bioethanol production than the single enzyme treatment when combined together with the fermenting microorganisms.

Moreover, a combination of co-cultured *S. cerevisiae* and *T. harzianum* was found to be better as the fermentative microorganisms than using the *S. cerevisiae* and *T. harzianum* independently in the SSF process. The combination of *S. cerevisiae* and *T. harzianum* was found to be the best compatible mixed culture for maximum bioethanol production using EFBs in the solid-state bioconversion process compared to other combinations [26,96]. In addition, *T. harzianum* is a prolific enzyme producer that aids in facilitating the saccharification of EFBs, as it is regarded as a potential cellulase enzyme producer [97,98]. The co-culture of ethanol-fermenting and amylolytic microorganisms has also shown great potential in making a cost-competitive SSF process for bioethanol production [99]. A study by Verma et al. [100] shows that the ethanol production by a co-culture of *S. diastaticus* and *S. cerevisiae* 21 (24.8 g/L) was higher than the monoculture of *S. diastaticus* (16.8 g/L) using raw, unhydrolyzed starch. According to Dey et al. [101], the co-cultivation of Baker's yeast *S. cerevisiae* and *P. stipitis* NCIM 3499 also resulted in a higher ethanol concentration of 42.34 g/L with 0.53 g/g yield from 18% (*w*/*w*) solid loading of pulp and paper sludge waste. Similarly, Izmirlioglu and Demirci [30], observed a maximum amount of bioethanol production at 35.9 g/L when *A. niger* and *S. cerevisiae* were co-cultured for the SSF of industrial waste potato mash. Liu et al. [102] obtained a 5.825 g/L ethanol yield (40.84% of theoretical yield) by using mixed cultures of *Trichoderma*, *S. cerevisiae*, and *Penicillium* for the bioethanol production of alkali-pretreated wheat bran.

3.5. SSF Optimization for Bioethanol Production

In simultaneous saccharification and fermentation, the enzymes and microbes will be simultaneously converted into ethanol [52]. Therefore, the optimization of the SSF process is important in order to achieve maximum bioethanol production from EFBs at a minimal cost [103]. Four parameters, including fermentation time, temperature, pH, and inoculum concentration, which have a significant influence on fermentation, were optimized using CCD-based RSM. The experimental design and response for the optimization of the SSF process of pretreated EFBs were as in Table S1. The interactive effect of the independent variables was studied in order to obtain optimum conditions for bioethanol production. A good correlation between the experimental and predicted bioethanol concentration from different parameters was observed. This indicates the high accuracy of the response surface model constructed in this experiment.

Further data analysis of the results obtained was performed using the RSM software to determine the suitable model that best fits the experimental data. A quadratic model was suggested as the model because the p-value was statistically significant with a *p*-value of <0.0001 (Table S2). The R2 value at 0.9633 was close to 1, hence indicating the high accuracy of this model and signifying a better correlation between the observed and predicted values [87]. The adjusted R2 of 0.9266 was in agreement with the predicted R2 of 0.7774. Adequate precision compares the average prediction error to the range of the predicted values at the design points [9]. Moreover, the lack of fit value of 2.95 implies that the lack of fit model was not significant relative to the pure error. There is a 15.41% (*p*-value of 0.1541) chance that a lack of fit value this large could occur due to noise. The experimental responses fit with the model when the lack of fit value obtained was not significant in the experiment and could be used to predict the optimum conditions accurately [18].

From the analysis of variance (ANOVA) (Table S3), the Model F-value of 26.25 implies that the model is significant. The ANOVA focused on the relationship between the independent and dependent variables based on the results and data obtained [66]. Based on the ANOVA, eight model terms, fermentation time (A), temperature (B), inoculum concentration (D), the interaction of fermentation time and inoculum concentration (AD), the interaction of temperature and pH (BC)), fermentation time (A2), temperature (B2) and inoculum concentration (D2), were found to be statistically significant with a p-value of less than 0.05 (<0.05), which affects the fermentation. The values of coefficient of variation (C.V. % = 8.37), standard deviation (SD = 0.59), and predicted residual sum of squares (PRESS = 29.69) were relatively low, which explained that the model had good precision and the experiments were reliable.

Final Equation in Terms of Coded Factors:

$$\begin{aligned}&\text{Bioethanol concentration (g/L)}\\&= 8.48 + 1.79\,(A) - 0.34\,(B) + 0.09\,(C) + 0.31\,(D) - 0.31\,(AB) - 0.074\,(AC) - 0.46\,(AD)\\&\quad 0.38\,(BC) - 0.043\,(BD) + 0.041\,(CD) - 1.14\,(A2) - 0.42\,(B2) - 0.021\,(C2) - O.26\,(D2)\end{aligned} \quad (4)$$

Note: A denotes the fermentation time (h), B is the temperature (°C), C is pH, and D is the inoculum concentration (% (v/v)).

Figure 7a–f shows the 3D response surface plots analysis of the CCD model for the optimized conditions during fermentation. Each figure represents the effect of two different variables on bioethanol production while the other conditions were kept constant at their optimum points [104]. The surface plots show the significant influences of each parameter on bioethanol production in this study. It is also used to investigate the interaction among the parameters and to determine the optimum concentration of each variable for maximum bioethanol production from EFBs [103]. The significant loss of EFBs during pretreatment, incomplete hydrolysis, inefficient fermentation conditions, and type has been identified as a major limitation that leads to poor yield in bioethanol production [105]. Hence, an optimization process was performed to improve the fermentation parameters which influence the bioethanol production efficiency of EFBs. In this study, the effects of fermentation time, temperature, pH, and inoculum concentration on bioethanol production were studied. From the 3D response surface plot analysis, the optimum predicted conditions for bioethanol production from EFBs were: 72 h fermentation time, temperature 30 °C, pH 4.8, and 10% (v/v) inoculum concentration. Under the above conditions, the maximum experimental bioethanol production was found to be 9.95 g/L, while the predicted response was 9.46 g/L.

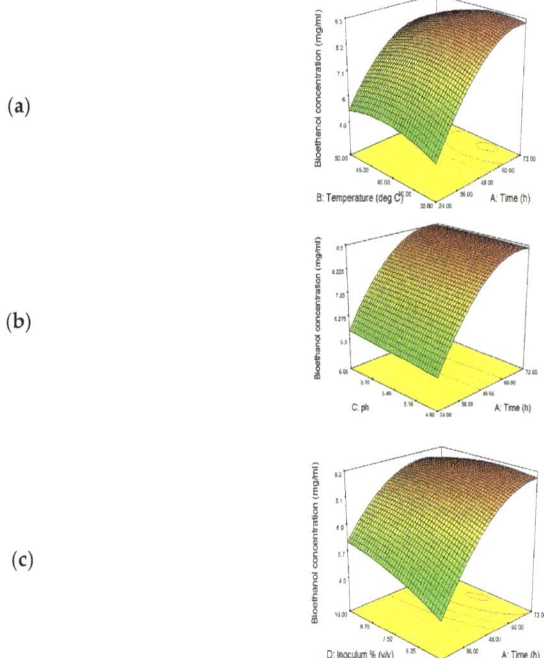

Figure 7. *Cont.*

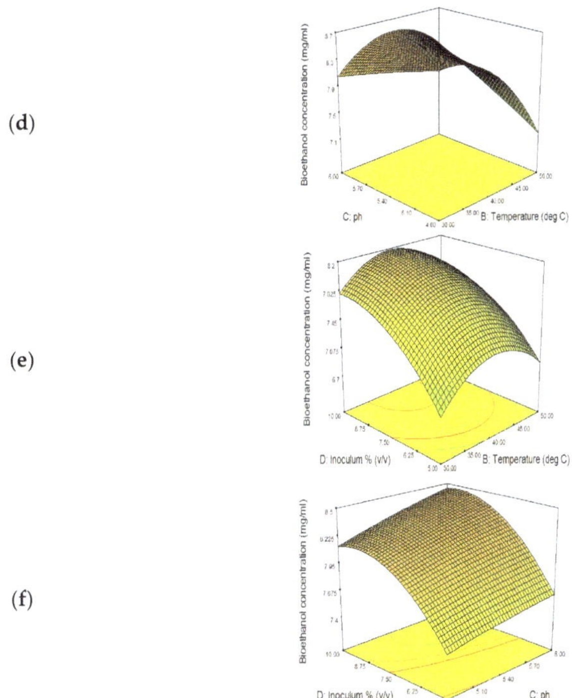

Figure 7. The three-dimensional (3D) surface plot of interaction in the fermentation process based on optimized conditions: (**a**) temperature and time; (**b**) pH and time; (**c**) inoculum concentration and time; (**d**) pH and temperature; (**e**) inoculum concentration and temperature; and (**f**) inoculum concentration and pH. (Note: the area of optimum conditions for fermentation is represented by the darker region).

3.5.1. Effect of Fermentation Time

From the studies, the highest bioethanol concentration obtained was 9.95 ± 0.41 g/L at 72 h while the lowest was 3.35 ± 0.56 g/L at 24 h. The highest bioethanol concentration was produced at a longer fermentation time of 72 h. Bioethanol production gradually increased from 24 to 72 h since the fermentable sugars were sufficient for the growth of microorganisms in order to digest the sugars into bioethanol. It can be seen that the bioethanol concentration tends to increase with the time of fermentation until all the fermentable sugars in the medium are completely utilized by the fermenting microbes. The bioethanol production was found to decrease slightly after 72 h [39].

In this current study, it can be seen that a fermentation time of 60 to 72 h shows a good correlation to the bioethanol production of EFBs. In [93], the bioethanol concentration improved with the increase in fermentation time using the co-culture of *T. harzianum* and *S. cerevisiae* of EFBs. The optimum fermentation time, 72 h was found as a suitable period to obtain higher bioethanol production. Similar results have also been reported by Syadiah et al. [106], in which the maximum ethanol production from sweet sorghum bagasse using a co-culture of *S. cerevisiae* and *Trichoderma reesei* was obtained at 72 h of fermentation with 6.60 g/L. Jambo et al. [107] revealed that the optimum fermentation time for bioethanol production from Eucheuma cottonii based on CCD was also 72 h.

3.5.2. Effect of Temperature

The highest bioethanol (9.95 ± 0.41 g/L) concentration was obtained at 30 °C. High temperature has been shown to lower bioethanol production [108]. Temperature has a major effect on bioethanol fermentation. The optimum temperature of the enzymes and yeast *S. cerevisiae* growth was 50 °C and 28 °C, respectively [44]. In this study, a new combination of enzymes (Cellulase and β-glucosidase) and microorganisms (*S. cerevisiae* and *T. harzianum*) were employed in the SSF fermentation process. Every enzyme and microorganism has its own optimum temperature for every process. Therefore, it is important to strike a balance between the optimal temperature for the enzymes and microorganisms used in the fermentation [109]. Thus, the optimization process for the parameter of temperature (30 °C to 50 °C) was carried out in the SSF fermentation.

In this study, the optimum temperature for the highest bioethanol production using co-cultures in EFBs was observed to be 30 °C. The cellulolytic enzyme activity shows a maximum at 30 °C in co-culture conditions in the present study, which might due to one of the enzymes; cellulase is derived from the microorganism *Trichoderma reesei*. Thus, the cellulolytic activity, which works best at 30 °C, influenced the glucose production from cellulase. This result is in accordance with the study of Ahmad et al. [110], in which the optimum temperature for maximum ethanol production was at a temperature of 30 °C. The ethanol yield was decreased significantly when temperature values were higher or lower than 30 °C. However, Verma et al. [100] suggested that a slight difference in temperature between 30 °C and 40 °C will not affect the ethanol fermentation of starch using a co-culture of *S. cerevisiae* and *S. diastaticus*. Research by Kassim et al. [111] reported that the lowest ethanol production rate was at 40 °C compared to other temperatures at 30 °C and 35 °C. This is because fermentation at higher temperatures can inhibit ethanol production. Moreover, a decrease in the viable cell number at temperatures above 30 °C would lower the bioethanol concentration and fermentation efficiency [33]. According to Park et al. [20], the optimal temperature for ethanol production using a fed-batch from the alkali-pretreated EFBs was 30 °C. Sahu et al. [112] attained the highest bioethanol production with 29.5 g/L at a 30 °C temperature for the fermentation process of glucose for rose petals.

3.5.3. Effect of pH

From the results, the highest bioethanol concentration of 9.95 ± 0.41 g/L was obtained at pH 4.8. In order to determine the effect of pH on the fermentation by the co-culture on bioethanol production, the citrate buffer pH was adjusted in the range of 4.8 to 6.0. In the present study, it was found that pH did not significantly affect the optimization of the fermentation, based on the ANOVA analysis. This occurred because the range of pH chosen for the optimization process was not wide enough to be used in the fermentation process. From the results, both the high (6.0) and low (4.8) pH values showed little difference in bioethanol production.

From the CCD optimization design, it was indicated that the optimum pH value for the fermentation process was 4.8. This shows that the co-culture preferred a slightly acidic condition to grow. Even though acids were required for the production of bioethanol, a highly acidic condition was not suitable for cell growth [113]. In Alam et al. [114], pH 5.5 was found as the optimum pH that led to a maximum bioethanol production of 7.4 g/L using co-cultured *S. cerevisiae* and *A. niger* for EFB fermentation. Meanwhile, Anu et al. [115] exhibited the best attribute for bioethanol production with 18.07 g/L for the enzymatic hydrolysate (20%) of pretreated rice straw at pH 6, 30 °C after 72 h. Meanwhile, the study by Chohan et al. [116] observed an increase in the ethanol yield from 0.14 g/g to 0.29 g/g after the pH was increased from 4.00 to 6.30. However, further increases in pH beyond 6.30 reduced the process yield. Hence, increasing the pH value significantly affected the production of ethanol and the rate of glucose consumption during the fermentation process.

3.5.4. Effect of Inoculum Concentration

According to the results, the highest bioethanol (9.95 ± 0.41 g/L) concentration was produced at 10% (*v*/*v*) inoculum concentration, respectively. These results were in line with the results obtained by Swain et al. [30] for the production of ethanol using sweet potato, in which the optimal inoculum size was 10%. Ansar et al. [117] described that the higher the percentage of inoculum used during fermentation, the higher the amount of ethanol produced. An increase in the inoculum concentration should increase the concentration of bioethanol. Different inoculum concentrations can be used to determine whether the ethanol yield and productivity were influenced [109]. In this study using the RSM approach, it was found that a high inoculum concentration increased the bioethanol yield.

The inoculum concentration used is one of the most critical factors which influences the industrial fermentation, lag phase duration, biomass yield, specific growth rate, and final product yield [118]. Kabbashi et al. [31] employed a 4% (*v*/*v*) inoculum size in the direct solid-state bioconversion of palm oil EFBs for bioethanol production with a maximum ethanol yield of 14.1% (*v*/*v*). The research by Neelakandan et al. [119] showed that the optimum inoculum concentration for cashew apple juice for bioethanol production was 8% (*v*/*v*) with a maximum bioethanol yield of 7.62% (*v*/*v*).

3.6. Bioethanol Production Using Optimized Conditions of Fermentation

The experimental analysis was performed to determine the optimized conditions for the fermentation process. Based on the optimization analysis of the experimental data, the suggested optimum levels of all the variables from the quadratic model of CCD in this study were 72 h of fermentation time, a temperature of 30 °C, pH 4.8, and an inoculum concentration of 6.79% (*v*/*v*). From these optimized conditions, the bioethanol concentration can reach up to 9.72 g/L with the desirability of 0.977.

A validation experiment was carried out to evaluate the conditions predicted by the CCD. The fermentation process was conducted under optimized conditions with 72 h of fermentation time, a temperature of 30 °C, pH 4.8, and inoculum concentration of 6.79% (*v*/*v*). The bioethanol concentration after the fermentation process was 9.65 g/L, which was in close agreement with the predicted value of 9.72 g/L. The difference between the predicted and experimental value was only 1.07%. Therefore, it can be concluded that the response surface from this study is reliable to be used to predict bioethanol production from the fermentation process.

4. Conclusions

Empty fruit bunches were treated with sequential acid-alkali pretreatment before being further used as the main feedstock in this study. A change in the physical characteristics and morphology of the EFBs before and after the pretreatment was confirmed by SEM and FTIR analysis. From the SEM analysis, the formation of pores and removal of silica was shown in the treated EFBs' structure. The FTIR spectra of EFBs showed a different graph pattern and peak between the raw and treated EFBs. The combination of enzymes and microorganisms in producing bioethanol was screened to determine the optimum concentration of this combination for the fermentation process of EFBs. It was found that enzyme combinations of cellulase and β-glucosidase with the microorganism combination of *S. cerevisiae* and *T. harzianum* had better results in the conversion of the EFBs into bioethanol production. From the GCMS analysis, this combination has the highest bioethanol concentration with 11.76 ± 0.79 g/L. The simultaneous saccharification and fermentation (SSF) optimization process was performed on pretreated EFBs by employing the central composite design of Response Surface Methodology. The effects of fermentation time, temperature, pH, and inoculum concentration on the fermentation were then analyzed. During fermentation, the highest bioethanol concentration was obtained at 72 h, 30 °C, pH 4.8, and an inoculum concentration of 10% (*v*/*v*). Based on the CCD analysis, the SSF of pretreated EFBs was repeated using the optimized conditions. From the results, the experimental data obtained were in close agreement with the RSM model prediction. Thus, it can be deduced that the RSM optimization of EFBs using SSF employed in this study is a promising tool for the better optimization of the fermentation process of bioethanol production in the future. Moreover, a new combination of enzymes and microbes was employed in the fermentation process. This combination has never been employed in other studies related to bioethanol production from EFBs using simultaneous saccharification and fermentation. Hence, this study can be a pioneer for the development of bioethanol production, as the results obtained were satisfactory with regard to bioethanol yield. Moreover, the employment of a central composite design from the RSM method for the optimization of the SSF process in this study showed a promising potential for the production of bioethanol using lignocellulosic biomass waste in the future. Thus, this study may contribute to future research for second-generation bioethanol using lignocellulosic biomass waste in Malaysia. In addition, the potential of EFBs as the main feedstock may contribute to the economic development of Malaysia by producing bioethanol, which is commercially valuable.

Supplementary Materials: The following are available online at https://www.mdpi.com/article/10.3390/fermentation8070295/s1, Table S1: Experimental design and responses for the simultaneous saccharification and fermentation process of pretreated EFBs for bioethanol production. Table S2: Model summary statistics of central composite design for the optimization of simultaneous saccharification and fermentation. Table S3: Analysis of variance table (ANOVA).

Author Contributions: E.D.; conceptualization, investigation, formal analysis, data curation, methodology, writing—original draft. R.A.; supervision, conceptualization, overall guidance, writing—review and editing, validation, revision of the paper. H.M.; writing—review and editing. M.K.S.; writing—review and editing. J.A.G.; writing—review and editing. P.R.; writing—review and editing. All authors have read and agreed to the published version of the manuscript.

Funding: This research received no external funding.

Institutional Review Board Statement: This study did not require any ethical approval.

Informed Consent Statement: Not applicable.

Data Availability Statement: Not applicable.

Acknowledgments: The authors would like to thank and acknowledge the Ministry of Education Malaysia (KPM) for financial support through the MyMaster Scholarship and Faculty of Science and Natural Resources, UMS.

Conflicts of Interest: The authors declare no conflict of interest.

References

1. Chen, J.; Zhang, B.; Luo, L.; Zhang, F.; Yi, Y.; Shan, Y.; Bianfang, L.; Yuan, Z.; Xin, W.; Xin, L. A review on recycling techniques for bioethanol production from lignocellulosic biomass. *Renew. Sustain. Energy Rev.* **2021**, *149*, 111370. [CrossRef]
2. Hosseinzadeh-bandbafha, H.; Nazemi, F.; Khounani, Z.; Ghanavati, H. Safflower-based biorefinery producing a broad spectrum of biofuels and biochemicals: A life cycle assessment perspective. *Sci. Total Environ.* **2022**, *802*, 149842. [CrossRef]
3. Hanif, M.; Mahlia, T.M.I.; Aditiya, H.B.; Chong, W.T. Techno-economic and environmental assessment of bioethanol production from high starch and root yield Sri Kanji 1 cassava in Malaysia. *Energy Rep.* **2016**, *2*, 246–253. [CrossRef]
4. Thakur, A.K.; Kaviti, A.K.; Mehra, R.; Mer, K.K.S. Progress in performance analysis of ethanol-gasoline blends on SI engine. *Renew. Sustain. Energy Rev.* **2017**, *69*, 324–340. [CrossRef]
5. Masum, B.M.; Masjuki, H.H.; Kalam, M.A.; Fattah, I.M.R.; Palash, S.M.; Abedin, M.J. Effect of ethanol—Gasoline blend on NOx emission in SI engine. *Renew. Sustain. Energy Rev.* **2013**, *24*, 209–222. [CrossRef]
6. Zabed, H.; Sahu, J.N.; Suely, A.; Boyce, A.N.; Faruq, G. Bioethanol production from renewable sources: Current perspectives and technological progress. *Renew. Sustain. Energy Rev.* **2017**, *71*, 475–501. [CrossRef]
7. Ahmad, A.; Buang, A.; Bhat, A.H. Renewable and sustainable bioenergy production from microalgal co-cultivation with palm oil mill effluent (POME): A review. *Renew. Sustain. Energy Rev.* **2016**, *65*, 214–234. [CrossRef]
8. Derman, E.; Abdulla, R.; Marbawi, H.; Sabullah, M.K. Oil palm empty fruit bunches as a promising feedstock for bioethanol production in Malaysia. *Renew. Energy* **2018**, *129*, 285–298. [CrossRef]
9. Khairil Anwar, N.A.K.; Hassan, N.; Mohd Yusof, N.; Idris, A. High-titer bio-succinic acid production from sequential alkalic and metal salt pretreated empty fruit bunch via simultaneous saccharification and fermentation. *Ind. Crop Prod.* **2021**, *166*, 113478. [CrossRef]
10. Anita, S.H.; Fitria; Solihat, N.N.; Sari, F.P.; Risanto, L.; Fatriasari, W.; Risanto, L.; Fatriasari, W.; Hermiati, E. Optimization of Microwave-Assisted Oxalic Acid Pretreatment of Oil Palm Empty Fruit Bunch for Production of Fermentable Sugars. *Waste Biomass Valorization* **2019**, *11*, 2673–2687. [CrossRef]
11. Chang, S.H. An overview of empty fruit bunch from oil palm as feedstock for bio-oil production. *Biomass Bioenergy* **2014**, *62*, 174–181. [CrossRef]

12. Tan, L.; Wang, M.; Li, X.; Li, H.; Zhao, J.; Qu, Y.; Choo, Y.M.; Loh, S.K. Fractionation of oil palm empty fruit bunch by bisulfite pretreatment for the production of bioethanol and high value products. *Bioresour. Technol.* **2016**, *200*, 572–578. [CrossRef] [PubMed]
13. Aditiya, H.B.; Chong, W.T.; Mahlia, T.M.I.; Sebayang, A.H.; Berawi, M.A.; Nur, H. Second generation bioethanol potential from selected Malaysia 's biodiversity biomasses: A review. *Waste Manag.* **2016**, *47*, 46–61. [CrossRef] [PubMed]
14. Piarpuzan, D.; Quintero, J.A.; Cardona, C.A. Empty fruit bunches from oil palm as a potential raw material for fuel ethanol production. *Biomass Bioenergy* **2011**, *35*, 1130–1137. [CrossRef]
15. Aguilar-reynosa, A.; Romaní, A.; Rodríguez-jasso, R.M.; Aguilar, C.N.; Garrote, G.; Ruiz, H.A. Comparison of microwave and conduction-convection heating autohydrolysis pretreatment for bioethanol production. *Bioresour Technol.* **2017**, *243*, 273–283. [CrossRef]
16. Abo-state, M.A.; Ragab, A.M.E.; El-gendy, N.S.; Farahat, L.A. Bioethanol Production from Rice Straw Enzymatically Saccharified by Fungal Isolates, *Trichoderma viride* F94 and *Aspergillus terreus* F98. *Soft* **2014**, *3*, 19–29. [CrossRef]
17. Javed, M.R.; Noman, M.; Shahid, M.; Ahmed, T.; Khurshid, M.; Rashid, M.H.; Ismail, M.; Sadaf, M.; Khan, F. Current situation of biofuel production and its enhancement by CRISPR /Cas9-mediated genome engineering of microbial cells. *Microbiol. Res.* **2019**, *219*, 1–11. [CrossRef]
18. Sharma, S.; Sharma, V.; Kuila, A. Simultaneous Saccharification and Fermentation of Corn Husk by Co-Culture Strategy. *J. Pet. Environ. Biotechnol.* **2019**, *9*, 360. [CrossRef]
19. Dahnum, D.; Octavia, S.; Triwahyuni, E.; Nurdin, M. Comparison of SHF and SSF processes using enzyme and dry yeast for optimization of bioethanol production from empty fruit bunch. *Energy Procedia* **2015**, *68*, 107–116. [CrossRef]
20. Park, E.Y.; Naruse, K.; Kato, T. One-pot bioethanol production from cellulose by co-culture of *Acremonium cellulolyticus* and *Saccharomyces cerevisiae*. *Biotechnol. Biofuels* **2012**, *5*, 64. [CrossRef]
21. Galbe, M.; Zacchi, G. A review of the production of ethanol from softwood. *Appl. Microbiol. Biotechnol.* **2002**, *59*, 618–628. [CrossRef] [PubMed]
22. Bhutto, A.W.; Qureshi, K.; Harijan, K.; Abro, R.; Abbas, T.; Bazmi, A.A. Insight into progress in pre-treatment of lignocellulosic biomass. *Energy* **2017**, *122*, 724–745. [CrossRef]
23. Rastogi, M.; Shrivastava, S. Recent advances in second generation bioethanol production: An insight to pretreatment, saccharification and fermentation processes. *Renew. Sustain. Energy Rev.* **2017**, *80*, 330–340. [CrossRef]
24. Cui, X.; Zhao, X.; Zeng, J.; Loh, S.K.; Choo, Y.M.; Liu, D. Robust enzymatic hydrolysis of Formiline-pretreated oil palm empty fruit bunches (EFB) for efficient conversion of cellulose to sugars and. *Bioresour. Technol.* **2014**, *166*, 584–591. [CrossRef] [PubMed]
25. Kim, S.; Kim, C.H. Bioethanol production using the sequential acid/alkali-pretreated empty palm fruit bunch fiber. *Renew. Energy* **2013**, *54*, 150–155. [CrossRef]
26. Jung, S.; Shetti, N.P.; Reddy, K.R.; Nadagouda, M.N.; Park, Y.; Aminabhavi, T.M.; Kwon, E.E. Synthesis of different biofuels from livestock waste materials and their potential as sustainable feedstocks—A review. *Energy Convers. Manag.* **2021**, *236*, 114038. [CrossRef]
27. Polprasert, S.; Choopakar, O.; Elefsiniotis, P. Bioethanol production from pretreated palm empty fruit bunch (PEFB) using sequential enzymatic hydrolysis and yeast fermentation. *Biomass Bioenergy* **2021**, *149*, 106088. [CrossRef]
28. Ali, I.W.; Aziz, K.K.; Syahadah, A.A.H. Bioethanol Production from Acid Hydrolysates of Date Palm Fronds Using a Co-culture of *Saccharomyces cerevisiae* and *Pichia stipitis*. *Int. J. Enhanc. Res. Sci. Technol. Eng.* **2014**, *3*, 35–44.
29. Mishra, A.; Ghosh, S. Saccharification of kans grass biomass by a novel fractional hydrolysis method followed by co-culture fermentation for bioethanol production. *Renew. Energy* **2020**, *146*, 750–759. [CrossRef]
30. Izmirlioglu, G.; Demirci, A. Simultaneous saccharification and fermentation of ethanol from potato waste by co-cultures of *Aspergillus niger* and *Saccharomyces cerevisiae* in biofilm reactors. *Fuel* **2017**, *202*, 260–270. [CrossRef]
31. Kabbashi, N.A.; Alam, M.Z.; Tompang, M.F. Direct Bioconversion of Oil Palm Empty Fruit Bunches for Bioethanol Production by Solid State Bioconversion. *IIUM Eng. J.* **2007**, *8*, 25–36. [CrossRef]
32. Adela, N.; Nasrin, A.B.; Loh, S.K.; Choo, Y.M. Bioethanol Production by Fermentation of Oil Palm Empty Fruit Bunches Pretreated with Combined Chemicals. *J. Appl. Environ. Biol. Sci.* **2014**, *4*, 234–242.
33. Swain, M.R.; Mishra, J.; Thatoi, H. Bioethanol Production from Sweet Potato (*Ipomoea batatas* L.) Flour using Co-Culture of *Trichoderma* sp. and *Saccharomyces cerevisiae* in Solid-State Fermentation. *Braz. Arch. Biol. Technol.* **2013**, *56*, 171–179. [CrossRef]
34. Bhadwal, A.S.; Le, A.; Le, T.T.; Shrivastava, S.; Bera, T. *Trichoderma koningii* assisted biogenic synthesis of silver nanoparticles and evaluation of their antibacterial activity. *Adv. Nat. Sci. Nanosci. Nanotechnol.* **2013**, *4*, 35005.
35. Bergman, L.W. Growth and Maintenance of Yeast. In *Methods in Molecular Biology, Two Hybrid Systems: Methods and Protocols*; MacDonald, P.N., Ed.; Humana Press Inc.: Totowa, NJ, USA, 2001; Volume 177, pp. 9–39.
36. Rodrigues, B.; Lima-Costa, M.E.; Constantino, A.; Raposo, S.; Felizardo, C.; Gonçalves, D.; Peinado, J.M. Growth kinetics and physiological behavior of co-cultures of *Saccharomyces cerevisiae* and *Kluyveromyces lactis*, fermenting carob sugars extracted with whey. *Enzym. Microb. Technol.* **2016**, *92*, 41–48. [CrossRef] [PubMed]
37. Paschos, T.; Xiros, C.; Christakopoulos, P. Simultaneous saccharification and fermentation by co-cultures of *Fusarium oxysporum* and *Saccharomyces cerevisiae* enhances ethanol production from liquefied wheat straw at high solid content. *Ind. Crop Prod.* **2015**, *76*, 793–802. [CrossRef]

38. Sasikumar, E.; Viruthagiri, T. Optimization of Process Conditions Using Response Surface Methodology (RSM) for Ethanol Production from Pretreated Sugarcane Bagasse: Kinetics and Modeling. *Bioenergy Resour.* **2008**, *1*, 239–247. [CrossRef]
39. Duangwang, S.; Sangwichien, C. Utilization of Oil Palm Empty Fruit Bunch Hydrolysate for Ethanol Production by Baker's Yeast and Loog-Pang. *Energy Procedia* **2015**, *79*, 157–162. [CrossRef]
40. Sudiyani, Y.; Styarini, D.; Triwahyuni, E.S.; Sembiring, K.C.; Aristiawan, Y.; Han, M.H. Utilization of biomass waste empty fruit bunch fiber of palm oil for bioethanol production using pilot—Scale unit. *Energy Procedia* **2013**, *32*, 31–38. [CrossRef]
41. Baharuddin, A.S.; Nor, A.A.R.; Umi, K.M.S.; Mohd, A.H.; Minato, W.; Yoshihito, S. Evaluation of pressed shredded empty fruit bunch (EFB)-palm oil mill effluent (POME) anaerobic sludge based compost using Fourier transform infrared (FTIR) and nuclear magnetic resonance (NMR) analysis. *Afr. J. Biotechnol.* **2011**, *10*, 8082–8089.
42. Sindhu, R.; Kuttiraja, M.; Prabisha, T.P.; Binod, P.; Sukumaran, R.K.; Pandey, A. Development of a combined pretreatment and hydrolysis strategy of rice straw for the production of bioethanol and biopolymer. *Bioresour. Technol.* **2016**, *215*, 110–116. [CrossRef] [PubMed]
43. Hamouda, H.I.; Nassar, H.N.; Madian, H.R.; Amr, S.S.A.; El-Gendy, N.S. Response Surface Optimization of Bioethanol Production from Sugarcane Molasses by *Pichia veronae* Strain HSC-22. *Biotechnol. Res. Int.* **2015**, *10*, 905792. [CrossRef] [PubMed]
44. Mansa, R.F.; Chen, W.F.; Yeo, S.J.; Farm, Y.Y.; Bakar, H.A.; Sipaut, C.S. Chapter 13: Fermentation Study on Macroalgae Eucheuma cottonii for Bioethanol Production via Varying Acid Hydrolysis. In *Advanced Biofuels*; Sarbatly, R.P.H., Ed.; Springer Science and Business Media: New York, NY, USA, 2013; pp. 219–240.
45. Tan, I.S.; Lee, K.T. Enzymatic hydrolysis and fermentation of seaweed solid wastes for bioethanol production: An optimization study. *Energy* **2014**, *78*, 53–62. [CrossRef]
46. Burhani, D.; Putri, A.M.H.; Waluyo, J.; Nofiana, Y.; Sudiyani, Y. The effect of two-stage pretreatment on the physical and chemical characteristic of oil palm empty fruit bunch for bioethanol production. *AIP Conf. Proc.* **2017**, *1904*, 20016.
47. Campioni, T.S.; Soccol, C.R.; Libardi Junior, N.; Rodrigues, C.; Woiciechowski, A.L.; Letti, L.A.J.; Vandenberghe, L.P.D.S. Sequential chemical and enzymatic pretreatment of palm empty fruit bunches for *Candida pelliculosa* bioethanol production. *Biotechnol. Appl. Biochem.* **2019**, *67*, 723–731. [CrossRef]
48. Akhtar, J.; Teo, C.L.; Lai, L.W.; Hassan, N.; Idris, A.; Aziz, R.A. Factors affecting delignification of oil palm empty fruit bunch by microwave-assisted dilute acid/alkali pretreatment. *BioResources* **2015**, *10*, 588–596. [CrossRef]
49. Ebrahimi, M.; Villaflores, O.B.; Ordono, E.E.; Caparanga, A.R. Effects of acidified aqueous glycerol and glycerol carbonate pretreatment of rice husk on the enzymatic digestibility, structural characteristics, and bioethanol production. *Bioresour. Technol.* **2017**, *228*, 264–271. [CrossRef]
50. Abdul, P.M.; Jahim, M.J.; Harun, S.; Markom, M.; Lutpi, N.A.; Hassan, O.; Mohd Nor, M.T. Effects of changes in chemical and structural characteristic of ammonia fibre expansion (AFEX) pretreated oil palm empty fruit bunch fibre on enzymatic saccharification and fermentability for biohydrogen. *Bioresour. Technol.* **2016**, *211*, 200–208. [CrossRef]
51. Taherzadeh, M.J.; Karimi, K. Pretreatment of Lignocellulosic Wastes to Improve Ethanol and Biogas Production: A Review. *Int. J. Mol. Sci.* **2008**, *9*, 1621–1651. [CrossRef]
52. Sari, A.A.; Ariani, N.; Muryanto; Kristiani, A.; Utomo, T.B.; Sudarno. Potential of Oil Palm Empty Fruit Bunches for Bioethanol Production and Application of Chemical Methods in Bioethanol Wastewater Treatment OPEFB for Bioethanol and Its Wastewater Treatment. In Proceedings of the International Conference on Sustainable and Renewable Energy Engineering, Hiroshima, Japan, 10–12 May 2017; pp. 1–4.
53. Triwahyuni, E.; Muryanto; Sudiyani, Y.; Abimanyu, H. The effect of substrate loading on simultaneous saccharification and fermentation process for bioethanol production from oil palm empty fruit bunches. *Energy Procedia* **2015**, *68*, 138–146. [CrossRef]
54. Fatriasari, W.; Ulwan, W.; Aminingsih, T.; Puspita, F.; Suryanegara, L.; Heri, A.; Ghozali, M.; Kholida, L.N.; Hussin, M.H. Optimization of maleic acid pretreatment of oil palm empty fruit bunches (OPEFB) using response surface methodology to produce reducing sugars. *Ind. Crop Prod.* **2021**, *171*, 113971. [CrossRef]
55. Baharuddin, A.S.; Sulaiman, A.; Kim, D.H.; Mokhtar, M.N.; Hassan, M.A.; Wakisaka, M.; Shirai, Y.; Nishida, H. Selective component degradation of oil palm empty fruit bunches (OPEFB) using high-pressure steam. *Biomass Bioenergy* **2013**, *55*, 268–275. [CrossRef]
56. Paramasivan, S.; Sankar, S.; Velavan, R.S.; Krishnakumar, T.; Sithara, R.; Batcha, I.; Muthuvelu, K.S. Assessing the potential of lignocellulosic energy crops as an alternative resource for bioethanol production using ultrasound assisted dilute acid pretreatment. *Mater. Today Proc.* **2021**, *45*, 3279–3285. [CrossRef]
57. Musatto, S.I.; Dragone, G.; Fernandes, M.; Milagres, A.M.F.; Roberto, I.C. The effect of agitation speed, enzyme loading and substrate concentration on enzymatic hydrolysis of cellulose from brewer's spent grain. *Cellulose* **2008**, *15*, 711–721. [CrossRef]
58. Ariffin, H.; Hassan, M.; Umi Kalsom, M.; Abdullah, N.; Shirai, Y.; Ariffin, H. Effect of physical, chemical and thermal pretreatments on the enzymatic hydrolysis of oil palm empty fruit bunch (OPEFB). *J. Trop. Agric. Food Sci.* **2008**, *36*, 1–10.
59. Kim, D.Y.; Kim, Y.; Kim, T.; Oh, K. Two-stage, acetic acid-aqueous ammonia, fractionation of empty fruit bunches for increased lignocellulosic biomass utilization. *Bioresour. Technol.* **2016**, *199*, 121–127. [CrossRef]
60. Alvira, P.; Tomás-Pejó, E.; Ballesteros, M.; Negro, M.J. Pretreatment technologies for an efficient bioethanol production process based on enzymatic hydrolysis: A review. *Bioresour. Technol.* **2010**, *101*, 4851–4861. [CrossRef]
61. Santosh, I.; Ashtavinayak, P.; Amol, D.; Sanjay, P. Enhanced bioethanol production from different sugarcane bagasse cultivars using co-culture of *Saccharomyces cerevisiae* and *Scheffersomyces* (*Pichia*) *stipitis*. *J. Environ. Chem. Eng.* **2017**, *5*, 2861–2868. [CrossRef]

62. Liong, Y.Y.; Halis, R.; Lai, O.M.; Mohamed, R. Conversion Of Lignocellulosic Biomass From Grass To Bioethanol Using Materials Pretreated With Alkali And The White Rot Fungus, *Phanerochaete Chrysosporium*. *BioResources* **2012**, *7*, 5500–5513. [CrossRef]
63. Thanapimmetha, A.; Saisriyoot, M.; Khomlaem, C.; Chisti, Y.; Srinophakun, P. A comparison of methods of ethanol production from sweet sorghum bagasse. *Biochem. Eng. J.* **2019**, *151*, 107352. [CrossRef]
64. Coral Medina, J.D. Woiciechowski, A.; Zandona Filho, A.; Noseda, M.D.; Kaur, B.S.; Soccol, C.R. Lignin preparation from oil palm empty fruit bunches by sequential acid/alkaline treatment—A biorefinery approach. *Bioresour. Technol.* **2015**, *194*, 172–178. [CrossRef]
65. Norul Izani, M.A.; Paridah, M.T.; Anwar, U.M.K.; Mohd Nor, M.Y.; Ng, P.S. Effects of fiber treatment on morphology, tensile and thermogravimetric analysis of oil palm empty fruit bunches fibers. *Compos. B Eng.* **2013**, *45*, 1251–1257. [CrossRef]
66. Chin, S.X.; Chia, H.C.; Zakaria, S.; Fang, Z.; Ahmad, S. Ball milling pretreatment and diluted acid hydrolysis of oil palm empty fruit bunch (EFB) fibres for the production of levulinic acid. *J. Taiwan Inst. Chem. Eng.* **2015**, *52*, 85–92. [CrossRef]
67. Tye, Y.Y.; Leh, C.P.; Wan Abdullah, W.N. Total glucose yield as the single response in optimizing pretreatments for *Elaeis guineensis* fibre enzymatic hydrolysis and its relationship with chemical composition of fibre. *Renew. Energy* **2017**, *114*, 383–393. [CrossRef]
68. Shamsudin, S.; Md Shah, U.K.; Zainudin, H.; Abd-Aziz, S.; Mustapa Kamal, S.M. Shirai, Y.; Hassan, M.A. Effect of steam pretreatment on oil palm empty fruit bunch for the production of sugars. *Biomass Bioenergy* **2012**, *36*, 280–288. [CrossRef]
69. Isroi; Cifriadi, A.; Panji, T.; Wibowo, N.A.; Syamsu, K. Bioplastic production from cellulose of oil palm empty fruit bunch. *IOP Conf. Ser. Earth Environ. Sci.* **2017**, *65*, 12011. [CrossRef]
70. Nurul Hazirah, C.H.; Masturah, M.; Osman, H.; Jamaliah, M.J.; Shuhaida, H. Preliminary Study on Analysis of the Chemical Compositions and Characterization of Empty Fruit Bunch (EFB) in Malaysia. *Adv. Mater. Res.* **2014**, *970*, 204–208. [CrossRef]
71. Eliza, M.Y.; Shahruddin, M.; Noormaziah, J.; Rosli, W.D.W. Carboxymethyl Cellulose (CMC) from oil palm empty fruit bunch (OPEFB) in the new solvent dimethyl sulfoxide (DMSO)/tetrabutylammonium fluoride (TBAF). *J. Phys. Conf. Ser.* **2015**, *622*, 12026. [CrossRef]
72. Zulkefli, S.; Abdulmalek, E.; Abdul Rahman, M.B. Pretreatment of oil palm trunk in deep eutectic solvent and optimization of enzymatic hydrolysis of pretreated oil palm trunk. *Renew. Energy* **2017**, *107*, 36–41. [CrossRef]
73. Hamzah, F.; Idris, A.; Shuan, T.K. Preliminary study on enzymatic hydrolysis of treated oil palm (Elaeis) empty fruit bunches fibre (EFB) by using combination of cellulase and β, 1-4 glucosidase. *Biomass Bioenergy* **2011**, *35*, 1055–1059. [CrossRef]
74. Abu Bakar, N.K.; Zanirun, Z.; Abd-Aziz, S.; Ghazali, F.M.; Hassan, M.A. Production of fermentable sugars from oil palm empty fruit bunch using crude cellulase cocktails with *Trichoderma asperellum* UPM1 and *Aspergillus fumigatus* UPM2 for bioethanol production. *BioResources* **2012**, *7*, 3627–3639.
75. Hossain, M.N.B.; Basu, J.K.; Mamun, M. The Production of Ethanol from Micro-Algae Spirulina. *Procedia Eng.* **2015**, *105*, 733–738. [CrossRef]
76. Yang, J.; Kim, J.E.; Kim, J.K.; Lee, S.H.; Yu, J.H.; Kim, K.H. Evaluation of commercial cellulase preparations for the efficient hydrolysis of hydrothermally pretreated empty fruit bunches. *BioResources* **2017**, *12*, 7834–7840. [CrossRef]
77. Bhatia, S.K.; Bhatia, R.K.; Choi, Y.K.; Kan, E.; Kim, Y.G.; Yang, Y.H. Biotechnological potential of microbial consortia and future perspectives. *Crit. Rev. Biotechnol.* **2018**, *38*, 1209–1229. [CrossRef]
78. Kusfanto, H.F.; Maggadani, B.P.; Suryadi, H. Fermentation of Bioethanol from the Biomass Hydrolyzate of Oil Palm Empty Fruit Bunch using selected yeast isolates. *Int. J. Appl. Pharm.* **2017**, *9*, 49–53. [CrossRef]
79. Nguyen, K.; Murray, S.; Lewis, J.A.; Kumar, P. Morphology, cell division, and viability of *Saccharomyces cerevisiae* at high hydrostatic pressure. *arXiv* **2017**, arXiv:170300547.
80. Evcan, E.; Tari, C. Production of bioethanol from apple pomace by using cocultures: Conversion of agro-industrial waste to value added product. *Energy* **2015**, *88*, 775–782. [CrossRef]
81. Ab Majid, A.H.; Zahran, Z.; Abd Rahim, A.H.; Ismail, N.A.; Abdul Rahman, W. Mohammad Zubairi, K.S.; Satho, T. Morphological and molecular characterization of fungus isolated from tropical bed bugs in Northern Peninsular Malaysia, *Cimex hemipterus* (Hemiptera: Cimicidae). *Asian Pac. J. Trop. Biomed.* **2015**, *5*, 707–713. [CrossRef]
82. Kannangara, S.; Dharmarathna, R.M.G.C.S. Jayarathna, D. Isolation, Identification and Characterization of *Trichoderma* Species as a Potential Biocontrol Agent against *Ceratocystis paradoxa*. *J. Agric. Sci.* **2017**, *12*, 51–62.
83. Shah, S.; Nasreen, S.; Sheikh, P. Cultural and Morphological Characterization of *Trichoderma* spp. Associated with Green Mold Disease of *Pleurotus* spp. in Kashmir. *Res. J. Microbiol.* **2012**, *7*, 139–144. [CrossRef]
84. Mustafa, A.; Khan, M.A.; Inam-ul-Haq, M.; Pervez, M.A.; Umar, U.D. Usefulness of Different culture media for in-vitro evaluation of *Trichoderma* spp. against seed borne fungi of economic importance. *J. Phytopathol.* **2009**, *21*, 83–88.
85. Prajankate, P.; Siwarasak, P. Co-culture of *Trichoderma reesei* RT-P1 with *Saccharomyces cerevisiae* RT-P2: Morphological Study. *Sci. Technol.* **2011**, *4*, 75–78.
86. De Azevedo, A.M.C.; De Marco, J.L.; Felix, C.R. Characterization of an amylase produced by a *Trichoderma harzianum* isolate with antagonistic activity against *Crinipellis perniciosa*, the causal agent of witches' broom of cocoa. *FEMS Microbiol. Lett.* **2000**, *188*, 171–175. [CrossRef]
87. Kumar, D.; Surya, K.; Verma, R. Bioethanol production from apple pomace using co-cultures with *Saccharomyces cerevisiae* in solid-state fermentation. *J. Microbiol. Biotechnol. Food Sci.* **2020**, *9*, 742–745. [CrossRef]
88. Aunsbjerg, S.D.; Andersen, K.R.; Knøchel, S. Real-time monitoring of fungal inhibition and morphological changes. *J. Microbiol. Methods* **2015**, *119*, 196–202. [CrossRef]

89. Alsuhaim, H.; Vojisavljevic, V.; Pirogova, E. Effects of Non-thermal Microwave Exposures on the Proliferation Rate of *Saccharomyces Cerevisiae* yeast. In Proceedings of the World Congress on Medical Physics and Biomedical Engineering, Beijing, China, 26–31 May 2012; pp. 1–5.
90. Jung, Y.H.; Kim, I.J.; Kim, H.K.; Kim, K.H. Dilute acid pretreatment of lignocellulose for whole slurry ethanol fermentation. *Bioresour. Technol.* **2013**, *132*, 109–114. [CrossRef]
91. Raman, J.K.; Gnansounou, E. Ethanol and lignin production from Brazilian empty fruit bunch biomass. *Bioresour. Technol.* **2014**, *172*, 241–248. [CrossRef]
92. Pandey, A.K.; Edgard, G.; Negi, S. Optimization of concomitant production of cellulase and xylanase from *Rhizopus oryzae* SN5 through EVOP-factorial design technique and application in Sorghum Stover based bioethanol production. *Renew Energy* **2016**, *98*, 51–56. [CrossRef]
93. Rajnish, K.N.; Samuel, M.S.; John, A.J.; Datta, S.; Chandrasekar, N.; Balaji, R.; Jose, S.; Selvarajan, E. Immobilization of cellulase enzymes on nano and micro-materials for breakdown of cellulose for biofuel production-a narrative review. *Int. J. Biol. Macromol.* **2021**, *182*, 1793–1802. [CrossRef] [PubMed]
94. Shokrkar, H.; Ebrahimi, S. Synergism of cellulases and amylolytic enzymes in the hydrolysis of microalgal carbohydrates. Biofuels. *Bioprod. Biorefining* **2018**, *12*, 749–755. [CrossRef]
95. Poornejad, N.; Karimi, K.; Behzad, T. Ionic Liquid Pretreatment of Rice Straw to Enhance Saccharification and Bioethanol Production. *J. Biomass Biofuel* **2014**, *1*, 8–15. [CrossRef]
96. Alam, M.Z.; Kabbashi, N.A.; Tompang, M.F. Development of single-step bioconversion process for bioethanol production by fungi and yeast using oil palm empty fruit bunches. In Proceedings of the 20th Symposium Malaysian Chememical Engineering, Kuala Lumpur, Malaysia, 19 December 2006; pp. 1–6.
97. Ferreira Filho, J.A.; Horta, M.A.C.; Beloti, L.L.; Dos Santos, C.A.; de Souza, A.P. Carbohydrate-active enzymes in *Trichoderma harzianum*: A bioinformatic analysis bioprospecting for key enzymes for the biofuels industry. *BMC Genom.* **2017**, *18*, 779. [CrossRef] [PubMed]
98. Wang, H.; Zhai, L.; Geng, A. Enhanced cellulase and reducing sugar production by a new mutant strain *Trichoderma harzianum* EUA20. *J. Biosci. Bioeng.* **2020**, *129*, 242–249. [CrossRef]
99. Qi, G.; Xiong, L.; Luo, M.; Huang, Q.; Huang, C.; Li, H.; Xuefang, C.; Xinde, C. Solvents production from cassava by co-culture of *Clostridium acetobutylicum* and *Saccharomyces cerevisiae*. *J. Environ. Chem. Eng.* **2018**, *6*, 128–133. [CrossRef]
100. Verma, G.; Nigam, P.; Singh, D.; Chaudhary, K. Bioconversion of starch to ethanol in a single-step process by coculture of amylolytic yeasts and *Saccharomyces cerevisiae* 21. *Bioresour. Technol.* **2000**, *72*, 261–266. [CrossRef]
101. Dey, P.; Rangarajan, V.; Nayak, J.; Bhusan, D.; Branden, S. An improved enzymatic pre-hydrolysis strategy for efficient bioconversion of industrial pulp and paper sludge waste to bioethanol using a semi-simultaneous saccharification and fermentation process. *Fuel* **2021**, *294*, 120581. [CrossRef]
102. Liu, Y.; Zhang, Y.; Xu, J.; Sun, Y.; Yuan, Z.; Xie, J. Consolidated bioprocess for bioethanol production with alkali-pretreated sugarcane bagasse. *Appl. Energy* **2015**, *157*, 517–522. [CrossRef]
103. Gonzales, R.R.; Kim, J.S.; Kim, S. Optimization of dilute acid and enzymatic hydrolysis for dark fermentative hydrogen production from the empty fruit bunch of oil palm. *Int. J. Hydrog. Energy* **2019**, *44*, 2191–2202. [CrossRef]
104. Zainan, N.H.; Alam, Z.; Al-khatib, M.F. Production of sugar by hydrolysis of empty fruit bunches using palm oil mill effluent (POME) based cellulases: Optimization study. *Afr. J. Biotechnol.* **2011**, *10*, 18722–18727.
105. Kamoldeen, A.A.; Keong, C.; Nadiah, W.; Abdullah, W.; Peng, C. Enhanced ethanol production from mild alkali-treated oil-palm empty fruit bunches via co-fermentation of glucose and xylose. *Renew. Energy* **2017**, *107*, 113–123. [CrossRef]
106. Syadiah, E.A.; Haditjaroko, L.; Syamsu, K. Bioprocess engineering of bioethanol production based on sweet sorghum bagasse by co-culture technique using *Trichoderma reesei* and *Saccharomyces cerevisiae*. *IOP Conf. Ser. Earth Environ. Sci.* **2018**, *209*, 12018. [CrossRef]
107. Jambo, S.A.; Abdulla, R.; Marbawi, H.; Gansau, J.A. Response surface optimization of bioethanol production from third generation feedstock—*Eucheuma cottonii*. *Renew. Energy* **2018**, *132*, 1–10. [CrossRef]
108. Khoja, A.H.; Ehsan, A.; Kashaf, Z.; Ansari, A.A.; Azra, N.; Muneeb, Q. Comparative study of bioethanol production from sugarcane molasses by using *Zymomonas mobilis* and *Saccharomyces cerevisiae*. *Afr. J. Biotechnol.* **2015**, *14*, 2455–2462.
109. Park, J.; Oh, B.; Seo, J.; Hong, W.K.; Yu, A.; Sohn, J.H.; Kim, C.H. Efficient Production of Ethanol from Empty Palm Fruit Bunch Fibers by Fed-Batch Simultaneous Saccharification and Fermentation Using *Saccharomyces cerevisiae*. *Appl. Biochem. Biotechnol.* **2013**, *170*, 1807–1814. [CrossRef] [PubMed]
110. Ahmad, F.U.; Bukar, A.; Usman, B. Production of bioethanol from rice husk using *Aspergillus niger* and *Trichoderma harzianum*. Bayero. *J. Pure Appl. Sci.* **2017**, *10*, 280.
111. Kassim, M.A.; Loh, S.K.; Bakar, N.A.; Aziz, A.A.; Mat Som, R. Bioethanol production from Enzymatically Saccharified Empty Fruit Bunches Hydrolysate using *Saccharomyces cerevisiae*. *Res. J. Env. Sci.* **2011**, *5*, 573.
112. Sahu, O. Appropriateness of rose (*Rosa hybrida*) for bioethanol conversion with enzymatic hydrolysis: Sustainable development on green fuel production. *Energy* **2021**, *232*, 120922. [CrossRef]
113. Ibrahim, M.; Abd-aziz, S.; Hassan, M.A. Oil Palm Empty Fruit Bunch as Alternative Substrate for Acetone—Butanol—Ethanol Production by *Clostridium butyricum* EB6. *Appl. Biochem. Biotechnol.* **2012**, *166*, 1615–1625. [CrossRef]

114. Alam, M.Z.; Al khatib, N.H.; Rashid, M.F. Optimization of bioethanol production from empty fruit bunches by co-culture of *Saccharomyces cerevisae* and *Aspergillus niger* using statistical experimental design. *J. Pure Appl. Microbiol.* **2014**, *8*, 731–740.
115. Anu; Singh, B.; Kumar, A. Process development for sodium carbonate pretreatment and enzymatic saccharification of rice straw for bioethanol production. *Biomass Bioenergy* **2020**, *138*, 105574. [CrossRef]
116. Chohan, N.A.; Aruwajoye, G.S.; Sewsynker-Sukai, Y.; Gueguim Kana, E.B. Valorisation of potato peel wastes for bioethanol production using simultaneous saccharification and fermentation: Process optimization and kinetic assessment. *Renew. Energy* **2019**, *146*, 1031–1040. [CrossRef]
117. Ansar; Nazaruddin; Azis, A.D.; Fudholi, A. Enhancement of bioethanol production from palm sap (*Arenga pinnata* (Wurmb) Merr) through optimization of *Saccharomyces cerevisiae* as an inoculum. *J. Mater. Res. Technol.* **2021**, *14*, 548–554. [CrossRef]
118. de Albuquerque Wanderley, A.C.; Soares, M.L.; Gouveia, E.R. Selection of inoculum size and *Saccharomyces cerevisiae* strain for ethanol production in simultaneous saccharification and fermentation (SSF) of sugar cane bagasse. *Afr. J Biotechnol* **2014**, *13*, 2762–2765.
119. Neelakandan, T.; Usharani, G.; Nagar, A. Optimization and Production of Bioethanol from Cashew Apple Juice Using Immobilized Yeast Cells by *Saccharomyces cerevisiae*. *Am. J. Sci. Res.* **2009**, *4*, 85–88.

Article

Preparation of Porous Biochar from Soapberry Pericarp at Severe Carbonization Conditions

Wen-Tien Tsai [1,*], Tasi-Jung Jiang [1], Yu-Quan Lin [1], Hsuan-Lun Chang [2] and Chi-Hung Tsai [3]

1 Graduate Institute of Bioresources, National Pingtung University of Science and Technology, Neipu Township, Pingtung 912, Taiwan; joybook528er@gmail.com (T.-J.J.); wsx55222525@gmail.com (Y.-Q.L.)
2 Department of Biological Science and Technology, National Pingtung University of Science and Technology, Neipu Township, Pingtung 912, Taiwan; hsuanlun98@gmail.com
3 Department of Resources Engineering, National Cheng Kung University, Tainan 701, Taiwan; ap29fp@gmail.com
* Correspondence: wttsai@mail.npust.edu.tw

Abstract: The residue remaining after the water extraction of soapberry pericarp from a biotechnology plant was used to produce a series of biochar products at pyrolytic temperatures (i.e., 400, 500, 600, 700 and 800 °C) for 20 min plant was used to produce a series of biochar products. The effects of the carbonization temperature on the pore and chemical properties were investigated by using N_2 adsorption–desorption isotherms, energy dispersive X-ray spectroscopy (EDS) and Fourier-transform infrared spectroscopy (FTIR). The pore properties of the resulting biochar products significantly increased as the carbonization temperature increased from 700 to 800 °C. The biochar prepared at 800 °C yielded the maximal BET surface area of 277 m^2/g and total pore volume of 0.153 cm^3/g, showing that the percentages of micropores and mesopores were 78% and 22%, respectively. Based on the findings of the EDS and the FTIR, the resulting biochar product may be more hydrophilic because it is rich in functional oxygen-containing groups on the surface. These results suggest that soapberry pericarp can be reused as an excellent precursor for preparing micro-mesoporous biochar products in severe carbonization conditions.

Keywords: soapberry pericarp; carbonization; biochar; pore property; surface chemistry

Citation: Tsai, W.-T.; Jiang, T.-J.; Lin, Y.-Q.; Chang, H.-L.; Tsai, C.-H. Preparation of Porous Biochar from Soapberry Pericarp at Severe Carbonization Conditions. *Fermentation* **2021**, *7*, 228. https://doi.org/10.3390/fermentation7040228

Academic Editors: Alessia Tropea and Gunnar Lidén

Received: 6 September 2021
Accepted: 8 October 2021
Published: 11 October 2021

1. Introduction

Biochar is a carbon-rich material which can be produced from a variety of lignocellulosic residues in a closed system under limited or no oxygen (or air). Due to its chemical and physical characteristics, biochar can be used as a product itself or as an ingredient within a mixed product for multiple objectives, including soil improvement, waste management, energy (or fuel) production, water pollution, and mitigation of climate change [1]. For example, biochar has been commonly used as a soil enhancer, thus making soils more fertile and also sequestering carbon in soils for a long time without greenhouse gas (GHG) emissions [2]. Concerning its porous structure and surface characterization, biochar has high adsorption potential for the removal of pollutants from water streams [3–7]. In recent years, there is an increasing interest in exploiting biochar as an excellent carbon material for environmental applications, or reusing different lignocellulosic feedstocks for producing biochar with high pore properties (e.g., specific surface area), which includes wood [8], oil palm shell [9], maize straw [10], cocoa pod husk [11], rice husk [12] and so on.

Soapberry (*Sapindus mukorossi*), also called soapnut, is a deciduous plant in the family *Sapindaceae*, which is commonly planted in tropical and sub-tropical Asian regions (including Taiwan) for its folk values [13]. The fruit is famous for the natural surfactants (i.e., saponins) present in its pericarp. Apart from its traditional use in detergents and shampoo for hair, skin and clothing, the pharmacological and biological actions of this plant have been recently exploited in the fields of medicine [14,15] and herbicides [16,17]. Because of the chemical characteristics of its saponins, it has been used in food applications as a natural preservative and emulsifier [18]. In order to extract the saponins from soapberry pericarp, the commonly used methods were to use the aqueous extraction solvent with the proper solid/liquid ratio at a mild temperature [14]. Furthermore, the residual biomass after the extraction of soapberry pericarps was thus generated without further utilization, causing problems of waste management and environmental pollution. In this regard, reusing the residual soapberry pericarp as a precursor in the production of carbon materials could be a promising route.

Similar to other lignocellulosic biomasses, the main components of soapberry pericarp are composed of cellulose, lignin and hemicellulose. However, there is limited literature on the use of soapberry pericarp for biochar production [19,20]. Zhang et al. [19] performed the oxidative torrefaction of three nutshells (including soapberry pericarp) at 250 and 300 °C for the production of biochar, which has potential for reuse as a solid fuel and for carbon storage. Velusamy et al. [20] prepared biochar from soapberry pericarp at 450 °C for about 2 h under a heating rate of 3 °C/min. The resulting biochar material (specific surface area = 2.2 m^2/g) was tested to determine its adsorption performance of antibiotic ciprofloxacin (one of emerging contaminants) in aqueous solution.

In view of the few studies on the preparation of biochar from soapberry pericarp in the literature, the aim of this work was to produce soapberry-based biochar products at severe carbonization temperatures (i.e., 400–800 °C), because this process parameter has the greatest influence on their physical structures [21,22]. The pore and chemical characteristics of the resulting biochar products were characterized as a function of carbonization temperature.

2. Materials and Methods

2.1. Materials

In this work, the residual soapberry pericarp as a starting feedstock for biochar production was obtained from a local biotechnology factory (Tainan city, Taiwan), which adopted the mild water system for extracting saponins from soapberry pericarp. The biomass sample first had its moisture removed under sunlight and was then dried by an air-circulation oven. The dry sample was shredded by a knifer and further sieved to a size in the range of mesh no. 80 (opening = 0.18 mm) and 40 (opening = 0.40 mm). The sample was subsequently used for the thermochemical analyses and the carbonization experiments.

2.2. Thermochemical Properties of Soapberry Pericarp

The characteristics of biomass feedstock greatly influence the performance of a thermochemical conversion system [23]. In this work, the thermochemical properties of soapberry pericarp included proximate analysis, ultimate analysis, calorific value, inorganic element analysis and thermogravimetric analysis (TGA). The operations and procedures for these thermochemical analyses have been reported previously [24,25].

2.3. Carbonization Experiments

In order to enhance the pore properties of the resulting biochar, the preparation of biochar from soapberry pericarp (about 5 g for each experiment) was carried out at higher carbonization temperatures (400–800 °C by an interval of 100 °C) for 20 min under the nitrogen gas flow (500 cm^3/min). Another carbonization experiment was performed at the highest temperature (i.e., 800 °C) for 80 min to evaluate the effect of residence time on the pore properties preliminarily. The operations of carbonization experiments and procedures at 10 °C/min for producing biochar products can refer to the previous studies [26,27]. Herein, the yield of the resulting biochar was obtained by the ratio of its weight to the weight of soapberry pericarp fed.

2.4. Analysis of Resulting Biochar Properties

The pore properties of the resulting biochar, including surface area, pore volume and pore size distribution, were determined by an accelerated surface area and porosimetry instrument (ASAP 2020; Micromeritics Co., Norcross, GA, USA), which was based on nitrogen (N_2) adsorption–desorption isotherms at −196 °C [28]. Prior to the measurement, the biochar sample was degassed at 250 °C in a vacuum for 3 h. The N_2 isotherms were measured over a relative pressure (P/Po) range between 10^{-5} and 0.999. The Brunauer–Emmett–Teller (BET) surface area and micropore volume (and micropore surface area) were determined by using the BET equation and *t*-plot analysis, respectively [29]. On the other hand, the elemental distributions and functional groups of the resulting biochar products with high pore properties were observed by energy-dispersive X-ray spectroscopy (EDS) (7021-H; HORIBA Co., Kyoto, Japan) and Fourier-transform infrared spectroscopy (FTIR) (FT/IR-4600; JASCO Co., Tokyo, Japan), respectively. The biochar sample preparation and analytical conditions have been stated in the previous study [11].

3. Results and Discussion

3.1. Thermochemical Characteristics of Soapberry Pericarp

Table 1 showed the main thermochemical properties of the dried soapberry pericarp (SP), including proximate analysis, ultimate analysis and calorific value. The data in Table 1 were very close to those in the literature [19]. For instance, the carbon content and calorific value in Table 1 were 52.96 wt% and 21.75 MJ/kg, respectively, in comparison with 53.24 wt% and 20.93 MJ/kg [19]. By contrast, the dried SP has a relatively lower ash content (2.28 wt%) than those of other biomass residues in the range from 1.41 to 20.26% (dry basis) [30]. The calorific value (i.e., 20.96 MJ/kg) was in accordance with its high contents of carbon (C, 52.96 wt%) and hydrogen (H, 7.29 wt%). It should be noted that the contents of nitrogen (N, 1.48 wt%) and sulfur (S, 0.36 wt%) for the dried biomass were obviously higher than those for other biomass husks [30], thus posing significant emissions of nitrogen oxides (NOx) and sulfur oxides (SOx) while it is burned without an installed control system. The main inorganic elements in the biomass sample (as listed in Table 2) are calcium (Ca), potassium (K), magnesium (Mg) and iron (Fe), which could be present in the forms of oxides and/or carbonates [30]. Furthermore, the contents of inorganic elements (i.e., Si, Al, Na, Cu, P and Ti) were not determined by the inductively coupled plasma-optical emission spectrometer (ICP-OES) because they were lower than their method detection levels (Table 2). Based on the data in Table 2, the total amounts (1.73 wt%) of inorganic elements by their oxides (i.e., CaO, K_2O, MgO and Fe_2O_3) were close to the low ash content (2.28 wt%) in Table 1.

Table 1. Thermochemical properties of soapberry pericarp (SP).

Properties [a]	Value
Proximate analysis [b]	
Ash (wt%)	2.28 ± 0.03
Volatile matter (wt%)	77.44 ± 1.33
Fixed carbon [c] (wt%)	20.28
Ultimate analysis [d]	
Carbon (wt%)	52.96 ± 0.01
Hydrogen (wt%)	7.29 ± 0.15
Oxygen (wt%)	36.94 ± 0.02
Nitrogen (wt%)	1.48 ± 0.12
Sulfur (wt%)	0.36 ± 0.10
Calorific value (MJ/kg) [b]	21.75 ± 0.18

[a] On a dry basis. [b] The mean ± standard deviation for three determinations. [c] By difference. [d] The mean ± standard deviation for two determinations.

Table 2. Contents of inorganic elements of soapberry pericarp (SP).

Inorganic Element	Value [a]	Method Detection Limit (ppm)
Ca (wt%)	0.544	
K (wt%)	0.527	
Mg (wt%)	0.121	
Fe (wt%)	0.092	
Si (wt%)	ND [a]	63.0
Al (wt%)	ND	11.4
Na (wt%)	ND	3.0
Cu (wt%)	ND	3.6
P (wt%)	ND	39.6
Ti (wt%)	ND	2.4

[a] Not detectable.

The thermogravimetric analysis (TGA) and derivative thermogravimetry (DTG) curves of the dried SP (about 0.2 g) were obtained at four different heating rates (i.e., 5, 10, 15 and 20 °C/min) under the nitrogen flow (50 cm^3/min), as depicted in Figure 1. Obviously, these curves revealed similar thermal behaviors according to the residual weight percentages as a function of temperature. Using the TGA curve at 10 °C/min as an example, the initial weight decline occurred in the range from 100 to 200 °C, which should be attributed to the losses of attached matters (e.g., moisture) and light volatiles due to the incipient decompositions of liable organic matters. Subsequently, significant mass loss was observed at the temperature range of 200 to 400 °C, which corresponded well to the decompositions of organic constituents such as hemicellulose and cellulose [18]. When the temperature was raised to above 400 °C, the continued mass loss was caused by the volatilization of complex lignocellulosic fractions (e.g., lignin) and inorganic carbonates/oxides. These data were consistent with those in the literature [31]. From the TGA curves (Figure 1), it was suggested that the dried SP could be an excellent precursor for producing biochar at higher carbonization temperatures. In this regard, this work adopted carbonization conditions in the range from 400 to 800 °C at the heating rate of 10 °C/min for producing biochar products with high pore properties.

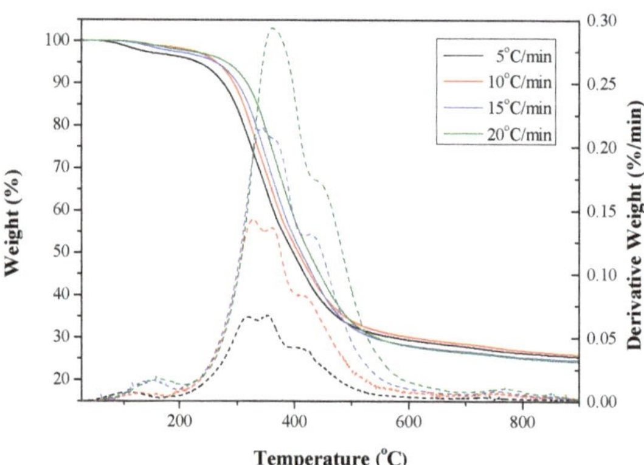

Figure 1. Thermogravimetric analysis (TGA, solid line)/derivative thermogravimetry (DTG, dotted line) curves of soapberry pericarp (SP) at various heating rates (5–20 °C/min).

3.2. *Pore Properties of Resulting Biochar*

The pore properties of porous materials commonly refer to specific surface area, pore volume, pore size and pore size distribution, which are closely related to their applications. In order to compare with the data clearly, the resulting products were coded by the "SP-BC-temperature" to mean the preparation of biochar (BC) from soapberry pericarp (SP) at different carbonization temperatures. The average yields of the resulting products prepared in duplicate at 400, 500, 600, 700 and 800 °C were 34.5, 26.5, 26.7, 23.7 and 21.0 wt%, respectively. At higher carbonization temperatures, the charring will cause the gradual mass decline by the formation of pyrolyzed gases such as carbon monoxide (CO). At this time, the chemical structure will be transformed from amorphous char to composite char with turbostratic crystallites [21]. Therefore, it was shown that the pore properties of biochar products would be enhanced with increasing temperature due to the development of porosity. The formations of micropores and mesopores were attributed to the fused ring structures of the aromatic carbon matrix.

Table 3 summarized the pore properties of the SP-BC products, which were prepared at 400–800 °C (an interval of 100 °C) for 20 min. Obviously, the BET surface areas of the resulting biochar products significantly rose from 0.3 to 277.1 m^2/g when the carbonization temperature increased from 400 to 800 °C, especially for the temperature in the range of 700 to 800 °C. As mentioned above, the carbonization at high temperature will lead to the formation of nanopores in the resulting turbostratic char, thus enhancing its pore properties significantly [21]. In order to test the effect of residence time on the pore properties of the resulting biochar, another carbonization experiment was performed at 800 °C for 80 min, showing a slight decline in the BET surface area from 277.1 to 240.8 m^2/g. This finding can be attributed to the severe carbonization reaction, leading to the structural breakdown of the resulting biochar and a reduction in its surface area [21,22,32]. On the other hand, the SP-BC-800 product was prepared in duplicate to assure its high pore properties consistently, showing that the BET surface area was about 300 ± 30 m^2/g. As listed in Table 3, other pore properties such as single point surface area and total pore volume also indicated an increasing trend. As summarized above, the temperature should be the most important process parameter for determining the pore properties of the biochar products as more pores were generated in severe carbonization conditions. The findings were consistent with those reported by the other feedstocks such as cocoa pod husk [11], rice husk [12], goat manure [21], biogas digestate [33] and dairy manure [34]. The maximal BET surface area of about 300 m^2/g can be obtained by using these feedstocks in the biochar production when the carbonization temperature reached 800 or 900 °C. In addition, the average pore diameter was obtained from the data on the BET surface area and the total pore volume assuming the pore is of cylindrical and uniform geometry, showing that the pore diameter of the SP-BC-800 product was close to the boundary limit (2.0 nm) between micropores and mesopores. In this regard, both the microporous and mesoporous structures could be presented in the optimal biochar SP-BC-800.

Table 3. Pore properties of SP-BC products.

Pore Property	SP-BC-400	SP-BC-500	SP-BC-600	SP-BC-700	SP-BC-800	SP-BC-800 [a]
Surface area (m^2/g)						
Single point surface area [b]	0.3	1.3	1.2	13.3	282.0	231.4
BET surface area [c]	0.3	1.4	1.5	14.1	277.1	240.8
Langmuir surface area	0.4	3.7	1.8	21.0	410.6	355.2
t-plot micropore area [d]	0.4	0.4	1.1	11.7	226.5	194.6
t-plot external surface area	0.0	1.0	0.4	2.3	50.6	46.2
Pore volume (cm^3/g)						
Total pore volume [e]	0.0005	0.005	0.002	0.010	0.153	0.130
t-plot micropore area [d]	0.0003	0.0002	0.001	0.006	0.119	0.096
Pore size (nm)						
Average pore width [f]	7.317	12.654	5.142	2.784	2.210	2.160

[a] This biochar product was prepared at 800 °C for 80 min. [b] Calculated by the single point BET method at relative pressure of 0.30. [c] Calculated by the BET method at relative pressure range of 0.06–0.30 (9 points). [d] Calculated by the *t*-plot method. [e] Calculated by the single point adsorption at relative pressure of 0.995 (pore diameter less than 38.17 nm. [f] Calculated from the ratio of the total pore volume (V_t) and the BET surface area (S_{BET}) if the pore is of cylindrical geometry (i.e., Average pore width = 4 × V_t/S_{BET}).

Figure 2 showed the N_2 adsorption/desorption isotherms of all resulting SP-BC products at −196 °C. Herein, the isotherms of the SP-BC-400 and SP-BC-500 products were not depicted in Figure 2 because of their relatively low adsorption/desorption amounts. From the isotherm shape, the SP-BC-800 product obviously exhibited the characteristics of Type I. Based on the definition by the International Union of Pure and Applied Chemistry (IUPAC) [28], carbon material with the Type I isotherm could be highly microporous because it has a high potential for adsorption at very low relative pressure (P/P_0) (<0.1) by micropore filling. However, typical microporous materials often contain pores over a wide range of sizes, including micropores (<2 nm) and mesopores (2–50 nm). When the values of P/P_0 increased from 0.1 to 1.0, the curves in Figure 2 indicated a low slope of the plateau, which was attributed to the multilayer adsorption on the pore surface of the resulting biochar products. Based on the capillary condensation of N_2 at a relative pressure of about 1.0, all pores were thus filled by liquid N_2, suggesting that total pore volume can be estimated by converting the saturated adsorption amount into liquid N_2 volume using its liquid density (i.e., 0.808 g/cm^3). It was also seen that the resulting SP-BC-800 product also exhibited a hysteresis loop with the Type IV isotherms [29], which showed the existence of a micro-mesoporous composite structure from the Type I and Type IV isotherms. Furthermore, this can be observed in its pore size distribution (Figure 3), which can be consistently linked to its pore properties, as listed in Table 3.

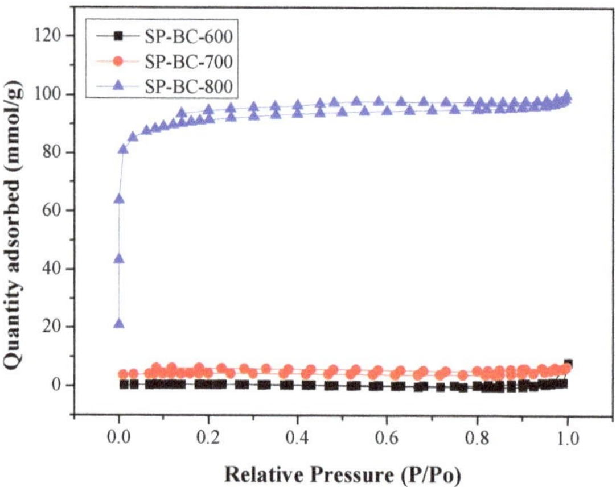

Figure 2. N_2 adsorption-desorption isotherms of SP-BC products prepared at different carbonization temperatures.

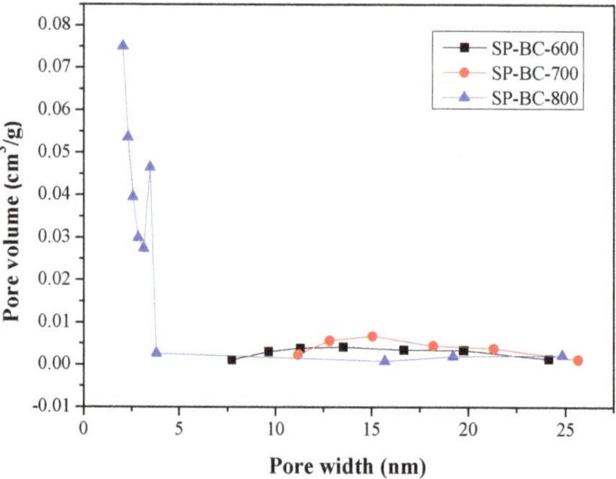

Figure 3. Pore size distribution curves of SP-BC products prepared at different carbonization temperatures.

3.3. Chemical Characteristics of Resulting Biochar

In this work, the changes in the elemental distributions on the surface of the biochar products were analyzed by energy dispersive X-ray spectroscopy (EDS). Figure 4 illustrated the EDS spectra of the resulting biochar products (i.e., SP-BC-600 and SP-BC-800) prepared at different temperatures (i.e., 600 and 800 °C). It clearly showed that the main elements on the surface of the biochar products included carbon and oxygen. In addition, the carbon contents of the biochar products slightly increased from 76.7 wt% to 80.2 wt% as the temperature increased from 600 °C to 800 °C. The high content of oxygen in the biochar should be indicative of the functional oxygen-containing groups on the surface such as carbonyl (C=O-). Furthermore, the presence of magnesium (Mg), potassium (K) and calcium (Ca) was observed in the biochar products, which could be associated with the forms of carbonates or oxides. It can be seen that the metal elements in the biochar products should be derived from the precursor SP, as listed in Table 2.

(a) SP-BC-600

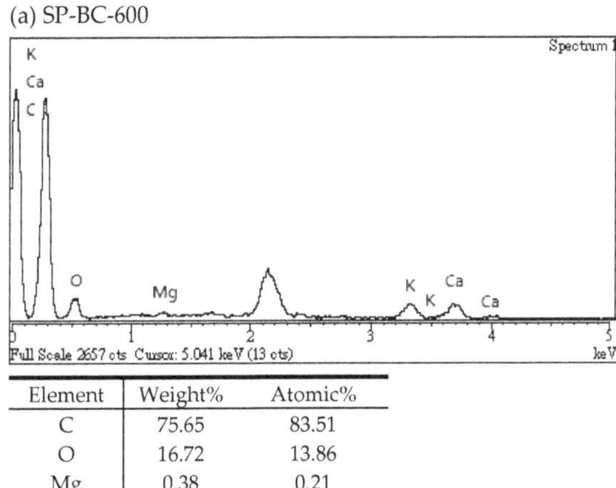

Element	Weight%	Atomic%
C	75.65	83.51
O	16.72	13.86
Mg	0.38	0.21
K	3.37	1.14
Ca	3.88	1.28

(b) SP-BC-800

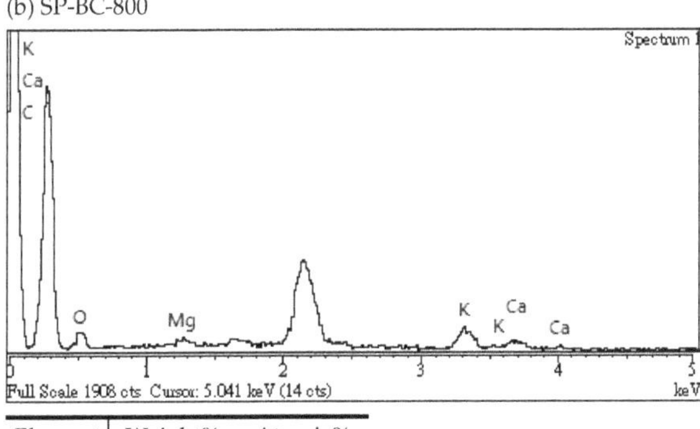

Element	Weight%	Atomic%
C	80.15	86.91
O	13.26	10.79
Mg	0.58	0.31
K	4.10	1.37
Ca	1.91	0.62

Figure 4. Energy-dispersive X-ray spectroscopy (EDS) spectra of (**a**) SP-BC product (SP-BC-600) and (**b**) SP-BC product (SP-BC-800).

The Fourier-transform infrared spectroscopy (FTIR) spectra of the biochar product with the highest pore properties (i.e., SP-BC-800) was recorded in the range of 400–4000 cm^{-1}, as shown in Figure 5. Clearly, there were four significant absorption peaks at about 3440, 1640, 1385 and 1115 cm^{-1}, which could be associated with functional oxygen-containing groups [35,36]. For example, the absorption peak at 3450 cm^{-1} in the biochar product could be assigned to the stretching vibration of hydroxyl group (O-H). In addition, the sharp peaks at about 1640, 1385 and 1115 cm^{-1} may be attributed to the carbonyl (C=O-), O-H bending (phenolic) and symmetric C-O stretching, respectively [36]. From the findings of the EDS (Figure 4) and the FTIR (Figure 5), the SP-BS-800 biochar product may be hydrophilic because its surface is rich in oxygen.

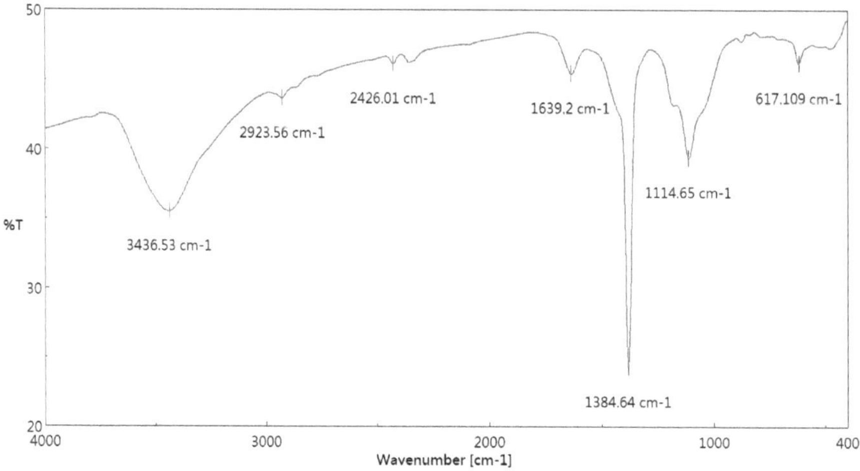

Figure 5. FTIR spectrum of the resulting biochar (SP-BC-800).

4. Conclusions

The preparation of porous biochar from soapberry pericarp has been carried out by a carbonization process at 400–800 °C for 20 min. Under the conditions examined, the process had a biochar yield of at least 20%. The pore properties of the resulting biochar products indicated an increasing trend when the carbonization temperature increased from 400 to 800 °C, especially for the temperature in the range of 700 to 800 °C. The biochar with a BET surface area of about 300 m^2/g was produced at 800 °C for 20 min. According to the data on N_2 isotherms and pore size distribution, the existence of micro-mesoporous composite structure in the optimal biochar was shown. From the findings of the EDS and the FTIR, the biochar product may be hydrophilic because its surface is rich in oxygen. These results suggest that soapberry pericarp can be reused as an excellent precursor for preparing micro-mesoporous biochar products in severe carbonization conditions. It would be helpful to study the adsorptive removal of cationic pollutants from the aqueous solution using the resulting biochar material with high pore properties.

Author Contributions: Conceptualization, W.-T.T.; methodology, C.-H.T.; validation, T.-J.J. and H.-L.C.; formal analysis, T.-J.J. and H.-L.C.; data curation, Y.-Q.L.; writing-original draft preparation, W.-T.T.; writing-review and editing, W.-T.T.; visualization, Y.-Q.L.; supervision, W.-T.T. All authors have read and agreed to the published version of the manuscript.

Funding: This research received no external funding.

Institutional Review Board Statement: Not applicable.

Informed Consent Statement: Not applicable.

Data Availability Statement: Data is contained within the article.

Acknowledgments: Sincere appreciation was expressed to acknowledge the National Pingtung University of Science and Technology for their assistances in the energy-dispersive X-ray spectroscopy (EDS) analysis.

Conflicts of Interest: The authors declare no conflict of interest.

References

1. Lehmann, J.; Joseph, S. Biochar for environmental management: An introduction. In *Biochar for Environmental Management*, 2nd ed.; Lehmann, J., Joseph, S., Eds.; Routledge: New York, NY, USA, 2015; pp. 1–13.
2. BIOCHAR (International Biochar initiative). Available online: https://biochar-international.org/biochar/ (accessed on 3 September 2021).
3. Mohan, D.; Sarswat, A.; Ok, Y.S.; Pittman, C.U., Jr. Organic and inorganic contaminants removal from water with biochar, a renewable, low cost and sustainable adsorbent—A critical review. *Bioresour. Technol.* **2014**, *160*, 191–202. [CrossRef] [PubMed]
4. Inyang, M.I.; Gao, B.; Yao, Y.; Xue, Y.; Zimmerman, A.; Mosa, A.; Pullammanappallil, P.; Ok, Y.S.; Cao, X. A review of biochar as a low-cost adsorbent for aqueous heavy metal removal. *Crit. Rev. Environ. Sci. Technol.* **2016**, *46*, 406–433. [CrossRef]
5. Yin, Q.; Zhang, B.; Wang, R.; Zhao, Z. Biochar as an adsorbent for inorganic nitrogen and phosphorus removal from water: A review. *Environ. Sci. Pollut. Res.* **2017**, *24*, 26297–26309. [CrossRef]
6. Xiang, W.; Zhang, X.; Chen, J.; Zou, W.; He, F.; Hu, X.; Tsang, D.C.W.; Ok, Y.S.; Gao, B. Biochar technology in wastewater treatment: A critical review. *Chemosphere* **2020**, *252*, 126539. [CrossRef]
7. Qiu, B.; Tao, X.; Wang, H.; Li, W.; Ding, X.; Chu, H. Biochar as a low-cost adsorbent for aqueous heavy metal removal: A review. *J. Anal. Appl. Pyrolysis* **2021**, *155*, 105081. [CrossRef]
8. Jiang, S.; Nguyen, T.A.H.; Rudolph, V.; Yang, H.; Zhang, D.; Ok, Y.S.; Huang, L. Characterization of hard- and softwood biochars pyrolyzed at high temperature. *Environ. Geochem. Health* **2017**, *39*, 403–415. [CrossRef]
9. Kong, S.H.; Lam, S.S.; Yek, P.N.Y.; Liew, R.K.; Ma, N.L.; Osman, M.S.; Wong, C.C. Self-purging microwave pyrolysis: An innovative approach to convert oil palm shell into carbon-rich biochar for methylene blue adsorption. *J. Chem. Technol. Biotechnol.* **2019**, *94*, 1397–1405. [CrossRef]
10. Guo, C.; Zou, J.; Yang, J.; Wang, K.; Song, S. Surface characterization of maize-straw derived biochar and their sorption mechanism for Pb^{2+} and methylene blue. *PLoS ONE* **2020**, *15*, e0238105. [CrossRef] [PubMed]
11. Tsai, W.T.; Lin, Y.Q.; Tsai, C.H.; Chen, W.S.; Chang, Y.T. Enhancing the pore properties and adsorption performance of cocoa pod husk (CPH)-derived biochars via post-acid treatment. *Processes* **2020**, *8*, 144. [CrossRef]
12. Tsai, W.T.; Lin, Y.Q.; Huang, H.J. Valorization of rice husk for the production of porous biochar materials. *Fermentation* **2021**, *7*, 70. [CrossRef]
13. Mondal, M.H.; Malik, S.; Garain, A.; Mandal, S.; Saha, B. Extraction of natural surfactant saponin from soapnut (*Sapindus mukorossi*) and its utilization in the remediation of hexavalent chromium from contaminated water. *Tenside Surfactants Deterg.* **2017**, *54*, 519–529. [CrossRef]
14. Liu, M.; Chen, Y.L.; Kuo, Y.H.; Lu, M.K.; Liao, C.H. Aqueous extract of *Sapindus mukorossi* induced cell death of A549 cells and exhibited antitumor property In Vivo. *Sci. Rep.* **2018**, *8*, 4831. [CrossRef] [PubMed]
15. Xu, Y.Y.; Gao, Y.; Chen, Z.; Zhao, G.C.; Liu, J.M.; Wang, X.; Gao, S.L.; Zhang, D.G.; Jia, L.M. Metabolomics analysis of the soapberry (*Sapindus mukorossi* Gaertn.) pericarp during fruit development and ripening based on UHPLC-HRMS. *Sci. Rep.* **2021**, *11*, 11657. [CrossRef] [PubMed]
16. Eddaya, T.; Boughdad, A.; Sibille, E.; Chaimbault, P.; Zaid, A.; Amechouq, A. Biological activity of *Sapindus mukorossi* Gaerten (Sapindaceae) aqueous extract against *Thysanoplusia orichalcea* (Lepidoptera- Noctuidae). *Ind. Crops Prod.* **2013**, *50*, 325–332. [CrossRef]
17. Dai, Z.Y.; Wang, J.; Ma, X.J.; Sun, J.; Tang, F. Laboratory and field evaluation of the phytotoxic activity of *Sapindus mukorossi* Gaertn pulp extract and identification of a phytotoxic substance. *Molecules* **2021**, *26*, 1318. [CrossRef]
18. Gonzalez, P.J.; Sorensen, P.M. Characterization of saponin foam from *Saponaria officinalis* for food applications. *Food Hydrocoll.* **2020**, *101*, 105541. [CrossRef]
19. Zhang, C.; Ho, S.H.; Chen, W.H.; Fu, Y.; Chang, J.S.; Bi, X. Oxidative torrefaction of biomass nutshells: Evaluations of energy efficiency as well as biochar transportation and storage. *Appl. Energy* **2019**, *235*, 428–441. [CrossRef]
20. Velusamy, K.; Periyasamy, S.; Kumar, P.S.; Jayaraj, T.; Krishnasamy, R.; Sindhu, J.; Sneka, D.; Subhashini, B.; Vo, D.V.N. Analysis on the removal of emerging contaminant from aqueous solution using biochar derived from soap nut seeds. *Environ. Pollut.* **2021**, *287*, 117632. [CrossRef]
21. Marco Keiluweit, M.; Nico, P.; Johnson, M.G.; Kleber, M. Dynamic molecular structure of plant biomass-derived black carbon (biochar). *Environ. Sci. Technol.* **2010**, *44*, 1247–1253. [CrossRef]
22. Chia, C.H.; Downie, A.; Munroe, P. Characteristics of biochar: Physical and structural properties. In *Biochar for Environmental Management*, 2nd ed.; Lehmann, J., Joseph, S., Eds.; Routledge: New York, NY, USA, 2015; pp. 89–109.
23. Basu, P. *Biomass Gasification, Pyrolysis and Torrefaction*, 2nd ed.; Academic Press: San Diego, CA, USA, 2013.

24. Tsai, W.T.; Huang, P.C. Characterization of acid-leaching cocoa pod husk (CPH) and its resulting activated carbon. *Biomass Convers. Biorefin.* **2018**, *8*, 521–528. [CrossRef]
25. Tsai, W.T.; Lin, Y.Q.; Tsai, C.H.; Chung, M.H.; Chu, M.H.; Huang, H.J.; Jao, Y.H.; Yeh, S.I. Conversion of water caltrop husk into biochar by torrefaction. *Energy* **2020**, *195*, 116967. [CrossRef]
26. Touray, N.; Tsai, W.T.; Chen, H.R.; Liu, S.C. Thermochemical and pore properties of goat-manure-derived biochars prepared from different pyrolysis temperatures. *J. Anal. Appl. Pyrolysis* **2014**, *109*, 116–122. [CrossRef]
27. Tsai, W.T.; Huang, C.N.; Chen, H.R.; Cheng, H.Y. Pyrolytic conversion of horse manure into biochar and its thermochemical and physical properties. *Waste Biomass Valoriz.* **2015**, *6*, 975–981. [CrossRef]
28. Condon, J.B. *Surface Area and Porosity Determinations by Phyisorption: Measurement and Theory*; Elsevier: Amsterdam, The Netherlands, 2006.
29. Lowell, S.; Shields, J.E.; Thomas, M.A.; Thommes, M. *Characterization of Porous Solids and Powders: Surface Area, Pore Size and Density*; Springer: Dordrecht, The Netherlands, 2006.
30. Jenkins, B.M.; Baxter, L.L.; Miles, T.R., Jr.; Miles, T.R. Combustion properties of biomass. *Fuel Process. Technol.* **1998**, *54*, 17–46. [CrossRef]
31. Mosek, O.; Johnston, C.T. Thermal analysis for biochar characterisation. In *Biochar: A Guide to Analytical Methods*; Singh, B., Camps-Arbestain, M., Lehmann, J., Eds.; CRC Press: Boca Raton, FL, USA, 2017; pp. 283–293.
32. Mukome, F.N.D.; Parikh, S.J. Chemical, physical, and surface characterization of biochar. In *Biochar: Production, Characterization, and Applications*; Ok, Y.S., Uchimiya, S.M., Chang, S.X., Bolan, N., Eds.; CRC Press: Boca Raton, FL, USA, 2016; pp. 67–96.
33. Hung, C.Y.; Tsai, W.T.; Chen, J.W.; Lin, Y.Q.; Chang, Y.M. Characterization of biochar prepared from biogas digestate. *Waste Manag.* **2017**, *66*, 53–60. [CrossRef]
34. Tsai, W.T.; Huang, C.P.; Lin, Y.Q. Characterization of biochars produced from dairy manure at high pyrolysis temperatures. *Agronomy* **2019**, *9*, 634. [CrossRef]
35. Cantrell, K.B.; Hunt, P.G.; Uchimiya, M.; Novak, J.M.; Ro, K.S. Impact of pyrolysis temperature and manure source on physicochemical characteristics of biochar. *Bioresour. Technol.* **2012**, *107*, 419–428. [CrossRef] [PubMed]
36. Johnston, C.T. Biochar analysis by Fourier-transform infra-red spectroscopy. In *Biochar: A Guide to Analytical Methods*; Singh, B., Camps-Arbestain, M., Lehmann, J., Eds.; CRC Press: Boca Raton, FL, USA, 2017; pp. 199–213.

fermentation

MDPI

Article

Residual Gas for Ethanol Production by *Clostridium carboxidivorans* in a Dual Impeller Stirred Tank Bioreactor (STBR)

Carolina Benevenuti [1,2], Marcelle Branco [3], Mariana do Nascimento-Correa [1,2], Alanna Botelho [1], Tatiana Ferreira [3] and Priscilla Amaral [1,*]

[1] Department of Biochemical Engineering, School of Chemistry, Universidade Federal do Rio de Janeiro, Rio de Janeiro 21941-909, RJ, Brazil; carolbenevenuti@hotmail.com (C.B.); marianamattosquim@gmail.com (M.d.N.-C.); lanna.mbotelho@gmail.com (A.B.)

[2] Technology Center for Chemical and Textile Industry, SENAI Innovation Institute for Biosynthetics, Rio de Janeiro 21941-909, RJ, Brazil

[3] Department of Organic Process, School of Chemistry, Universidade Federal do Rio de Janeiro, Rio de Janeiro 21941-909, RJ, Brazil; cellebranco@gmail.com (M.B.); tatiana@eq.ufrj.br (T.F.)

* Correspondence: pamaral@eq.ufrj.br; Tel.: +55-21-3938-7623

Citation: Benevenuti, C.; Branco, M.; do Nascimento-Correa, M.; Botelho, A.; Ferreira, T.; Amaral, P. Residual Gas for Ethanol Production by *Clostridium carboxidivorans* in a Dual Impeller Stirred Tank Bioreactor (STBR). *Fermentation* **2021**, *7*, 199. https://doi.org/10.3390/fermentation7030199

Academic Editor: Alessia Tropea

Received: 26 August 2021
Accepted: 17 September 2021
Published: 21 September 2021

Abstract: Recycling residual industrial gases and residual biomass as substrates to biofuel production by fermentation is an important alternative to reduce organic wastes and greenhouse gases emission. *Clostridium carboxidivorans* can metabolize gaseous substrates as CO and CO_2 to produce ethanol and higher alcohols through the Wood-Ljungdahl pathway. However, the syngas fermentation is limited by low mass transfer rates. In this work, a syngas fermentation was carried out in serum glass bottles adding different concentrations of Tween®80 in ATCC®2713 culture medium to improve gas-liquid mass transfer. We observed a 200% increase in ethanol production by adding 0.15% (*v/v*) of the surfactant in the culture medium and a 15% increase in biomass production by adding 0.3% (*v/v*) of the surfactant in the culture medium. The process was reproduced in stirred tank bioreactor with continuous syngas low flow, and a maximum ethanol productivity of 0.050 g/L.h was achieved.

Keywords: synthesis gas fermentation; volumetric mass transfer coefficient; Tween 80® surfactant

1. Introduction

Global energy consumption has increased over the last decades, with demand forecasted to be 248 quadrillions BTU of liquid fuels by 2050, which represents an increase of 50% compared to 2021 [1]. The use of fossil-based energy has been declining since its use drives climate changes and air pollution [2]. Additionally, the constant fluctuation of oil prices caused by political and economic instability around the world brings insecurity to this industrial sector [3].

In this scenario, there is an increasing demand for renewable and carbon-neutral fuels, especially those produced through microbial fermentation, such as ethanol and butanol [4,5]. Initially, ethanol was the main focus, which can either be a stand-alone fuel or a gasoline-ethanol blend [4]. However, its low caloric value and hygroscopicity limit the use and transportation of ethanol in the current infrastructure. Therefore, the interest in butanol as a liquid fuel, which is less hygroscopic and provides higher caloric value in comparison to ethanol, has increased in recent years [6]. Currently, those alcohols are produced through direct fermentation of sugars extracted from food or energy crops, with pretreatment steps to hydrolyze carbohydrate polymers, increasing costs and byproduct formation [7,8].

The indirect fermentation, or hybrid process, consists of the conversion of a wide variety of carbonaceous compounds to synthesis gas, also named syngas, through gasification, followed by its fermentation to desired products by specific biocatalysts [9,10]. Syngas, mainly composed of CO (carbon monoxide), CO_2 (carbon dioxide), and H_2 (hydrogen), can be obtained from biomass, coal, animal or municipal solid waste, and industrial CO-rich off-gases [11]. The hybrid process uses whole biomass components, including lignin, is not dependent on feedstock composition, and eliminates complex pretreatments and high enzyme costs [12].

Several *Clostridium* species are known to produce biofuels, but only a few of them use syngas as sole carbon and energy sources [13]. *Clostridium carboxidivorans* is an acetogenic bacteria capable of producing ethanol, butanol, and hexanol—valuable as fuels or even as platform chemicals in the pharmaceutical, perfume, and textile industries—from syngas [5,14–17].

Although syngas fermentation is a promising technology, it faces several challenges: low product yield, high separation cost, inhibitory compounds in syngas (i.e., tar, sulfur, and ash), and, mainly, low gas-liquid mass transfer [11,12,18–20]. Metabolic engineering, culture medium formulation, and different bioreactor designs have been proposed to overcome those challenges [11,21].

Efforts to increase the gas-liquid mass transfer usually include the study of different reactor designs such as stirred tank reactor (STR) [14,22], bubble column reactor (BCR) [23], hollow fiber membrane reactor (HFMR) [24], monolithic biofilm reactor (MBR) [23], trickle bed reactor (TBR) [25], and horizontal rotating packed bed biofilm reactor (h-RPB) [26]. STR is the most usual bioreactor used for biotechnology due to its good mixing and simple operation. A widely used approach to enhance mass transfer in STR is to increase the agitation speed and the gas flow rate. However, these strategies are not economically feasible to scale up due to the high energy consumption and microbial shear stress [11].

Another feasible approach to enhance gas-liquid mass transfer is to add some chemical agents, such as surfactants, or some vibration techniques in the culture medium to promote fine gas bubbles in the liquid phase [27–29]. The addition of surfactants in the culture medium enables the stabilization of microbubbles, avoiding coalescence. These agents can reduce interfacial free energy, reducing the liquid surface tension [30]. Coelho et al. [20] reported a significant increase (120%) in carbon monoxide mass transfer coefficient when Tween® 80 and/or PFC (perfluorocarbon) were added to water. Carbon monoxide fermentation by *Butyribacterium methylotrophicum* using Tween and Brij surfactants showed that only the Tween surfactants did not affect bacterial growth, and Tween® 80 showed a higher growth rate, comparatively [29]. Tween surfactants seem to be non-toxic and do not inhibit cell growth [29,31,32]. This approach can maintain the simplicity of STR with the advantage of high mass transfer coefficients, typical of HFMR that are difficult to operate and scale up [33].

In this study, we evaluated the effect of different concentrations of Tween® 80 in ATCC® 2713 culture medium for *Clostridium carboxidovorans* syngas fermentation in 100 mL serum bottles, and the best condition was validated in a stirred tank bioreactor (STBR).

2. Materials and Methods

2.1. Materials

Peptone, sodium pyruvate, tryptone, yeast extract, sodium dithionite, glucose, hemin, L-arginine, and menadione were obtained from Sigma-Aldrich (São Paulo, Brazil). Sodium chloride was obtained from Vetec (Rio de Janeiro, Brazil) and Tween® 80 from Isofar (Rio de Janeiro, Brazil). Syngas was provided by White Martins Praxair Inc. (Rio de Janeiro, Brazil) in a pressurized cylinder with a pre-established composition, based on gas obtained from pyrolysis of urban wastes (MAIM/INNOVA technology, [34]): 25% CO, 43.9% H_2, 10.02% CO_2, 10.05% N_2, and 11.01% CH_4.

2.2. Strain, Culture Medium, and Inoculum Preparation

Clostridium carboxidivorans DSM15243 was obtained from Deutsche Sammlung von Mikroorganismen und Zellkulturen GmbH (DSMZ, Braunschweig, Germany). The cells were activated, stored, and grown under anaerobic condition in 50 mL serum bottles containing 30 mL of TPYarg (Tryptone, Peptone, Yeast extract, and arginine) medium, containing the following composition (per liter): tryptone, 12 g; peptone, 12 g; yeast extract, 7 g; L-arginine, 1.2 g [5]. Syngas was flushed in the liquid phase for 5 min, and then all serum glass bottles were sealed with gas impermeable butyl rubber septum stoppers and aluminum seals. These bottles were autoclaved at 121 °C for 20 min for sterilization, inoculated (0.05 g dry weight cell/L) after cooling, followed by syngas addition in the headspace for 1 min. For both activation and growth, bottles were incubated for 48 and 24 h, respectively, in a horizontal position [35] at 37 °C and 150 rpm in Infors HT-Multitron Pro shaker.

2.3. *Syngas Fermentation in Serum Glass Bottles*

All fermentations were performed in 100 mL serum glass bottles containing 50 mL of ATCC® 2713 (tryptone, 10 g/L; gelatin peptone, 10 g/L; yeast extract, 5 g/L; glucose, 1 g/L; sodium chloride, 5 g/L; L-arginine, 1 g/L; sodium pyruvate, 1 g/L; menadione, 0.5 mg/L and hemin, 5 mg/L) culture medium with different concentrations of Tween® 80 (0, 0.07%, 0.15% and 0.3% *v/v*). After mediums preparation, nitrogen was flushed in the liquid phase for 30 min, and then syngas was flushed in the liquid phase for 5 min. The glass bottles were sealed with gas impermeable butyl rubber septum stoppers and aluminum seals and sterilized in an autoclave at 0.5 atm for 20 min. After sterilization, seed culture was aseptically inoculated in all glass bottles to achieve 0.05 g dry weight of cells/L. Syngas was aseptically added in the headspace, and the bottles were incubated horizontally at 37 °C and 150 rpm in Infors HT Multitron shaker. Cell growth was measured in real-time through non-invasive technology using Cell Growth Quantifier (CGQ) sensors from Aquila Biolabs, collecting biomass concentration data every 30 s. Biomass concentration was measured through an equipment particular optical unit (backscatter), which is converted to optical density at 600 nm (OD_{600}) by a standard curve previously obtained in CGQ. Fermented culture mediums were sampled for high-performance liquid chromatography (HPLC) analysis.

2.4. *Syngas Fermentation in Stirred Tank Bioreactor*

Syngas fermentation was conducted in a 1-L cylindrical stirred tank reactor (TEC-BIO-1.5, Tecnal Scientific Equipment Co., Piracibada, SP, Brazil) with an internal diameter of 9 cm and a maximum working volume of 1.0 L. The production medium (0.75 L) was the ATCC® 2713 medium (tryptone, 10 g/L; gelatin peptone, 10 g/L; yeast extract, 5 g/L; glucose, 1 g/L; sodium chloride, 5 g/L; L-arginine, 1 g/L; sodium pyruvate, 1 g/L; menadione, 0.5 mg/L and hemin, 5 mg/L) with Tween® 80, when its effect was validated. The bioreactor containing the production medium was autoclaved at 121 °C for 20 min, and, after cooling (room temperature), an inert gas (N_2) was flushed in liquid phase for 60 min. Then, syngas was flushed in the liquid for 30 min, and seed culture was inoculated just after under aseptic conditions to an initial cell concentration of 0.05 g dry weight of cells/L. The temperature was set at 37 °C, and medium was recirculated through a peristaltic pump coupled to the bioreactor at each sampling. Samples were withdrawn from the recycle line using an infusion set (Wiltex, 0.64 mm × 19 mm) and a 3.0 mL syringe (BD Plastipak).

Agitation speed was set at 300 rpm with a six-bladed Smith impeller (radial flow impeller 4.0 cm above the vessel bottom) and a six-bladed Rushton impeller (radial flow impeller 11.5 cm above the vessel bottom). Syngas was continuously supplied at the bottom of the bioreactor with a gas flow rate of 0.5 L/min controlled by a rotameter (Matheson, model FM-1000 VIH). The schematic diagram of the stirred tank reactor (STR) is shown in Figure 1.

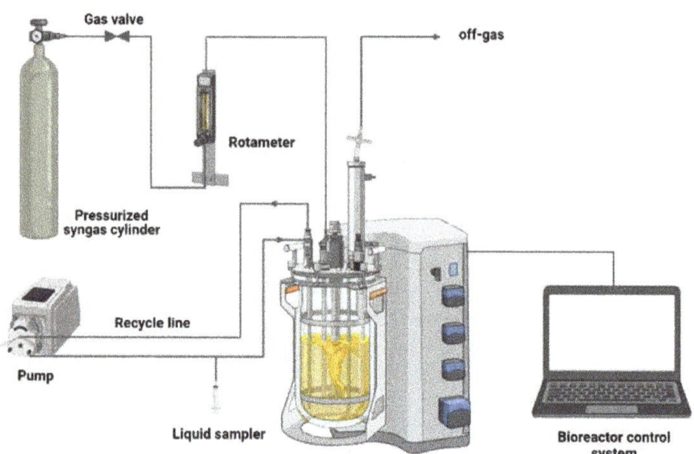

Figure 1. Schematic diagram of STBR used for *C. carboxidivorans* syngas fermentation. Created in Biorender.com (accessed on 13 July 2021).

Cell dry weight concentration (g dry weight of cells/L) was estimated by optical density measurement at 600 nm (OD_{600}). The OD_{600} was measured using a UV-VIS spectrophotometer (Shimadzu UV-1800). Cell dry weight concentration was determined using a standard curve previously obtained.

2.5. Analytical Methods

2.5.1. Dry Weight Cell

The direct dry weight cell was obtained through filtration and drying protocol. Five-milliliter samples of fermented culture medium were filtered using a 0.22 μm membrane and dried to constant weight at 60 °C using an incubator from Memmert IF55. The dry weight cell per liter was calculated using the cell weight and the sample volume.

2.5.2. Metabolites Analyses

Acetic acid, ethanol, and butanol were analyzed by HPLC (High-Performance Liquid Chromatography) from Shimadzu equipped with Aminex® HPX-87 H, 300 × 7.8 mm (Bio-Rad Laboratories Ltd., Mississauga, ON, Canada) column and RI (refractive index) detector (Shimadzu®). The mobile phase was H_2SO_4 5 mM at 0.6 mL/min flow rate. The column temperature was set at 55 °C. 20 μL of centrifuged and filtered samples were automatically injected into the equipment. The quantification of each metabolite was performed through an external standard (ESTD) curve previously obtained at specific retention times (acetic acid, 14.911 min; ethanol, 22.080 min and butanol, 37.074 min).

3. Results and Discussion

3.1. Serum Bottles Fermentation

3.1.1. Cell Growth

Syngas fermentation by *C. carboxidivorans* in serum bottles with ATCC®2713 medium and different Tween®80 concentrations (0, 0.07%, 0.15%, and 0.3% (*v/v*)) was monitored during 120 h. Cell dry weight per liter obtained by the CGQ equipment is depicted in Figure 2. A short lag phase was observed for all medium tested, probably due to the presence of glucose in the culture medium. This carbohydrate is quickly metabolized by *C. carboxidivorans* as a preferential substrate for heterotrophic growth.

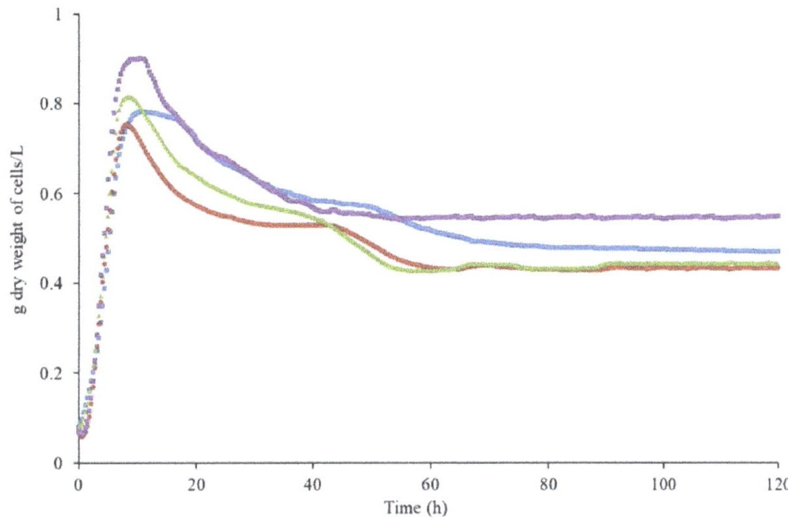

Figure 2. Cell growth of *Clostridium carboxidivorans* during syngas fermentation in the following media: ATCC®2713 (blue square), ATCC®2713 with 0.07% (*v/v*) Tween®80 (red circle), ATCC®2713 with 0.15% (*v/v*) Tween®80 (green triangle), and ATCC®2713 with 0.3% (*v/v*) Tween®80 (purple square).

C. carboxidivorans growth profiles in ATCC®2713 medium with different concentrations of Tween®80 were similar. The maximum biomass concentration for all media was detected after about 10 h due to fast glucose consumption, followed by an accented drop until approximately 50 h. After that, it stabilized at approximately 0.5–0.6 g/L. This growth profile has been shown for syngas batch fermentations with *C. carboxidivorans* in other works [22,36,37]. Since syngas is not continuously fed in serum bottles, cell growth reaches the stationary phase, when there is a balance between growth and death of cells. Fernández-Naveira et al. [36] showed that ethanol causes inhibition of cell growth. So, without substrate supply and with a toxic compound being produced, it is possible that cell death overlaps cell growth, and autolysis may occur, reducing turbidity [38].

Higher biomass concentration was obtained in medium with 0.3% Tween® 80 (0.9 g dry weight of cells/L after 10 h), 15% more compared to pure ATCC® 2713 medium. After 120 h of fermentation, the medium with 0.3% Tween® 80 also showed the highest final biomass concentration (0.52 g dry weight of cells/L) among all other conditions. This might be related to the beneficial effect of Tween® 80 in CO and CO_2 assessment after glucose exhaustion.

The specific growth rates of the fermentations with different Tween® 80 concentrations were also very similar (Table 1), with a higher value for the medium with 0.3% Tween® 80.

Table 1. Specific *Clostridium carboxidivorans* growth rate in ATCC®2713 medium with different concentrations of Tween®80.

Culture Medium	μ (h^{-1})
ATCC® 2713	0.310 ± 0.13
ATCC® 2713 + 0.07% Tween® 80	0.359 ± 0.12
ATCC® 2713 + 0.15% Tween® 80	0.350 ± 0.06
ATCC® 2713 + 0.3% Tween® 80	0.414 ± 0.04

Considering that Tween®80 can physically interact with dispersed bubbles in the liquid, causing emulsion formation, we did not know if CGQ sensors would generate wrong OD_{600} measurements due to the different concentrations of surfactant in the medium, leading to false biomass concentration data. To confirm the final OD measured by CGQ sensors, cell dry weight direct determination was performed after 120 h of fermentation. The results obtained showed that the OD_{600} measured by CGQ sensors were reliable since the conditions that led to lower or higher cell concentration were the same with both measurements, as shown in Table 2.

Table 2. *Clostridium carboxidivorans* cell concentration (g dry weight of cells/L) after 120 h of syngas fermentation obtained by CGQ measurement and calculated by direct cell dry weight measurement for different media.

Culture Medium	CGQ Measurement	Direct Cell Dry Weight
ATCC® 2713	0.477 ± 0.023	0.433 ± 0.156
ATCC® 2713 + 0.07% Tween® 80	0.390 ± 0.003	0.413 ± 0.065
ATCC® 2713 + 0.15% Tween® 80	0.475 ± 0.054	0.443 ± 0.075
ATCC® 2713 + 0.3% Tween® 80	0.520 ± 0.019	0.477 ± 0.015

3.1.2. Metabolites Production

Gaseous substrates are assimilated by acetogenic bacteria through the Wood-Ljungdahl pathway producing acetyl-CoA, an important intermediate to acids and alcohols production. Most acetogenics show a defined pattern of metabolites production, in which acids are produced in a first stage called acetogenesis, followed by the conversion of these acids into the respective alcohols, called solventogenesis. Ethanol production by *Clostridium carboxidivorans* has its particularities. It can be produced either directly from acetyl-CoA in a two-step reaction via acetaldehyde, requiring 4 molecules of NADH, or via acetate and subsequent reduction to acetaldehyde, producing 1 molecule of ATP and consuming 4 NADH per molecule of ethanol produced. Therefore, acetic acid produced during syngas fermentation by *Clostridium carboxidivorans* is an important indicator of potential ethanol production [39].

In our previous studies, we have identified that ethanol production by this strain in ATCC® 2713 medium increases gradually until 24 h, then stabilizes. In other culture mediums, it starts to increase again after 70 h [5] So, we decided to sample serum bottle fermentations at strategic points (24 h—the first peak, 96 h—the second peak, and then 120 h, to verify final stabilization) to avoid volume reduction. Higher acetic acid concentration was obtained in pure ATCC®2713 medium (4.44 g/L) after 120 h of syngas fermentation, which is 85% more than the amount obtained in ATCC®2713 medium containing Tween®80 (2.3 g/L) (Figure 3).

Despite this higher acetic acid concentration in the medium without the surfactant, higher ethanol production was detected after 96 h of syngas fermentation in ATCC®2713 medium with 0.15% (*v/v*) Tween®80 (1.90 g/L) (Figure 4). This value is 3.2 fold higher than that obtained using pure ATCC®2713 medium (0.58 g/L) at the same fermentation time. After 120 h, ethanol concentration using 0.15% and 0.3% Tween®80 were 1.79 g/L and 1.83 g/L, respectively, representing an increase of approximately 200% compared to pure ATCC®2713 medium (0.58 g ethanol/L). The addition of Tween®80 in ATCC®2713 medium resulted in less acetic acid accumulation and higher ethanol production, probably due to the greater availability of inorganic carbon (CO and CO_2) and protons (NADH) generated by important Wood-Ljungdahl enzymes as hydrogenases (HYA) and carbon monoxide dehydrogenases (CODH). The surfactant could improve carbon monoxide (CO) and carbon dioxide (CO_2) availability, resulting not only in more carbon fixation in the pathway but also more proton generation to be consumed in the following steps.

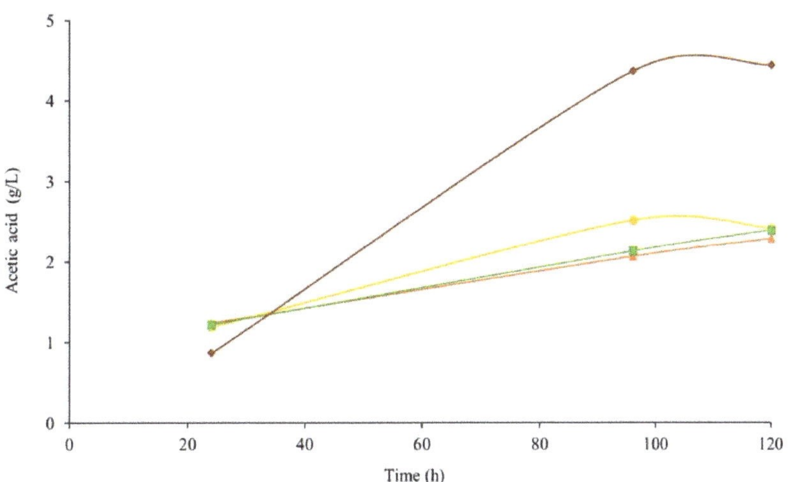

Figure 3. Acetic acid production by *Clostridium carboxidivorans* during syngas fermentation in ATCC®2713 (red diamond), ATCC®2713 with 0.07% (*v/v*) Tween®80 (orange triangle), ATCC®2713 with 0.15% (*v/v*) Tween®80 (yellow circle), and ATCC®2713 with 0.3% (*v/v*) Tween®80 (green square).

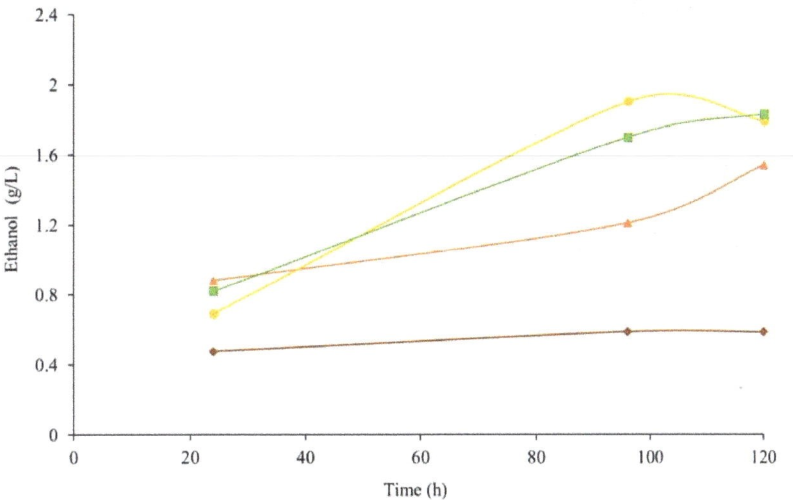

Figure 4. Ethanol production by *Clostridium carboxidivorans* during syngas fermentation in ATCC®2713 (red diamond), ATCC®2713 with 0.07% (*v/v*) Tween®80 (orange triangle), ATCC®2713 with 0.15% (*v/v*) Tween®80 (yellow circle), and ATCC®2713 with 0.3% (*v/v*) Tween®80 (green square).

The highest ethanol productivity was also obtained using 0.15% Tween®80 (*v/v*), which was 0.02 g/L.h after 96 h of syngas fermentation. The critical micelle concentration (CMC) of Tween®80 as informed by the supplier (Sigma-Aldrich) is 0.012 mM, which is equivalent to 0.15% (*v/v*). At CMC, the lowest superficial tension and, therefore, the largest interfacial area between gas and aqueous phase is attained. Probably, better results of ethanol production using 0.15% Tween®80 are related to the increase in mass transfer of the substrates (CO and CO_2) from syngas to the aqueous phase for microbial assimilation. There was no butanol production during 120 h of syngas fermentation by *Clostridium carboxidivorans* in serum bottles.

Since the use of ATCC®2713 medium with 0.15% (*v/v*) Tween®80 Led to higher ethanol production and represented lower cost compared to the medium with 0.30% Tween®80 (*v/v*), we decided to use 0.15% Tween®80 for the validation experiment in stirred tank bioreactor.

3.2. Stirred Tank Bioreactor Fermentation

3.2.1. Cell Growth

C. carboxidivorans growth in ATCC®2713 medium with 0.15% (*v/v*) Tween®80 was monitored during 120 h of syngas fermentation in STBR (Figure 5). As observed in serum bottles, the maximum biomass concentration was also obtained at the beginning of the experiment, followed by a reduction in cell concentration. However, the exponential growth phase was longer in bioreactor fermentation, taking more than 20 h, and the decrease in cell concentration was less deep.

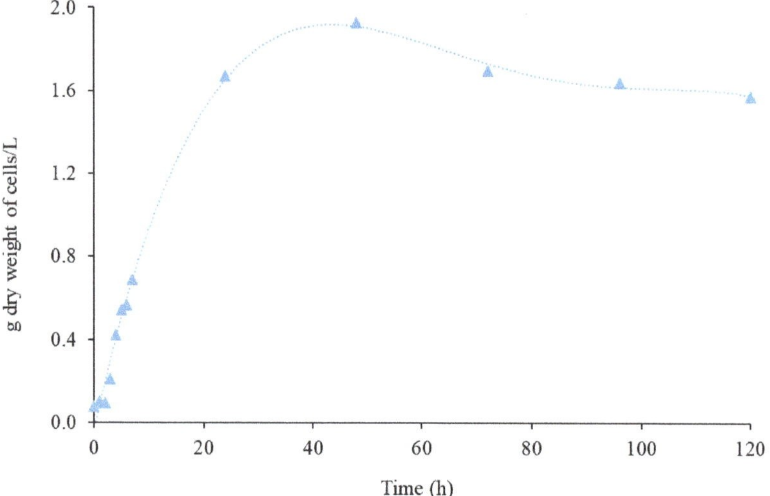

Figure 5. *Clostridium carboxidivorans* growth during syngas fermentation in ATCC®2713 medium with 0.15% (*v/v*) Tween®80 in stirred tank bioreactor.

The lag phase lasted less than 2 h, in accordance to observed in serum bottles fermentation experiments. The maximum biomass concentration and the specific growth rate were 1.93 g dry weight of cells/L and 0.377 h^{-1}, respectively. The biomass production after 120 h of fermentation in ATCC®2713 with 0.15% Tween®80 (1.67 g dry weight of cells/L) was 106% higher than the maximum biomass achieved with the same medium in serum bottle fermentation (0.81 g dry weight of cells/L). This might be related to the greater availability of gaseous substrates in bioreactor configuration since the syngas was fed continuously at a low flow rate, and the better system agitation which probably promoted an increase in mass transfer.

3.2.2. Metabolites Production

Unlike most acetogenic bacteria, for which solvents are only detected after the acidogenic phase, during the stationary growth phase [40], ethanol production started along with cell growth for *C. carboxidivorans* syngas fermentation (Figure 6). According to Shen et al. [41], despite ethanol being considered a non-growth-associated metabolite in *C. carboxidivorans* syngas fermentation, there is evidence which indicates that it is produced in both growth- and non-growth-associated phases. This can happen because the bacteria can use glucose for growth and the Wood-Ljungdahl pathway to metabolize CO as a carbon source for ethanol production simultaneously if the diffusion of the gaseous substrates is efficient [41]. Therefore, it is possible that the contribution of Tween®80 to mass transfer induces ethanol production in the exponential growth phase, which continues to increase even after cell growth has stopped.

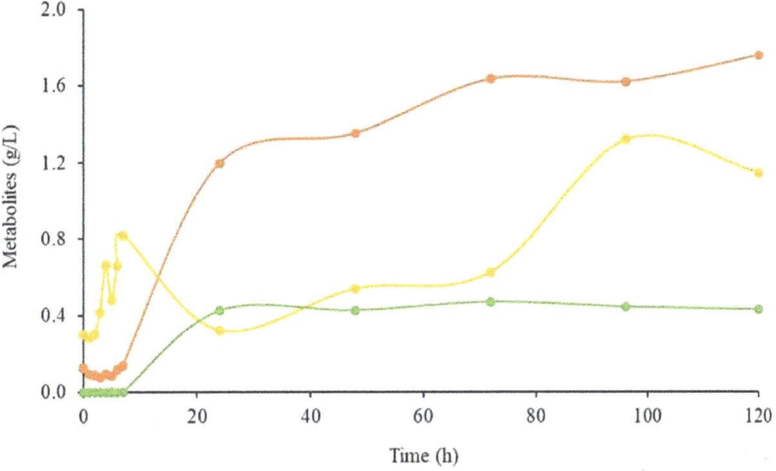

Figure 6. Acetic acid (yellow), ethanol (orange), and butanol (green) production during *Clostridium carboxidivorans* syngas fermentation in ATCC®2713 with 0.15% (*v/v*) Tween®80 in bioreactor.

The maximum ethanol concentration obtained in STBR after 120 h in ATCC®2713 medium with 0.15% Tween®80 was 1.76 g/L. Although biomass production in ATCC®2713 with 0.15% Tween®80 in the bioreactor was much higher (140%) than the obtained in serum bottle fermentation, an increase in ethanol production in the bioreactor experiment was not observed. However, productivity was higher because, at 24 h of fermentation, 1.2 g/L of ethanol had already been produced, resulting in 0.050 g/L.h, while only 0.69 g/L had been produced in a serum bottle, which resulted in a 44% lower productivity (0.028 g/L.h).

The maximum acetic acid concentration was obtained after 96 h in STBR (1.32 g/L), 43% lower than the obtained in serum bottles fermentation (2.3 g/L). However, in STBR, butanol production was detected (0.43 g/L after 24 h), which was null in serum bottles experiments

The increase in biomass, ethanol, and butanol productions is probably a result of an enhancement in the gas-liquid mass transfer coefficient due to the bioreactor configuration as well as the addition of Tween® 80 in the culture medium as observed in the serum bottles fermentation. Studies have concluded that low concentrations of surface-active additives can affect gas-liquid mass transfer parameters such as the volumetric mass transfer coefficient (kLa) [42,43]. Belo et al. [44] studied the influence of Tween®80 in hydrodynamic parameters and mass transfer of carbon dioxide (CO_2) in aqueous solution. The presence of Tween® 80 generated an important increase in the gas-liquid interfacial area caused by a decrease in the bubble diameter. As reported by Coelho et al. [20], the addition of 0.15% (*v/v*) of Tween®80 in water resulted in an increase of 120% in the carbon monoxide (CO) kLa. This result was obtained using the same bioreactor design and operational conditions as described herein.

Regarding solvent production during cell growth, Shen et al. [12] also reported a similar mixotrophic scenario with *C. carboxidivorans* using a monolithic biofilm reactor in which about 1.5 g/L of biomass was produced after 48 h of fermentation. In the mentioned study, a mineral medium with 10 g/L of fructose and a synthetic syngas (20% CO, 5% H_2, 15% CO_2, 60% N_2) was used, which explains the fast biomass production when compared to processes that use only inorganic carbon as substrate. In another study using CO and

CO_2 as carbon sources in a batch fermentation with continuous syngas feed, 0.42 g/L of *C. carboxidivorans* biomass was achieved in 750 h of processing [45].

4. Discussion

The results obtained in the present investigation were compared with similar recent studies reported in the literature (Table 3). Ethanol productivity using 0.15% Tween®80 was 0.050 g/L.h, a superior value than was obtained by Doll et al. [14] using two CSTR in series after 200 h of fermentation, under similar conditions. Fernández-Naveira et al. [36] reported an autotrophic fermentation process with pure carbon monoxide (CO) as substrate in a CSTR with 1.2 L of working volume. After 245 h of fermentation, ethanol concentration was 5.6 g/L, representing lower ethanol productivity than obtained in the present study. Shen et al. [26] reported the highest ethanol productivity using a horizontal rotating packed bed bioreactor (0.279 g/L.h) with pressurized headspace at 29.7 psi, which requires high energy consumption and special equipment to support high pressures.

Table 3. Ethanol production from syngas fermentation in different systems.

Biocatalysts	Reactor [a]	$CO:H_2:CO_2:N_2:CH_4$	Ethanol (g/L)	Ethanol Productivity (g/L.h)	Fermentation Period (h)	References
Clostridium ragsdalei	CSTR	40:30:30:0:0	13.2	0.044	300	[11]
	TBR	38:28:28:5:0	5.7	0.003	1662	[46]
	TBR	38:28:28:5:0	13.2	0.158	84	[25]
Clostridium carboxidivorans	CSTR	20:10:20:50:0	2.7	0.008	340	[45]
	CSTR	100:0:0:0:0	5.6	0.023	245	[36]
	CSTR	20:10:20:50:0	2.34	0.011	210	[45]
	CSTR	30:20:10:40:0	5.9	0.032	185	[47]
	CSTR	20:5:15:60:0	2.1	0.082	25	[26]
	h-RPB	20:5:15:60:0	7	0.279	25	[26]
	CSTR	80:0:20:0:0	6	0.03	200	[14]
	STBR	25:44:10:10:11	1.2	0.050	24	This study
Clostridium ljungdahli	CSTR	65:30:5:0:0	3.8	0.005	730	[48]
	HFM	25:15:25:40:0	1.09	0.005	216	[49]

[a] CSTR: continuous stirred tank reactor; TBR: trickle bed reactor; BCR: bubble column reactor; h-RPB: horizontal rotating packed bed biofilm reactor; HFM: hollow fiber membrane; STBR: stirred tank bioreactor.

The highest ethanol productivities from syngas fermentations found in the literature are generally related to sophisticated bioreactor designs (TBR [25], h-RPB [26]), which are difficult to scale up and operate, especially because of the preliminary step of film formation needed for these bioreactors. Considering that in this study, we proposed a continuous syngas feed at a very low flow rate and a simple reactor configuration, with no pressure or pH control. The ethanol productivity obtained was promising compared to studies reported in the literature using similar substrates, operational conditions, and microorganisms. Tween®80 is a relatively cheap input (US $30/L at MilliporeSigma website—https://www.sigmaaldrich.com/US/en, accessed on 30 July 2021), and a small amount is needed to increase ethanol production (0.0015 L per liter of culture medium). Moreover, it is suitable for microbial culture, without toxicity to bacterial cells [29]. Although Bredweel et al. [29] have tested Tween®80 in carbon monoxide fermentation, this is the first report in the literature concerning the effect of different concentrations of Tween®80 in syngas fermentations and its validation in bioreactor scale. Tween®80 can cause foam depending on concentration, medium composition, mechanical agitation, and gas flow rate, and serum bottles do not evaluate this problem. We have validated the use of Tween®80 in syngas fermentation in a bioreactor since we detected product formation and no foam was observed. Besides, the gas composition used herein is much more realistic when compared to syngas obtained from waste material pyrolysis [50]. Further research is needed to evaluate syngas composition after fermentation to verify the variability of CO, CO_2, and H_2 consumption with and without Tween®80 to provide in-depth understanding of its effect.

5. Conclusions

The effect caused by the addition of Tween®80 to ATCC®2713 medium was evidenced by an increase in biomass and ethanol production during *Clostridium carboxidivorans* syngas fermentation in serum bottles and validated in a stirred tank bioreactor. The presence of this surfactant probably led to the reduction of bubble size, increasing the gas-liquid interfacial area, which resulted in the increase of CO and CO_2 mass transfer coefficients. The biomass and ethanol productions increased by 15% and 200% using Tween®80 in the culture medium, respectively, compared to pure ATCC®2713 medium. In a bioreactor, 106% more biomass was produced compared to serum bottle fermentation, but the same ethanol concentration was achieved.

Author Contributions: C.B., T.F., and P.A. conceived and planned the experiments. C.B., A.B., M.B., and M.d.N.-C. carried out the experiments. C.B. and M.B. were responsible for analyses. C.B. wrote the manuscript with support from T.F. and P.A. T.F. and P.A. supervised the project. All authors provided critical feedback and helped shape the research, analysis, and manuscript. All authors have read and agreed to the published version of the manuscript.

Funding: This research was funded by Fundação Carlos Chagas Filho de Amparo à Pesquisa do Estado do Rio de Janeiro, grant number FAPERJ-E-26/010.002984/2014, and Conselho Nacional de Desenvolvimento Científico e Tecnológico, grant number CNPq–PQ/2 308626/2019-2.

Institutional Review Board Statement: Not applicable.

Informed Consent Statement: Not applicable.

Acknowledgments: The group would like to acknowledge Nei Pereira Jr. and Roberta Ribeiro for intellectual guidance. We thank SENAI's Innovation Institute for Biosynthetic for the use of Cell Growth Quantifier (CGQ) sensors from Aquila Biolabs for cell growth measurement in real-time through non-invasive technology.

Conflicts of Interest: The authors declare no conflict of interest.

References

1. EIA with Projections to 2050. Available online: https://www.eia.gov/outlooks/aeo/ (accessed on 12 August 2021).
2. Gildemyn, S.; Molitor, B.; Usack, J.G.; Nguyen, M.; Rabaey, K.; Angenent, L.T. Upgrading syngas fermentation effluent using Clostridium kluyveri in a continuous fermentation. *Biotechnol. Biofuels* **2017**, *10*, 1–15. [CrossRef] [PubMed]
3. Mohammadi, M.; Najafpour, G.D.; Younesi, H.; Lahijani, P.; Uzir, M.H.; Mohamed, A.R. Bioconversion of synthesis gas to second generation biofuels: A review. *Renew. Sustain. Energy Rev.* **2011**, *15*, 4255–4273. [CrossRef]
4. Ukpong, M.N.; Atiyeh, H.K.; De Lorme, M.J.M.; Liu, K.; Zhu, X.; Tanner, R.S.; Wilkins, M.R.; Stevenson, B.S. Physiological response of Clostridium carboxidivorans during conversion of synthesis gas to solvents in a gas-fed bioreactor. *Biotechnol. Bioeng.* **2012**, *109*, 2720–2728. [CrossRef]
5. Benevenuti, C.; Botelho, A.; Ribeiro, R.; Branco, M.; Pereira, A.; Vieira, A.C.; Ferreira, T.; Amaral, P. Experimental Design to Improve Cell Growth and Ethanol Production in Syngas Fermentation by Clostridium carboxidivorans. *Catalysts* **2020**, *10*, 59. [CrossRef]
6. Munasinghe, P.C.; Khanal, S.K. Biomass-derived syngas fermentation into biofuels. *Biofuels* **2011**, *101*, 79–98.
7. Sun, Y.; Cheng, J. Hydrolysis of lignocellulosic materials for ethanol production: A review. *Bioresour. Technol.* **2002**, *83*, 1–11. [CrossRef]
8. Boboescu, I.Z.; Gélinas, M.; Beigbeder, J.B.; Lavoie, J.M. A two-step optimization strategy for 2nd generation ethanol production using softwood hemicellulosic hydrolysate as fermentation substrate. *Bioresour. Technol.* **2017**, *244*, 708–716. [CrossRef] [PubMed]
9. Datar, R.P.; Shenkman, R.M.; Cateni, B.G.; Huhnke, R.L.; Lewis, R.S. Fermentation of biomass-generated producer gas to ethanol. *Biotechnol. Bioeng.* **2004**, *86*, 587–594. [CrossRef]
10. Lewis, R.S.; Tanner, R.S.; Huhnke, R.L. Huhnke Indirect or Direct Fermentation of Biomass to Fuel Alcohol. U.S. Patent No. 11/441,392, 25 May 2006.
11. Sun, X.; Atiyeh, H.K.; Zhang, H.; Tanner, R.S.; Huhnke, R.L. Enhanced ethanol production from syngas by Clostridium ragsdalei in continuous stirred tank reactor using medium with poultry litter biochar. *Appl. Energy* **2019**, *236*, 1269–1279. [CrossRef]
12. Shen, Y.; Brown, R.; Wen, Z. Syngas fermentation of Clostridium carboxidivoran P7 in a hollow fiber membrane biofilm reactor: Evaluating the mass transfer coefficient and ethanol production performance. *Biochem. Eng. J.* **2014**, *85*, 21–29. [CrossRef]

13. Liberato, V.; Benevenuti, C.; Coelho, F.; Botelho, A.; Amaral, P.; Pereira, N.; Ferreira, T. Chemicals production in a biorefinery context. *Catalysts* **2019**, *9*, 962. [CrossRef]
14. Doll, K.; Rückel, A.; Kämpf, P.; Wende, M.; Weuster-Botz, D. Two stirred-tank bioreactors in series enable continuous production of alcohols from carbon monoxide with Clostridium carboxidivorans. *Bioprocess. Biosyst. Eng.* **2018**, *41*, 1403–1416. [CrossRef]
15. Dürre, P.; Eikmanns, B.J. C1-carbon sources for chemical and fuel production by microbial gas fermentation. *Curr. Opin. Biotechnol.* **2015**, *35*, 63–72. [CrossRef]
16. Kennes, D.; Abubackar, H.N.; Diaz, M.; Veiga, M.C.; Kennes, C. Bioethanol production from biomass: Carbohydrate vs. syngas fermentation. *J. Chem. Technol. Biotechnol.* **2016**, *91*, 304–317. [CrossRef]
17. Zhang, J.; Taylor, S.; Wang, Y. Effects of end products on fermentation profiles in Clostridium carboxidivorans P7 for syngas fermentation. *Bioresour. Technol.* **2016**, *218*, 1055–1063. [CrossRef] [PubMed]
18. Kundiyana, D.K.; Huhnke, R.L.; Wilkins, M.R. Syngas fermentation in a 100-L pilot scale fermentor: Design and process considerations. *J. Biosci. Bioeng.* **2010**, *109*, 492–498. [CrossRef]
19. Richter, H.; Martin, M.E.; Angenent, L.T. A two-stage continuous fermentation system for conversion of syngas into ethanol. *Energies* **2013**, *6*, 3987–4000. [CrossRef]
20. Coelho, F.M.B.; Botelho, A.M.; Ivo, O.F.; Amaral, P.F.F.; Ferreira, T.F. Volumetric mass transfer coefficient for carbon monoxide in a dual impeller stirred tank reactor considering a perfluorocarbon–water mixture as liquid phase. *Chem. Eng. Res. Des.* **2019**, 160–169. [CrossRef]
21. Phillips, J.R.; Huhnke, R.L.; Atiyeh, H.K. Syngas fermentation: A microbial conversion process of gaseous substrates to various products. *Fermentation* **2017**, *3*, 28. [CrossRef]
22. Fernández-Naveira, Á.; Veiga, M.C.; Kennes, C. Effect of salinity on C1-gas fermentation by Clostridium carboxidivorans producing acids and alcohols. *AMB Express* **2019**, *9*, 1–11. [CrossRef] [PubMed]
23. Shen, Y.; Brown, R.; Wen, Z. Enhancing mass transfer and ethanol production in syngas fermentation of Clostridium carboxidivorans P7 through a monolithic biofilm reactor. *Appl. Energy* **2014**, *136*, 68–76. [CrossRef]
24. Coelho, F.M.B.; Nele, M.; Ribeiro, R.R.; Ferreira, T.F.; Amaral, P.F.F. Clostridium carboxidivorans' surface characterization using contact angle measurement (CAM). In Proceedings of the IConBM2016 2nd International Conference on BIOMASS, Sicily, Italy, 19–22 June 2016; Volume 50, pp. 277–282.
25. Devarapalli, M.; Lewis, R.S.; Atiyeh, H.K. Continuous ethanol production from synthesis gas by Clostridium ragsdalei in a trickle-bed reactor. *Fermentation* **2017**, *3*, 23. [CrossRef]
26. Shen, Y.; Brown, R.C.; Wen, Z. Syngas fermentation by Clostridium carboxidivorans P7 in a horizontal rotating packed bed biofilm reactor with enhanced ethanol production. *Appl. Energy* **2017**, *187*, 585–594. [CrossRef]
27. Aldric, J.M.; Gillet, S.; Delvigne, F.; Blecker, C.; Lebeau, F.; Wathelet, J.P.; Manigat, G.; Thonart, P. Effect of surfactants and biomass on the gas / liquid mass transfer in an aqueous-silicone oil two-phase partitioning bioreactor using Rhodococcus erythropolis T902. 1 to remove VOCs from gaseous effluents. *J. Chem. Technol. Biotechnol.* **2009**, *84*, 1274–1283. [CrossRef]
28. Rehman, F.; Medley, G.J.D.; Bandulasena, H.; Zimmerman, W.B.J. Fluidic oscillator-mediated microbubble generation to provide cost effective mass transfer and mixing ef fi ciency to the wastewater treat- ment plants. *Environ. Res.* **2015**, *137*, 32–39. [CrossRef] [PubMed]
29. Bredwell, M.D.; Telgenhoff, I.M.D.; Barnard, I.S.; Worden, R.M.I. Effect of surfactants on carbon monoxide fermentations by butyribacterium methylotroph icum. *Appl. Biochem. Biotechnol.* **1997**, *63–65*, 637–647. [CrossRef]
30. Myung, J.; Kim, M.; Pan, M.; Criddle, C.S.; Tang, S.K.Y. Bioresource technology low energy emulsion-based fermentation enabling accelerated methane mass transfer and growth of poly (3-hydroxybutyrate) -accumulating methanotrophs. *Bioresour. Technol.* **2016**, *207*, 302–307. [CrossRef] [PubMed]
31. Nielsen, C.K.; Kjems, J.; Mygind, T.; Snabe, T.; Meyer, R.L. Effects of tween 80 on growth and biofilm formation in laboratory media. *Front. Microbiol.* **2016**, *7*, 1–10. [CrossRef] [PubMed]
32. Eskandani, M.; Hamishehkar, H.; Dolatabadi, J.E.N. Cyto/genotoxicity study of polyoxyethylene (20) sorbitan monolaurate (tween 20). *DNA Cell Biol.* **2013**, *32*, 498–503. [CrossRef] [PubMed]
33. Yasin, M.; Park, S.; Jeong, Y.; Lee, E.Y.; Lee, J.; Chang, I.S. Effect of internal pressure and gas/liquid interface area on the CO mass transfer coefficient using hollow fibre membranes as a high mass transfer gas diffusing system for microbial syngas fermentation. *Bioresour. Technol.* **2014**, *169*, 637–643. [CrossRef]
34. Rodrigues, V.; Alberto, C.; Cosenza, N.; Barros, C.F.; Krykhtine, F.; Netto, E.; Fortes, S.; Alberto, C.; Cosenza, N.; Barros, C.F. Tratamento de resíduos sólidos urbanos e produção de energia: Análise de legislação para viabilidade econômica de soluçoes. In Proceedings of the XI Simpósio de Excelência em Gestão e Tecnologia—SEGeT, Rio de Janeiro, Brazil, 22–24 October 2014.
35. Ribeiro, R.R.; Coelho, F.; Ferreira, T.F.; Amaral, P.F.F. A new strategy for acetogenic bacteriacell growth and metabolites production using syngas in lab scale. *IOSR J. Biotechnol. Biochem.* **2017**, *03*, 27–30. [CrossRef]
36. Fernández-Naveira, Á.; Abubackar, H.N.; Veiga, M.C.; Kennes, C. Carbon monoxide bioconversion to butanol-ethanol by Clostridium carboxidivorans: Kinetics and toxicity of alcohols. *Appl. Microbiol. Biotechnol.* **2016**, *100*, 4231–4240. [CrossRef] [PubMed]
37. Fernández-Naveira, Á.; Abubackar, H.N.; Veiga, M.C.; Kennes, C. Efficient butanol-ethanol (B-E) production from carbon monoxide fermentation by Clostridium carboxidivorans. *Appl. Microbiol. Biotechnol.* **2016**, *100*, 3361–3370. [CrossRef]

38. Ogata, S. Morphological changes during conversion of clostridium saccharoperbutylacetonicum to protoplasts by sucrose-induced autolysis. *Microbiol. Immunol.* **1980**, *24*, 393–400. [CrossRef]
39. Daniell, J.; Köpke, M.; Simpson, S.D. Commercial biomass syngas fermentation. *Energies* **2012**, *5*, 5372–5417. [CrossRef]
40. Fernández-Naveira, Á.; Veiga, M.C.; Kennes, C. Glucose bioconversion profile in the syngas-metabolizing species Clostridium carboxidivorans. *Bioresour. Technol.* **2017**, *244*, 552–559. [CrossRef]
41. Shen, S.; Gu, Y.; Chai, C.; Jiang, W.; Zhuang, Y.; Wang, Y. Enhanced alcohol titre and ratio in carbon monoxide-rich off-gas fermentation of Clostridium carboxidivorans through combination of trace metals optimization with variable-temperature cultivation. *Bioresour. Technol.* **2017**, *239*, 236–243. [CrossRef]
42. Vasconcelos, J.M.T.; Rodrigues, J.M.L.; Orvalho, S.C.P.; Alves, S.S.; Mendes, R.L.; Reis, A. Effect of contaminants on mass transfer coefficients in bubble column and airlift contactors. *Chem. Eng. Sci.* **2003**, *58*, 1431–1440. [CrossRef]
43. Blanco, A.; García-Abuín, A.; Gómez-Díaz, D.; Navaza, J.M. Hydrodynamic and absorption studies of carbon dioxide absorption in aqueous amide solutions using a bubble column contactor. *Brazilian J. Chem. Eng.* **2013**, *30*, 801–809. [CrossRef]
44. Belo, I.; García-Abuín, A.; Gómez-Díaz, D.; Navaza, J.M.J.M.; Vidal-Tato, I. Effect of tween 80 on bubble size and mass transfer in a bubble contactor. *Chem. Eng. Technol.* **2011**, *34*, 1790–1796. [CrossRef]
45. Fernández-Naveira, Á.; Veiga, M.C.; Kennes, C. Effect of pH control on the anaerobic H-B-E fermentation of syngas in bioreactors. *J. Chem. Technol. Biotechnol.* **2017**, *92*, 1178–1185. [CrossRef]
46. Devarapalli, M.; Atiyeh, H.K.; Phillips, J.R.; Lewis, R.S.; Huhnke, R.L. Ethanol production during semi-continuous syngas fermentation in a trickle bed reactor using Clostridium ragsdalei. *Bioresour. Technol.* **2016**, *209*, 56–65. [CrossRef] [PubMed]
47. Fernández-naveira, Á.; Veiga, M.C.; Kennes, C. Selective anaerobic fermentation of syngas into either C_2–C_6 organic acids or ethanol and higher alcohols. *Bioresour. Technol.* **2019**, *280*, 387–395. [CrossRef] [PubMed]
48. Acharya, B.; Dutta, A.; Basu, P. Ethanol production by syngas fermentation in a continuous stirred tank bioreactor using Clostridium ljungdahlii. *Biofuels* **2019**, *10*, 221–237. [CrossRef]
49. Anggraini, I.D.; Kery, K.; Bandung, P.N.; Kresnowati, P.; Purwadi, R. Bioethanol production via syngas fermentation of clostridium ljungdahlii in a hollow fiber membrane supported bioreactor. *Chem. Eng.* **2019**, *10*, 481–490. [CrossRef]
50. Das, P.; Chandramohan, V.P.; Mathimani, T.; Pugazhendhi, A. Recent advances in thermochemical methods for the conversion of algal biomass to energy. *Sci. Total Environ.* **2021**, *766*, 144608. [CrossRef]

fermentation

MDPI

Article

Multi-Objective Sustainability Optimization of Biomass Residues to Ethanol via Gasification and Syngas Fermentation: Trade-Offs between Profitability, Energy Efficiency, and Carbon Emissions

Elisa M. de Medeiros [1,2], Henk Noorman [1,3], Rubens Maciel Filho [2] and John A. Posada [1,*]

1 Department of Biotechnology, Delft University of Technology, van der Maasweg 9, 2629 ZH Delft, The Netherlands; demedeiros.elisa@gmail.com (E.M.d.M.); henk.noorman@dsm.com (H.N.)
2 School of Chemical Engineering, University of Campinas, Av. Albert Einstein 500, Campinas 13084-852, Brazil; rmaciel@unicamp.br
3 DSM Biotechnology Center, P.O. Box 1, 2600 MA Delft, The Netherlands
* Correspondence: J.A.PosadaDuque@tudelft.nl

Abstract: This work presents a strategy for optimizing the production process of ethanol via integrated gasification and syngas fermentation, a conversion platform of growing interest for its contribution to carbon recycling. The objective functions (minimum ethanol selling price (MESP), energy efficiency, and carbon footprint) were evaluated for the combinations of different input variables in models of biomass gasification, energy production from syngas, fermentation, and ethanol distillation, and a multi-objective genetic algorithm was employed for the optimization of the integrated process. Two types of waste feedstocks were considered, wood residues and sugarcane bagasse, with the former leading to lower MESP and a carbon footprint of 0.93 USD/L and 3 g CO_2eq/MJ compared to 1.00 USD/L and 10 g CO_2eq/MJ for sugarcane bagasse. The energy efficiency was found to be 32% in both cases. An uncertainty analysis was conducted to determine critical decision variables, which were found to be the gasification zone temperature, the split fraction of the unreformed syngas sent to the combustion chamber, the dilution rate, and the gas residence time in the bioreactor. Apart from the abovementioned objectives, other aspects such as water footprint, ethanol yield, and energy self-sufficiency were also discussed.

Keywords: gasification; multi-objective optimization; bioethanol; syngas fermentation; modeling; sustainability

Citation: de Medeiros, E.M.; Noorman, H.; Maciel Filho, R.; Posada, J.A. Multi-Objective Sustainability Optimization of Biomass Residues to Ethanol via Gasification and Syngas Fermentation: Trade-Offs between Profitability, Energy Efficiency, and Carbon Emissions. *Fermentation* **2021**, 7, 201. https://doi.org/10.3390/fermentation7040201

Academic Editor: Alessia Tropea

Received: 23 July 2021
Accepted: 21 September 2021
Published: 23 September 2021

1. Introduction

In recent years, significant progress has been achieved in the field of biobased production, especially regarding ethanol production from lignocellulosic materials such as sugarcane bagasse, corn stover, and wood residues—the so-called 2nd-generation (2G) ethanol [1]. However, 2G ethanol is still hardly competitive with sugar-based or 1st-generation ethanol, and despite the existence of several commercial-scale plants based on 2G technologies, the actual throughput remains mostly below the installed capacity [2]; most of these production routes are based on hydrolysis and sugar fermentation [3]. In contrast, gasification-based pathways are considered promising due to the alleged feedstock flexibility and the potential to convert all parts of the biomass (including lignin) [4].

Gasification has a long history of applications with different purposes (heat, electricity, chemicals, or fuels), but most large-scale gasifiers operate with coal, while biomass gasification has been applied on a far more limited scale and has mostly been used for heat and power generation as an alternative to natural gas and biomass combustion [5]. Regarding biomass-to-fuel via gasification, there are currently only eight facilities with a technology readiness level (TRL) above six that are operational or under construction/commissioning, with five of them targeting ethanol production (two operational) and only one at a commercial scale: the Enerkem plant in Alberta, Canada, which converts municipal solid waste (MSW) to syngas, with further chemical conversion to ethanol and other chemicals [6].

Syngas can also be converted to ethanol via fermentation (i.e., using microbes instead of chemical catalysts). Among the abovementioned projects, only one (by LanzaTech/Aemetis) is projected for use in a gasification–fermentation route. This plant, which is still expected to begin construction, will first convert agricultural waste to syngas via plasma gasification [6], a relatively new technology with the ability to convert nearly any type of carbonaceous material yet at still high costs and limited process understanding [7]. Gas fermentation technology is a challenge despite significant developments in the past few years, which include the construction and operation of several demonstration plants to convert basic oxygen furnace (BOF) gas, a CO-rich gas, into ethanol [8].

The integration of biomass gasification and syngas fermentation (i.e., thermo-biochemical route), has been advocated as a promising and versatile contribution to the biobased economy [9]. The understanding of this linkage may be achieved through experiments at laboratory scale, designed with the perspective of the large-scale [10]. Since data sharing regarding the operation of existing large-scale lignocellulose gasification and syngas fermentation plants is still constrained by knowledge protection laws and agreements, the use of mathematical models is necessary to gather insights regarding the large-scale. The modeling strategies for the gasification process are currently more advanced than those developed for the fermentation process. Strategies used for simulating syngas-fermenting bacteria inside large-scale bioreactors include the use of black-box models [11] and largely complex genome-scale models [12] as well as combinations of the two strategies and thermodynamics [13,14]. As for the detailed simulation of mass transfer in large-scale bioreactors that are appropriate for syngas fermentations, there is only study that is currently known, which is at an early development stage [15].

Some of the aforementioned fermentation models have been included in process simulations aimed at assessing the link between gasification and fermentation [16,17], yet little research has been conducted to explore the simultaneous effects of the process conditions and design choices of different units on the performance of the whole process or to optimize it in terms of multiple objectives [18,19]. At the same time, integrated optimization may be indispensable for the commercialization of thermo-biochemical processes. As highlighted by Ramachandriya et al. [20], different challenges arise when integrating both conversion steps (e.g., low product yield, energy requirements in the gasifier, and inhibition caused by syngas impurities), but most studies in this field have focused on the microbial physiology of syngas fermenting bacteria [21–23]. On the other hand, research on biomass gasification has unveiled a complex relationship between the performance of different gasifier systems and multiple process conditions (steam to biomass ratio, temperature, air equivalence ratio, feedstock moisture, etc.) [24].

In this context, the main goals of this work are (i) the development of a framework for modeling and optimizing the integrated process for ethanol production from biomass via the thermo-biochemical route by considering two types of feedstock (sugarcane bagasse and wood residues); (ii) the holistic impact analysis of the operating conditions and design parameters; (iii) the analysis of the optimal trade-offs between economic, energy, and environmental performance; and (iv) the analysis of the Pareto-optimal conditions of the multiple units involved in the process by taking into account their interactions.

2. Materials and Methods

2.1. Modeling Framework

The ethanol production process is divided into five main units, as presented in Figure 1. In unit A100, the biomass feed is dried and gasified, after which the syngas is sent to a reformer. Hot streams from this unit are then cooled in A200, recovering its heat for steam and power generation, after which the cold syngas (~60 °C) is passed through a scrubber to remove contaminants. In A300, the syngas is compressed to the pressure at the bottom of the bioreactor, cooled to 37 °C, and mixed with recycled gas before being fed to the bioreactor. Cells are separated in a microfiltration membrane and recycled with a small purge, and the product stream (dilute ethanol with traces of acetic acid) is sent to A400 for ethanol recovery and purification using distillation and molecular sieves. Unit A500 produces cooling water and chilled water for the whole plant.

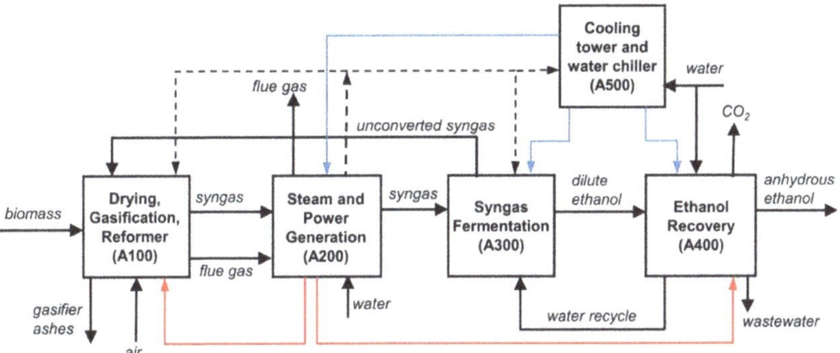

Figure 1. Block flow diagram of the thermo-biochemical route for ethanol production from biomass. Dashed lines: electricity streams; blue lines: cooling or chilled water; red lines: steam.

This section provides details about the operation of areas A100 and A200, while information about A300 and A400 can be found elsewhere [18].

Our approach to modeling syngas fermentation has been described elsewhere [17,18]. Previously we demonstrated the application of surrogate modeling and machine learning (specifically, artificial neural networks) as tools to simplify the evaluation of the responses originally obtained by rigorous models of the bioreactor and the distillation columns [18]. This strategy is repeated in the present work and is applied to the gasification model, which is described next.

2.1.1. Drying, Gasification, and Tar Reformer (A100)

As in de Medeiros et al. [25], the gasification process consists of a dual fluidized bed gasifier with the circulation of the char and bed material between the two beds, as schematized in Figure 2. Hot flue gas from the combustion zone (CZ) is used in the air pre-heater and the biomass dryer. Since char formation is regulated by the temperature in the gasification zone (T_{GZ}) and char is the main fuel in the combustion zone, the system in Figure 2 will reach an equilibrium point for T_{GZ} and T_{CZ} (temperature in the combustion zone), therefore making T_{GZ} an output of the process instead of an input. To transform T_{GZ} into an independent variable, we propose that other variables (namely, air flow rate and additional fuel fed to CZ) can be tuned to satisfy the energy balance for the desired T_{GZ}, which is not necessarily at the aforementioned equilibrium point. Therefore, the gasification model proposed here comprises an optimization routine in which, for a given T_{GZ}, we wish to minimize the square difference of the heat duty between GZ and CZ, here named Q_{diff}, by finding the corresponding values of three variables: AE (air excess fed to CZ), DT (temperature difference, T_{CZ}–T_{GZ}), and f (split fraction of biomass that is diverted to CZ instead of GZ). The energy difference Q_{diff} also considers a 2% loss of the lower heating value (LHV) of the biomass. Since we also wish to minimize the resources input for the whole process, for a given TGZ, the objective function becomes:

$$\min\left[\left(Q_{diff}\right)^2 + (AE) + (f) + (DT)\right] = f(T_{GZ}, AE, f, DT) \quad (1)$$

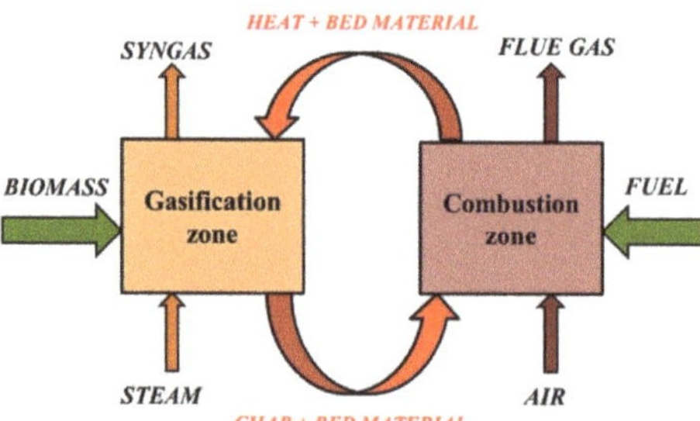

Figure 2. Schematic representation of dual fluidized bed reactor in A100.

The calculation of the energy difference Q_{diff} starts by estimating the outcomes of the gasification zone (syngas and char yields, and compositions), for which we use temperature-dependent correlations, which were previously adopted by NREL [26]. These correlations are second-degree polynomial functions of T_{GZ} that predict the yield of the syngas (scf/lb maf biomass) and the mass fractions (dry basis) of its main components (i.e., CO, CO_2, H_2, CH_4, C_2H_4, C_2H_6, C_2H_6, C_2H_2, C_6H_6). Although there is a correlation for the char yield, we follow NREL's recommendation of instead using the following algorithm based on the elemental balances: (i) for carbon, determine the total amount of C in the syngas from the results of the correlations and consider any remaining C to be in the form of char; (ii) for oxygen, assume that at least 4% of O in biomass ends in the char; then, if the O balance results in a deficit of this element, the water is decomposed. If there is an excess of O, then the exceeding amount is assumed to also be in the char; (iii) for sulfur, assume that at least 8.3% of the S in biomass is in char and that the remaining S is converted to H_2S in the syngas; (iv) for nitrogen, assume that at least 6.6% of the N in biomass goes into the char and that the remaining is converted to NH_3 in the syngas; (v) for hydrogen, determine the total amount of H in all of the components of the syngas and consider the remaining H to be in the char. To be coherent with the correlations, other conditions were assumed to be fixed and equal to the experiments described by the correlations, i.e., biomass moisture entering the GZ equal to 10% and steam to biomass ratio SBR = 0.4 kg/kg dry biomass.

To calculate Q_{diff}, the gasification unit was simulated in Aspen Plus following the flowsheet presented in Figure 3. The Aspen flowsheet is presented in Figure S1 in the Supplementary Materials. Biomass was specified as a non-conventional component described by its heating value and composition given by proximate and ultimate analyses. These can be found in de Medeiros et al. [25] and Capaz et al. [27], for bagasse and wood residues, respectively (see also Table S1 in the Supplementary Materials). For each temperature, the results of the GZ algorithm explained above were used as input in the yield reactor representing the GZ (R-01). A combustion reactor (R-02) is a stoichiometric reactor that is fed with the char generated in GZ, as well as the biomass that may be diverted for this use in the splitter (SP-01). In the simulation, there was also a yield reactor (not depicted in Figure 3) to transform the non-conventional component biomass into conventional components that could participate in combustion reactions. The dryer (D-01) was modeled in Aspen with a stoichiometric reactor and flash operation: the former converts the non-conventional biomass stream into a stream containing biomass and H_2O, which is later separated in the flash operation. The amount of H_2O generated in this stage is the difference between the initial moisture of the wet biomass and the final desired moisture of 10%. The output Q_{diff} is then the sum of the three heat streams related to these operations: the decomposition of nonconventional biomass, the gasification reactions R-01, and the combustion reactions R-02. The tar reformer was simulated as a stoichiometric reactor where the conversions of CH_4, C_2H_6, C_2H_4, and tars into CO and H_2 take place in fixed amounts, i.e., 80%, 99%, 90%, and 99%, respectively (the same as those adopted by NREL [26]). The heat duty was calculated in Aspen, and it was assumed to be provided by the combustion of the unconverted syngas from A300 as well as by a fraction of the unreformed syngas from the gasifier. The latter can be adjusted to not only meet the requirements of the tar reformer but to also increase the amount of energy available for steam/electricity production in A200.

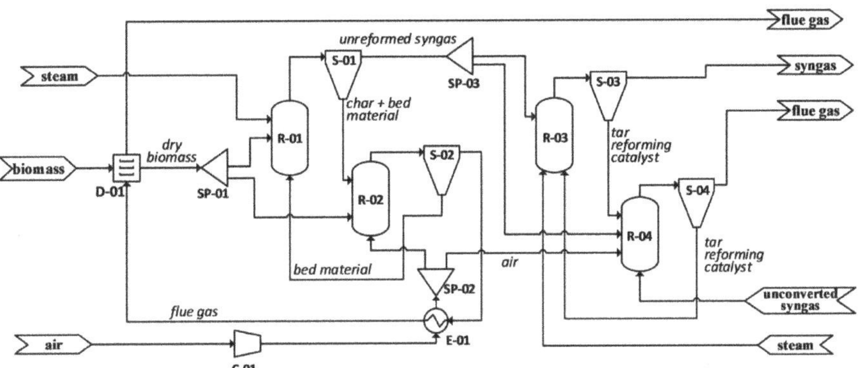

Figure 3. Simplified process flow diagram of A100: drying, gasification, and tar reformer. D-01: biomass dryer; SP-01 to SP-03: stream splitters; R-01 and R-02: gasification (GZ) and combustion (CZ) zones of dual bed gasifier; S-01 to S-04: cyclones; R-03 and R-04: tar reformer and catalyst regenerator; C-01: air blower; E-01: air pre-heater.

The minimization problem (Equation (1)) was solved in MATLAB for a range of T_{GZ}. Since the calculation of Q_{DIFF} and other the outputs require the Aspen simulation, one possible approach is to link both programs and to run the simulation every time the objective function needs to be evaluated. However, to make the framework more robust and to reduce the number of simulation runs, we instead decided to train artificial neural networks (ANNs) with the data generated in Aspen for multiple combinations of inputs (T_{GZ}, AE, f, DT). This procedure was previously explained in a different case [18]. These surrogate models were then used in the optimization problem, which was solved with *fmincon* in MATLAB. The ranges used to obtain the data were T_{GZ} between 700–1000 °C; AE between 10–150%; f between 0–0.5; and DT between 30–100 °C.

2.1.2. Heat Recovery and Power Generation (A200)

Energy is recovered from three streams of hot gases: syngas from the tar reformer, flue gas from the char combustor, and flue gas from the tar reformer/combustor (catalyst regenerator). These hot gases are used as energy inputs in a Rankine cycle with reheat (Figure 4) to produce electricity. In this cycle, there are two expansion stages (ST-01 and ST-02) with an intermediate re-heating operation (E-02) to increase the energy efficiency. In the 2nd stage, a slipstream is extracted to provide steam for the gasification and as process heat (for the distillation). The specifications of inlet/outlet pressure and temperature at the turbine were considered the same as those reported by NREL [26]. Since the properties (mass flow and temperature) of the hot streams are not fixed (i.e., they depend on the conditions of the process), the heat exchanger network (represented in the flowsheet by the exchangers E-01, E-02, and E-05) is designed with an algorithm that roughly maximizes the sensible heat that can be transferred from the hot to the cold streams. In this unit, the mass flow rate of the water/steam circulating in the Rankine cycle is set to meet the plant targets of electricity and steam consumption, but if heat is still available, then more water is provided to increase electricity production. This is achieved by a small optimization routine to maximize the amount of water while respecting the 1st and 2nd laws of thermodynamics.

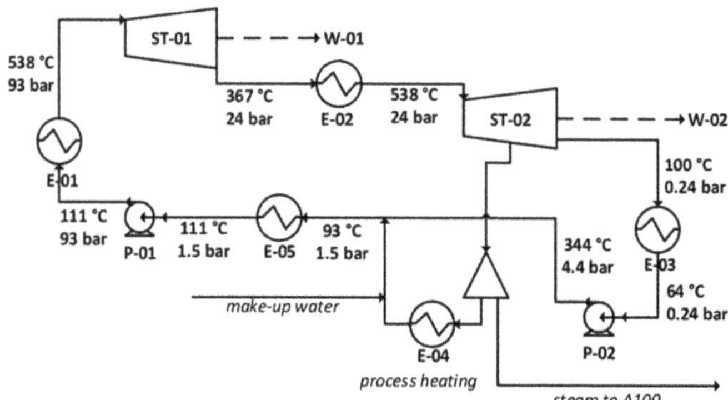

Figure 4. Rankine cycle with reheat. E-01 to E-05: heat exchangers (units represent series of exchangers); ST-01 and ST-02: steam turbine, 1st and 2nd stages; P-01 and P-02: water pumps.

Electricity generated in unit A200 is used to supply the gas compressors, air blowers, and pumps in the entire plant, as well as the water chiller (which produces chilled water for the bioreactor that must be kept at 37 °C). After heat recovery, the reformed syngas stream is further cooled to 60 °C using cooling water, and is fed to a scrubbing system following the same specifications as adopted by Dutta et al. [26], i.e., comprising a venturi scrubber, cyclone separator, and a quench water circulation system with a small purge and freshwater makeup.

2.2. Evaluation of Model Outputs and Multi-Objective Optimization

The modeling framework considers nine decision variables for the overall process optimization: in A100, (i) the T_{GZ} (temperature in the gasification zone of the gasifier), and (ii) f_s (fraction of unreformed syngas sent to combustion) are considered; in A300, (iii) the D_{rate} (dilution rate in the bioreactor), (iv) GRT (gas residence time, defined as the volume of liquid divided by fresh gas volumetric flow), (v) GRR (gas recycle ratio), (vi) L (column height), and (vii) V_R (volume of bioreactor) are considered; and in A400, (viii) the SF_{C1} (mass ratio of side stream to feed stream in the first distillation column) and (ix) RR_{C2} (molar reflux ratio in the second distillation column) are considered. The sustainability performance is measured by four types of responses: (i) economic; (ii) energetic; (iii) carbon footprint; and (iv) water footprint. The variable f_s is used to regulate the amount of energy (electricity and heat) that is produced inside the plant: if f_s is too high, the process exports energy and produces less ethanol; if it is too low, then the energy must be imported, which therefore increases the carbon footprint of the process and the utility cost. There is, of course, a point at which the process becomes self-sufficient, but it does not necessarily correspond to optimal process in terms of all of the sustainability criteria. The optimization was conducted for two feedstocks: sugarcane bagasse and wood (eucalyptus) residues.

The capital costs were calculated following the bare module costing technique detailed in Turton et al. [28]. For the gasification unit and steam turbine, the base costs were taken from NREL [26] and were corrected for inflation to the year 2019. The capacity was considered the same for both case studies: 2000 tonnes of dry biomass per day. The costs of heat exchangers, pumps, air blowers, and towers were calculated with the purchase–cost correlations available in Turton et al. [28]. For all types of equipment, the capacity ranges were respected by dividing the equipment into more units if necessary (for example, if the calculated heat exchanger area is greater than 1000 m^2). The economic performance indicator used for the optimization is the minimum ethanol selling price ($MESP$), i.e., the price to achieve NPV = 0. Other economic assumptions were the same as those in de Medeiros et al. (2020).

Table 1 presents the considerations of the prices and carbon footprint (emission factors) associated with raw materials and utilities used in the process. The costs of other raw materials, such as olivine and the tar reformer catalyst, were taken from NREL [26] and were assumed to have negligible carbon footprint contribution. It is worth mentioning that the fermentation nutrients were excluded from the analysis since they could not be calculated with our model, but they were not expected to have a significant impact on either *MESP* or CO_2 emissions, as shown in [16]. In an LCA study using data from LanzaTech, Handler et al. [29] reported that inputs such as nutrients, water, and chemicals together amounted to 9–20% of the CO_2eq emissions related to feedstock procurement (corn stover, switchgrass, or forest residue). Regarding the carbon footprint of lignocellulosic feedstocks (sugarcane bagasse or eucalyptus residues), these are considered here as co-products instead of waste, i.e., a fraction of the impacts associated with the production of sugarcane/ethanol or eucalyptus are allocated to the residual biomass according to their economic value [30].

Table 1. Prices and carbon footprint considered in this study.

Raw Material	Price	Carbon Footprint
Sugarcane bagasse	USD 45/t (db) (Bonomi et al., 2016)	0.042 kg CO_2eq/kg (db) (Capaz et al., 2020)
Wood residues	USD 11.3/t (db) (SEAB, 2019)	0.0189 kg CO_2eq/kg (db) (Capaz et al., 2020)
Electricity	USD 0.14/kWh (CPFL Energia, 2019)	0.17 kg CO_2eq/kWh (Capaz et al., 2020)
Steam	variable (Ulrich and Vasudevan, 2006)	70 kg CO_2eq/GJ (Ecoinvent)
Natural gas	USD 0.274/kg	2.63 kg CO_2eq/kg (Ecoinvent)

The energy efficiency considered here reflects how much of the energy input from biomass and heat/power (if these are not produced inside the plant) is available in the final product (anhydrous ethanol). If there is an excess of electricity production, for example, the carbon footprint of the process will be lower, but so will the energy efficiency. Finally, the water footprint is the total water consumed in the process divided by the production rate of ethanol. Cooling water make-up due to losses from evaporation, drift, and blowdown were assumed to be 0.4% of the total cooling water consumption.

Prior to the multi-objective optimization, a sensitivity analysis was conducted to determine the correlations between the input and output variables as well as the correlations between different responses. For the latter, principal component analysis was applied to a set of responses obtained under different combinations of input variables (4000 points), and the values of the main component coefficients (also called loadings) were used to interpret the correlations between the responses and thus to reduce the number of objectives. With the final set of objectives, the multi-objective optimization was then conducted in MATLAB using a genetic algorithm. The search ranges of the decision variables are shown in the Results section together with the ranges of the optimal Pareto results in Table 2 (Section 3.4).

Table 2. Multi-objective optimization of thermo-biochemical route: ranges of the Pareto-optimal solutions.

Decision Variables	Search Space	Bagasse	Wood Residues
T_{GZ} (°C)	700–1000	839–989	909–983
f_s	0–0.35	0.00182–0.280	0.111–0.330
D_{rate} (h^{-1})	0.05–0.15	0.0568–0.080	0.0560–0.0644
GRT (min)	5–40	21.6–32.1	21.7–33.0
GRR	0–0.5	0.0990–0.293	0.124–0.304
L (m)	30–50	43.1–47.2	40.4–48.9
V_R (m^3)	400–900	455–600	418–596
SF_{C1}	0.06–0.13	0.0894–0.0940	0.0886–0.0950
RR_{C2}	3–6	4.84–5.95	4.75–5.87

3. Results and Discussion

3.1. Gasification

As explained in Section 2.1.1, the gasification model expands the NREL algorithm [26] by tuning other process conditions to maintain a desired temperature in the gasification zone. The main results are presented in Figures 5 and 6. In Figure 5, the compositions are shown for bagasse only, but since the model uses temperature-dependent correlations for the dry molar fractions in the gas phase, there are virtually no differences between the dry composition obtained for the two feedstocks. This is certainly a limitation of the model because it means the feedstock composition has no effect on the dry gas composition; however since the differences are small (e.g., the bagasse has lower carbon content, 46.96% against 50.89%, as shown in Table S1 in the Supplementary Materials), we can assume that in view of the whole process and by recalling that the moisture at the entrance of the gasifier is the same (i.e., 10%), the main distinctive aspects of the feedstocks will be the initial moisture (50% for bagasse and 12% for wood), price, and carbon footprint. It is worth mentioning that the composition correlations were developed for different types of wood; hence, it is safe to affirm that the gasifier model is more accurate for eucalyptus residues.

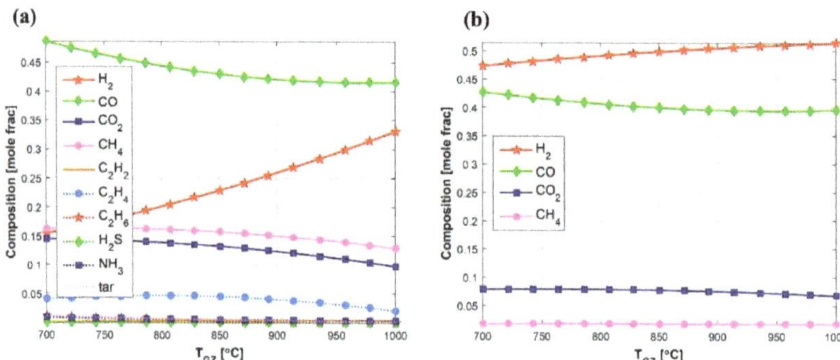

Figure 5. Molar composition of syngas (dry basis): (**a**) after gasifier; (**b**) after tar reformer (other species are not shown due to negligible concentrations).

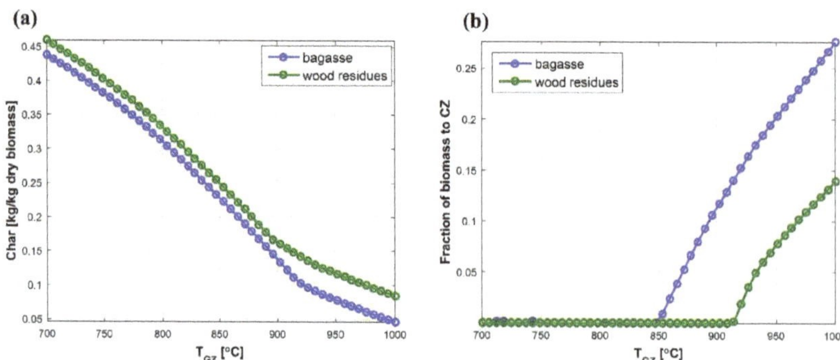

Figure 6. Main differences between the predictions of the gasifier model for bagasse and wood residues: (**a**) char yield; (**b**) required fraction of biomass sent to the combustion zone.

Differences in feedstock composition are compensated in the char yield, which is therefore lower for bagasse (Figure 6a). Another difference can be observed in the fraction of biomass that must be diverted to the combustion zone to maintain the desired temperature in the gasification zone (Figure 6b): in the case of bagasse, the fraction rises from zero at a lower temperature than the wood resides, not only due to the feedstock composition but also due to the heating value, which is lower for bagasse (16.05 MJ/kg against 18.61 MJ/kg, dry basis).

Another limitation of the model is the inability to predict the formation of toxic HCN. Although it is produced in much lower amounts than NH_3 [31], HCN has been reported to be the main reason behind the shutdown of the INEOS Bio gasification–fermentation plant in Florida [32]. On the other hand, recent studies have suggested that the syngas-fermenting microbe *Clostridium ljungdahlii* can adapt to the presence of cyanide and can achieve similar growth performance as it can without the contaminant [33]. Moreover, HCN removal from syngas can be accomplished through different cleaning processes, such as absorption into an aqueous solution followed by alkaline chlorination or oxidation or even direct through decomposition using heterogeneous catalysts during the gasification process [34]. It may be hypothesized that INEOS Bio underestimated the amount of HCN that would be produced in the gasifier and then, with the plant already constructed, it might have been too problematic to include further cleaning stages.

3.2. Bubble Column Bioreactor

A bubble column bioreactor is affected by several variables. For the optimization study, five variables were direct inputs of this unit (D_{rate}, GRT, GRR, L, V_R), but other variables were fixed (e.g., cell recycle ratio, at 0.85), or they were outcomes from other units (e.g., syngas composition). In a previous study [11], we showed how the syngas composition affects the gas conversion and ethanol productivity predicted by the biokinetic model. Figure 7 presents the main performance indicators of the bubble column reactor for different values of D_{rate} and GRT, with the syngas molar composition fixed at (CO:H_2:CO_2) = (0.4:0.5:0.1), column height L (m) = 40, volume V_R (m^3) = 500, and no gas recycling (GRR = 0). The responses presented in Figure 7 are conflicting and cannot be optimized simultaneously: for example, the highest ethanol titers are achieved under very low D_{rate} (<0.075 h^{-1}) and GRT, while the highest CO conversions are achieved with high GRT. The energy efficiency η_{LHV} is also favored under high GRT (due to higher conversion), but the productivity is favored at low GRT, achieving a maximum close to D_{rate} = 0.1 h^{-1}.

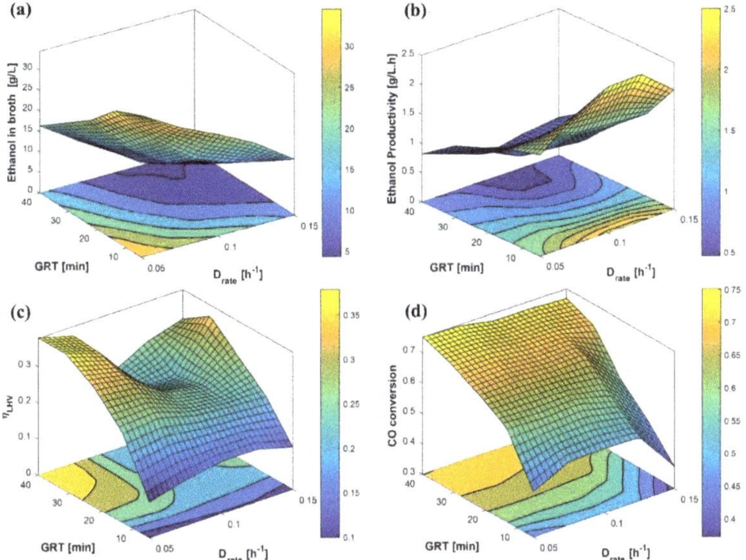

Figure 7. Bubble column reactor, sensitivity of D_{rate}, and GRT on model outcomes: (**a**) ethanol concentration ($g{\cdot}L^{-1}$) in the liquid phase; (**b**) ethanol productivity ($g{\cdot}L^{-1}{\cdot}h^{-1}$); (**c**) energy efficiency η_{LHV}; (**d**) CO conversion.

3.3. Global Effects of Input Variables and Correlations between Responses

Within the framework of the entire production process, the model was first used to predict the relevant responses to a set of combinations of decision variables. The results were then used to calculate the correlation coefficients between the decision variables and each of the responses, which are presented in Figure 8. First, it is worth noting that all of the decision variables have absolute correlation coefficients greater than 0.1 for at least one of the responses; for this reason, all of them are kept in the optimization problem. Second, T_{GZ}, f_s, and GRT dominate, with the highest correlation coefficients for all of the responses. Moreover, a few interpretations can be highlighted.

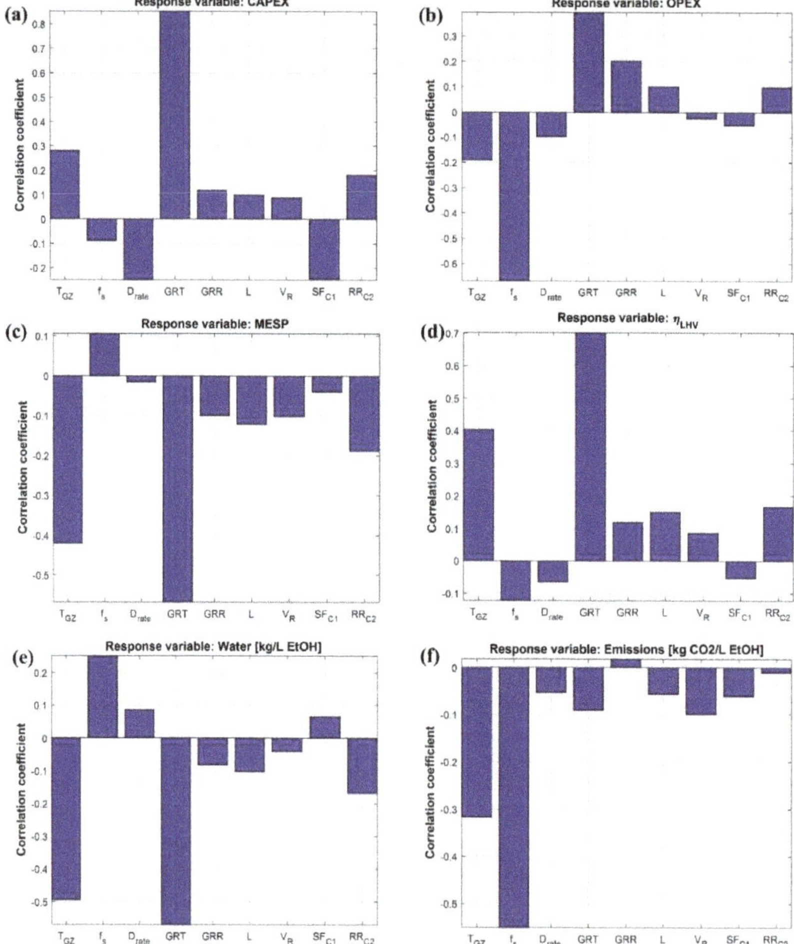

Figure 8. Correlation coefficients between decision variables and responses: (**a**) CAPEX, (**b**) OPEX, (**c**) MESP, (**d**) η_{LHV}, (**e**) Water use and (**f**) CO_{2eq} emissions. Results are presented for sugarcane bagasse only (results for wood residues are presented in Figure S2 in the Supplementary Materials).

GRT is a measure of the amount of fresh syngas fed to the bioreactor: for a fixed reactor volume, the higher the value of *GRT*, the lower the fresh gas volumetric flow rate fed to each vessel, which means that for the same syngas production rate (an outcome of the gasification unit), the number of reactor vessels must be increased, hence the large positive effect on *CAPEX*. The effect on *OPEX* is not straightforward because as seen in Figure 7, increasing the *GRT* increases the gas conversion but also decreases the ethanol titer (which means that more resources are used downstream). *MESP* and η_{LHV} show similar correlation coefficients but with opposite signs, also meaning that lower values of *MESP* are an indication of higher energy efficiency. The effect on the water use is approximately opposite to the energy efficiency, corroborating that higher energy use per liter of the product also prompts a higher cooling water requirement and therefore, more water make-up.

The split fraction of the unreformed syngas that was diverted to combustion (f_s) has large negative effects on both, the *OPEX* and the carbon footprint, since increasing f_s implies decreasing the input of external energy to the plant, hence lower costs and CO_{2eq} emissions. However, as seen in Figure 8c, f_s also has a small positive effect on MESP, meaning that the abovementioned gains are overshadowed by the reduced ethanol production.

Although increasing the temperature in the gasification zone (T_{GZ}) means sacrificing more biomass to combustion (Figure 6b), this loss is compensated by the reduced formation of char (Figure 6a), thus a higher syngas yield, plus a higher production of H_2 (Figure 5), which favors ethanol production during fermentation. The small increase in *CAPEX* (probably due to higher gas flow rates) is therefore repaid by these gains, as observed with the correlation coefficients of this variable compared to other responses.

To conduct the sustainability optimization, *MESP* was elected as the main economic indicator, while the other responses shown in Figure 8, apart from *CAPEX* and *OPEX*, were initially considered as objectives. The results of the correlation analysis described above also indicated existing correlations between the responses (e.g., between *MESP*, $\eta_{LHV,}$ and water); such hypothesis was verified using principal component analysis (PCA). PCA takes a set of multidimensional data and reduces the dimensions by creating new variables (principal components) that are linear combinations of the original variables. The values of these linear coefficients (sometimes called loadings) can then be compared to find correlations among the variables. In the present case, two principal components were found to explain more than 90% of the variance in the original data set; therefore, the coefficients of the first two components provide an accurate overview of these correlations, as depicted in Figure 9. As expected, *MESP*, $-\eta$, and water use are clustered in the same region with similar coordinates. Based on these results, we decided to exclude the water footprint from the multi-objective optimization and to proceed with three minimization objectives only: (i) *MESP*, (ii) $-\eta$ (because one of the goals is to maximize the energy efficiency), and (iii) carbon footprint.

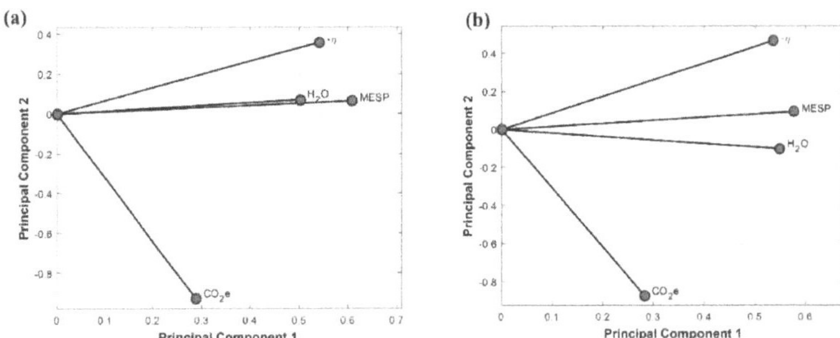

Figure 9. PCA of model responses showing the principal component coefficients (loadings) of the first two principal components: (**a**) sugarcane bagasse; (**b**) wood residues.

3.4. Multi-Objective Sustainability Optimization

Figure 10 presents the Pareto fronts and their respective interpolant surfaces that were obtained for the two feedstocks. Significantly lower carbon footprint and MESP values can be obtained with wood residues (0.93 USD/L against 1 USD/L and 3g CO_2eq/MJ against 10 g CO_2eq/MJ). The main reasons behind these results are the lower feedstock price, the lower feedstock-related emissions, and the initial moisture of the wood residues. The energy efficiency however approached 32% in both cases, a result that is lower than a previous estimation (η = 38%) [25] that considered a much more simplistic bioreactor model. Indeed, as demonstrated in de Medeiros et al. [18], an optimistic estimation of the gas–liquid mass transfer coefficient (k_La) can lead to a substantial improvement in energy efficiency and a reduction of *MESP*. Considering the high values of *MESP*, even under optimal conditions, and its dependence on the energy efficiency, the results presented here and in de Medeiros et al. [18] corroborate the need for improvement in the bioreactor, be it with novel reactor designs that facilitate gas–liquid mass transfer while keeping low cost or with genetic improvement of the microorganisms. These changes must, however, be followed by new optimization studies to re-evaluate the optimal process conditions.

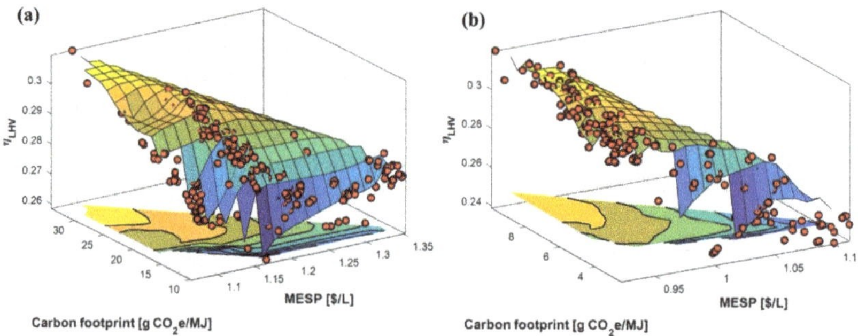

Figure 10. Pareto surfaces obtained with three-objective optimization of the thermo-biochemical route (Pareto-optimal solutions represented by red spheres): (**a**) sugarcane bagasse; (**b**) wood residues.

For both feedstocks, the optimal *MESP* can be decreased at the cost of higher GHG emissions; however, even at the lowest *MESP* values, the process still represents a significant emission reduction compared to gasoline (94 gCO_{2eq}/MJ) [35] and 1st-generation ethanol (38.5–44.9 g CO_{2eq}/MJ) [36]; however, it should be mentioned that our calculations do not take into account the emissions related to the distribution of ethanol. The results are comparable to other combinations of 2G technology and feedstock, for example, the biochemical route using wheat straw (16 g CO_{2eq}/MJ) [2] or sugarcane residues (17.5 gCO_{2eq}/MJ) [27]. Similarly, Handler et al. [29] reported GHG emissions from gas fermentation between 8.0 gCO_{2eq}/MJ for corn stover and 31.4 for basic oxygen furnace gas.

There are different sources of uncertainty in the modeling framework. First, those associated with the process models: for example, in the correlations used to predict syngas composition as a function of temperature or in the equations and parameters used for the calculation of the gas–liquid mass transfer coefficient (k_La) and reaction rates in the bioreactor model. These are uncertainties that can be attenuated with research to deliver more experimental data, either laboratory or industrial, to validate and improve the models. The other type of uncertainty is related to economic and environmental parameters and assumptions that are unrelated to process models, such as the price of raw materials, capital cost correlations, and emission factors. For example, biomass residues are not traditional materials with established market prices, but they acquire a so-called opportunity price as second-generation technologies or as other types of biomass valorization processes gain popularity. Similarly, one can expect that values of CO_{2eq} emissions due to feedstock procurement to depend not only on the location and type of biomass but also on the impact assessment methodology and database used for the calculation of these emission factors. In this context, Figure 11 presents the two-dimensional projections of the Pareto fronts from Figure 10 along with the uncertainty intervals obtained when four economic and environmental assumptions are varied within ±30% ranges: (i) feedstock price; (ii) *CAPEX* calculation; (iii) feedstock emission factor; and (iv) electricity emission factor. The points A, B, C, and D were selected as the most desirable candidates, as discussed further in Section 3.4.

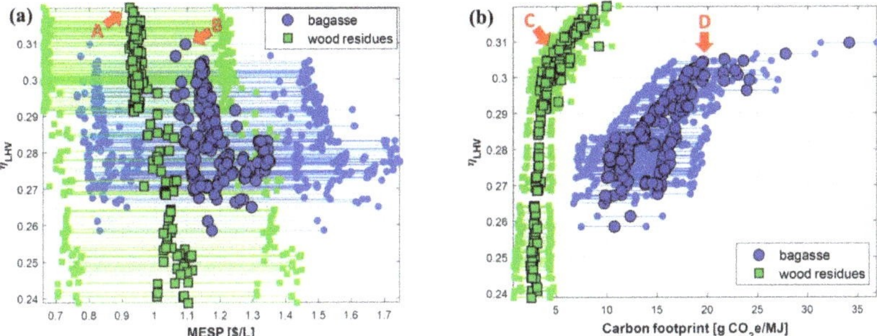

Figure 11. Projections of three-objective Pareto fronts in pairs, including intervals of ±30% uncertainty in economic and environmental assumptions: (**a**) MESP; (**b**) carbon footprint.

The large uncertainty intervals demonstrate the importance of being transparent about the assumptions and limitations of techno-economic and environmental assessments. Nevertheless, the main contribution of this paper is not the calculation of *MESP*, energy efficiency, and carbon footprint, but rather the strategies presented for sustainability optimization and the insights regarding the effects of interconnected input variables and their behavior at optimal solutions. This is illustrated in Figure 12 for the most relevant variables: T_{GZ}, f_s, D_{rate}, and *GRT*. As seen in Section 3.3, these variables showed the strongest correlations with the responses, which is why they are also more dispersed along the Pareto fronts. Other variables, however, were limited to more narrow ranges of optimal values when compared to their original search space. Ranges of Pareto-optimal values obtained for all of the decision variables are shown in Table 2, together with their original search space.

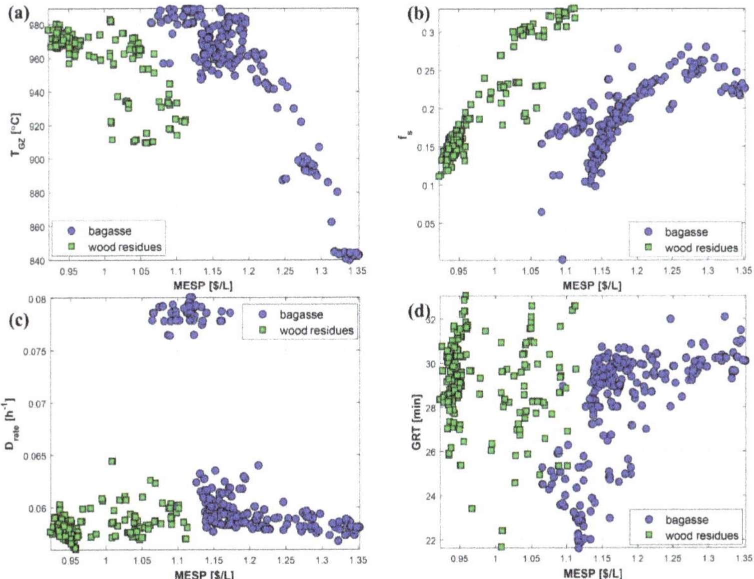

Figure 12. Pareto-optimal values of most relevant decision variables: (**a**) T_{GZ}; (**b**) f_s; (**c**) D_{rate}; (**d**) *GRT*.

The optimal trends presented in Figure 12 reinforce, to some extent, the correlations discussed in Section 3.3 (Figure 8). For example, lower *MESP* (and higher efficiency) can be achieved with a higher gasification temperature, while the opposite is observed for the variable f_s (fraction of unreformed syngas that is sent to combustion). The optimal values of D_{rate} are constrained to the range 0.055–0.08 h^{-1}, similar to what was observed in de Medeiros et al. [18]. Finally, *GRT* is spread over the range 22–32 min, but although its patterns are not as evident as seen for T_{GZ} and f_s, there seems to be a rough tendency of a higher *GRT* leading to a higher *MESP* (and lower η), which is, at first sight, in contrast to the results presented in Figure 8. However, when considering the entire *GRT* search space (see Table 2), the optimal values are closer to the upper bound than to the lower bound, therefore confirming that higher *GRT* is better for both *MESP* and η. It is when the data set is limited to the Pareto fronts that this pattern is not clear anymore, demonstrating that other input variables also exert strong effects on the optimal results.

Although the Pareto-optimal solutions are, by definition, equally optimal, points A, B, C, and D from Figure 11 can be selected as the best candidates according to the following criteria: first, given the current context, in which profitability is still the prevailing standard, points A and B are those for which both profitability and energy efficiency are maximized. It should be noted that it is not always the case that these two targets can be optimized at the same time (for example, see de Medeiros et al. [18]). Points C and D take into account the carbon footprint but do not consider it the most crucial target: beyond these points, minor improvements in the energy efficiency are followed by a proportionally larger increase in carbon emissions. Table 3 presents the values of the decision variables at these four solutions along with the corresponding values of the three targets. The main differences between the two types of solutions (A and B against C and D) are related to the gasification temperature (slightly lower in the second case), the bioreactor volume (also lower in the second case), and, more notably, the syngas fraction f_s, which is much higher when the carbon footprint is taken into account.

Table 3. Multi-objective optimization of thermo-biochemical route: selected optimal points.

Decision Variables	A (Wood)	B (Bagasse)	C (Wood)	D (Bagasse)
MESP (USD·L^{-1})	0.934	1.09	0.958	1.14
η	0.319	0.310	0.305	0.304
g CO_2eq/MJ	8.60	34.1	4.11	19.4
T_{GZ} (°C)	974	974	961	962
f_s	0.119	0.00182	0.186	0.119
D_{rate} (h^{-1})	0.0572	0.060	0.058	0.058
GRT (min)	30.3	28.9	31.8	29.8
GRR	0.245	0.248	0.247	0.283
L (m)	45.8	46.0	47.4	45.1
V_R (m^3)	503	554	485	551
SF_{C1}	0.0940	0.0920	0.0930	0.0921
RR_{C2}	5.11	5.13	5.10	5.00

Water footprint was also included in the analysis as a measure of direct water use (i.e., excluding the water footprint to produce the feedstock and raw materials), but as explained in Section 3.3, it was excluded from the multi-objective optimization due to its high correlation with both *MESP* and η. In Figure 13a the water footprint of the Pareto-optimal points is plotted against the corresponding results of *MESP*, with the minimum values being around 5 kg of water per liter of ethanol for both the bagasse and wood residues. As a comparison, Dutta et al. [26] reported 2.0 kg/L for ethanol production from wood via gasification and mixed alcohol synthesis, yet the LanzaTech process is expected to consume around 8.5 kg/L [29]. The ethanol yields (Figure 13b) are also comparable to other 2G processes found to be in the range 205–330 L/ton dry biomass [25,26,37].

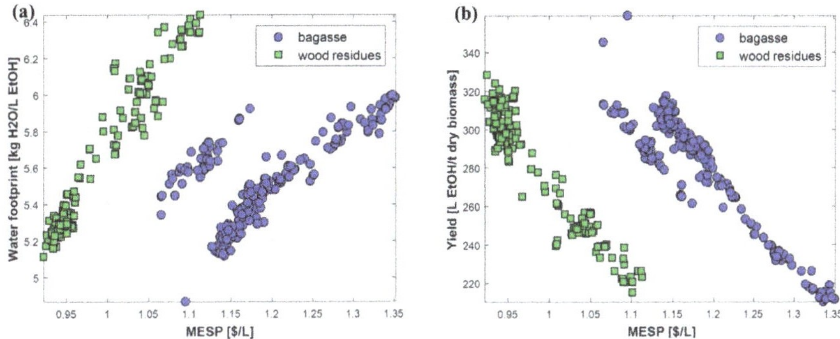

Figure 13. Pareto-optimal values of other performance indicators: (**a**) water footprint; (**b**) ethanol yield.

Finally, Figure 14 illustrates the trade-off between energy efficiency and self-sufficiency. The results indicate that energy self-sufficiency is not necessarily beneficial, as higher values of efficiency can be achieved when energy is purchased (in the form of steam and electricity) instead of produced entirely inside the plant, which sacrifices syngas that could be converted into ethanol. Though this conclusion may seem counterintuitive, it can be clarified by comparing Figure 14 with Figure 13b: as the energy demand increases with η, so does the ethanol yield, with gains that outweigh the extra energy requirement (*MESP* and η go in different directions, as seen in Figures 10 and 11a).

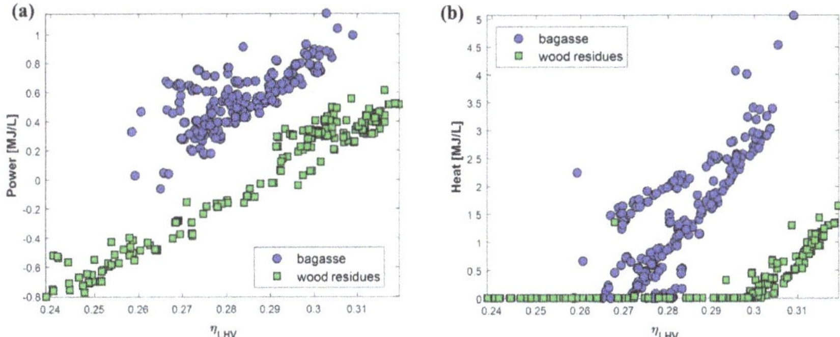

Figure 14. Pareto-optimal values of global energy balance (consumed minus produced in the plant): (**a**) power; and (**b**) heat (steam).

4. Conclusions

This work shows how the sustainability of a gasification–fermentation route can be improved and optimized by tuning the process conditions and design parameters related to different units of the process. The modeling framework of the whole process, from biomass to ethanol fuel, and the interconnected effects of input variables on multiple outcomes are discussed. The correlation coefficients among various decision variables and each of the responses were obtained from the parametric studies of gasification and fermentation models. A multi-objective optimization was applied as a tool for sustainability optimization that does not rely on assigning weights to goals of different natures (e.g., economic and environmental) and optimal trade-offs, were also discussed. Wood residue feedstock was found to be better in terms of a lower MESP and carbon footprint (0.93 USD/L, and 3 g CO_2eq/MJ) compared to sugarcane bagasse (1 USD/L, and 10 g CO_2eq/MJ). This is due to the lower price of the wood, lower feedstock-related emissions, and lower initial moisture. The optimal energy efficiency was found to be the same (32%) in both cases.

Further, the Pareto-optimal solutions and uncertainties in the economic and environmental factors that were used were illustrated. Although early stage economic calculations bear large uncertainties, the optimization results indicate the low competitiveness of this technology against current ethanol production from sugarcane or corn, unless improvements are made to increase the efficiency of bioreactors or if other actions are considered, such as subsidy schemes or carbon taxes; however, this was not evaluated in the present work. The water usage of this integrated process was shown to be lower than that of the current existing production processes, and the purchasing energy was found to be the better option over the energy self-sufficient process at the expense of syngas.

Supplementary Materials: The following are available online at https://www.mdpi.com/article/10.3390/fermentation7040201/s1, Figure S1: Aspen flowsheet of the gasification unit. Figure S2: Correlation coefficients between decision variables and responses (a) CAPEX, (b) OPEX, (c) MESP, (d) η_{LHV}, (e) Water use and (f) CO_{2eq} emissions. Results shown only for wood residues. Table S1: Elemental analysis (% dry basis) and moisture (%) of sugarcane bagasse and wood residues.

Author Contributions: Conceptualization, E.M.d.M., H.N., R.M.F., and J.A.P.; methodology, E.M.d.M., H.N., and J.A.P.; software, E.M.d.M.; validation, E.M.d.M., H.N., and J.A.P.; formal analysis, E.M.d.M.; investigation, E.M.d.M.; resources, H.N., R.M.F., and J.A.P.; data curation, E.M.d.M.; writing—original draft preparation, E.M.d.M.; writing—review and editing, H.N., R.M.F., and J.A.P.; visualization, E.M.d.M.; supervision, H.N., and J.A.P.; project administration, H.N., R.M.F., and J.A.P.; funding acquisition, H.N., R.M.F., and J.A.P. All authors have read and agreed to the published version of the manuscript.

Funding: This research received financial support from CAPES-BRAZIL, CTBE/CNPEM (Brazilian Bioethanol Science and Technology Laboratory), DSM, and the BE-Basic Foundation. This work was conducted as part of a Dual Degree PhD project under an agreement between UNICAMP and TU-DELFT.

Institutional Review Board Statement: Not applicable.

Informed Consent Statement: Not applicable.

Acknowledgments: The authors acknowledge the support from the Dual Degree PhD Programme and also acknowledge the efforts of Haneef Shijaz and Eduardo Almeida Benalcázar for their contributions to the review process.

Conflicts of Interest: The authors declare no conflict of interest.

References

1. Ayodele, B.V.; Alsaffar, M.A.; Mustapa, S.I. An Overview of Integration Opportunities for Sustainable Bioethanol Production from First- and Second-Generation Sugar-Based Feedstocks. *J. Clean. Prod.* **2020**, *245*, 118857. [CrossRef]
2. Padella, M.; O'Connell, A.; Prussi, M. What Is Still Limiting the Deployment of Cellulosic Ethanol? Analysis of the Current Status of the Sector. *Appl. Sci.* **2019**, *9*, 4523. [CrossRef]
3. Polprasert, S.; Choopakar, O.; Elefsiniotis, P. Bioethanol Production from Pretreated Palm Empty Fruit Bunch (PEFB) Using Sequential Enzymatic Hydrolysis and Yeast Fermentation. *Biomass Bioenergy* **2021**, *149*, 106088. [CrossRef]
4. Lak Kamari, M.; Maleki, A.; Nazari, M.A.; Sadeghi, M.; Rosen, M.A.; Pourfayaz, F. Assessment of a Biomass-Based Polygeneration Plant for Combined Power, Heat, Bioethanol and Biogas. *Appl. Therm. Eng.* **2021**, *198*, 117425. [CrossRef]
5. Kirkels, A.F.; Verbong, G.P.J. Biomass Gasification: Still Promising? A 30-Year Global Overview. *Renew. Sustain. Energy Rev.* **2011**, *15*, 471–481. [CrossRef]
6. Hrbek, J. *Status Report on Thermal Gasification of Biomass and Waste 2019: IEA Bioenergy Task 33 Special Report*; IEA Bioenergy: Vienna, Austria, 2019; ISBN 9781910154656.
7. Munir, M.T.; Mardon, I.; Al-Zuhair, S.; Shawabkeh, A.; Saqib, N.U. Plasma Gasification of Municipal Solid Waste for Waste-to-Value Processing. *Renew. Sustain. Energy Rev.* **2019**, *116*, 109461. [CrossRef]
8. Latif, H.; Zeidan, A.A.; Nielsen, A.T.; Zengler, K. Trash to Treasure: Production of Biofuels and Commodity Chemicals via Syngas Fermenting Microorganisms. *Curr. Opin. Biotechnol.* **2014**, *27*, 79–87. [CrossRef]
9. Posada, J.A.; Osseweijer, P. Socio-economic and environmental considerations for sustainable supply and fractionation of lignocellulosic biomass in a biorefinery context. In *Biomass Fractionation Technologies for a Lignocellulosic Feedstock Based Biorefinery*; Mussatto, S.L., Ed.; Elsevier: Amsterdam, The Netherlands, 2016; p. 674.
10. Noorman, H.J.; Heijnen, J.J. Biochemical Engineering's Grand Adventure. *Chem. Eng. Sci.* **2017**, *170*, 677–693. [CrossRef]

11. De Medeiros, E.M.; Posada, J.A.; Noorman, H.; Filho, R.M. Dynamic Modeling of Syngas Fermentation in a Continuous Stirred-Tank Reactor: Multi-Response Parameter Estimation and Process Optimization. *Biotechnol. Bioeng.* **2019**, *116*, 2473–2487. [CrossRef]
12. Chen, J.; Gomez, J.A.; Höffner, K.; Barton, P.I.; Henson, M.A. Metabolic Modeling of Synthesis Gas Fermentation in Bubble Column Reactors. *Biotechnol. Biofuels* **2015**, *8*, 89. [CrossRef]
13. Benalcázar, E.A.; Noorman, H.; Maciel Filho, R.; Posada, J.A. Modeling Ethanol Production through Gas Fermentation: A Biothermodynamics and Mass Transfer-Based Hybrid Model for Microbial Growth in a Large-Scale Bubble Column Bioreactor. *Biotechnol. Biofuels* **2020**, *13*, 59. [CrossRef]
14. Norman, R.O.J.; Millat, T.; Schatschneider, S.; Henstra, A.M.; Breitkopf, R.; Pander, B.; Annan, F.J.; Piatek, P.; Hartman, H.B.; Poolman, M.G.; et al. Genome-Scale Model of C. Autoethanogenum Reveals Optimal Bioprocess Conditions for High-Value Chemical Production from Carbon Monoxide. *Eng. Biol.* **2019**, *3*, 32–40. [CrossRef]
15. Abrahamson, B. Conceptual Design and Evaluation of a Commercial Syngas Fermentation Process. Master's Thesis, Technische Universiteit Delft, Delft, The Netherlands, 2019.
16. Benalcázar, E.A.; Gevers-Deynoot, B.; Noorman, H.; Osseweijer, P.; Posada, J.A. Production of Bulk Chemicals from Lignocellulosic Biomass via Thermochemical Conversion and Syngas Fermentation: A Comparative Techno-Economic and Environmental Assessment of Different Site-Specific Supply Chain Configurations: Techno-Economic and Environmental Assessment of Bulk Chemicals Production Though Biomass Gasification and Syngas Fermentation. *Biofuels Bioprod. Biorefining* **2017**, *11*, 861–886. [CrossRef]
17. De Medeiros, E.M.; Posada, J.A.; Noorman, H.; Filhob, R.M. Modeling and Multi-Objective Optimization of Syngas Fermentation in a Bubble Column Reactor. *Comput. Aided Chem. Eng.* **2019**, *46*, 1531–1536. [CrossRef]
18. De Medeiros, E.M.; Noorman, H.; Maciel Filho, R.; Posada, J.A. Production of Ethanol Fuel via Syngas Fermentation: Optimization of Economic Performance and Energy Efficiency. *Chem. Eng. Sci. X* **2020**, *5*, 100056. [CrossRef]
19. Almeida Benalcázar, E.; Noorman, H.; Maciel Filho, R.; Posada, J. Assessing the Sensitivity of Technical Performance of three Ethanol Production Processes based on the Fermentation of Steel Manufacturing Offgas, Syngas and a 3:1 Mixture Between H2 and CO_2. In *Computer Aided Chemical Engineering*; Elsevier: Amsterdam, The Netherlands, 2020; Volume 48, pp. 589–594. ISBN 978-0-12-823377-1.
20. Ramachandriya, K.D.; Kundiyana, D.K.; Sharma, A.M.; Kumar, A.; Atiyeh, H.K.; Huhnke, R.L.; Wilkins, M.R. Critical Factors Affecting the Integration of Biomass Gasification and Syngas Fermentation Technology. *AIMS Bioeng.* **2016**, *3*, 188–210. [CrossRef]
21. Bertsch, J.; Müller, V. Bioenergetic Constraints for Conversion of Syngas to Biofuels in Acetogenic Bacteria. *Biotechnol. Biofuels* **2015**, *8*, 210. [CrossRef] [PubMed]
22. Pinto, T.; Flores-Alsina, X.; Gernaey, K.V.; Junicke, H. Alone or Together? A Review on Pure and Mixed Microbial Cultures for Butanol Production. *Renew. Sustain. Energy Rev.* **2021**, *147*, 111244. [CrossRef]
23. Yang, C.; Dong, L.; Gao, Y.; Jia, P.; Diao, Q. Engineering Acetogens for Biofuel Production: From Cellular Biology to Process Improvement. *Renew. Sustain. Energy Rev.* **2021**, *151*, 111563. [CrossRef]
24. Liakakou, E.T.; Infantes, A.; Neumann, A.; Vreugdenhil, B.J. Connecting Gasification with Syngas Fermentation: Comparison of the Performance of Lignin and Beech Wood. *Fuel* **2021**, *290*, 120054. [CrossRef]
25. De Medeiros, E.M.; Posada, J.A.; Noorman, H.; Osseweijer, P.; Filho, R.M. Hydrous Bioethanol Production from Sugarcane Bagasse via Energy Self-Sufficient Gasification-Fermentation Hybrid Route: Simulation and Financial Analysis. *J. Clean. Prod.* **2017**, *168*, 1625–1635. [CrossRef]
26. Humbird, D.; Davis, R.; Tao, L.; Kinchin, C.; Hsu, D.; Aden, A.; Schoen, P.; Lukas, J.; Olthof, B.; Worley, M.; et al. Process Design and Economics for Conversion of Lignocellulosic Biomass to Ethanol. *NREL Tech. Rep.* **2011**, *303*, 275–3000.
27. Junqueira, T.L.; Chagas, M.F.; Gouveia, V.L.R.; Rezende, M.C.A.F.; Watanabe, M.D.B.; Jesus, C.D.F.; Cavalett, O.; Milanez, A.Y.; Bonomi, A. Techno-Economic Analysis and Climate Change Impacts of Sugarcane Biorefineries Considering Different Time Horizons. *Biotechnol. Biofuels* **2017**, *10*, 50. [CrossRef]
28. Turton, R.; Bailie, R.C.; Whiting, W.B.; Shaeiwitz, J.A. *Analysis, Design and Synthesis of Chemical Processes*; Prentice Hall: Boston, MA, USA, 2008; ISBN 9780135129661.
29. Handler, R.M.; Shonnard, D.R.; Griffing, E.M.; Lai, A.; Palou-Rivera, I. Life Cycle Assessments of Ethanol Production via Gas Fermentation: Anticipated Greenhouse Gas Emissions for Cellulosic and Waste Gas Feedstocks. *Ind. Eng. Chem. Res.* **2016**, *55*, 3253–3261. [CrossRef]
30. Capaz, R.S.; de Medeiros, E.M.; Falco, D.G.; Seabra, J.E.A.; Osseweijer, P.; Posada, J.A. Environmental Trade-Offs of Renewable Jet Fuels in Brazil: Beyond the Carbon Footprint. *Sci. Total Environ.* **2020**, *714*, 136696. [CrossRef]
31. Wilk, V.; Hofbauer, H. Conversion of Fuel Nitrogen in a Dual Fluidized Bed Steam Gasifier. *Fuel* **2013**, *106*, 793–801. [CrossRef]
32. Lane, J. Digest Feedback May Help Explain INEOS Bio's High Levels of HCN Gas. 2014, pp. 1–2. Available online: https://www.biofuelsdigest.com/bdigest/2014/09/08/feedback-may-help-explain-ineos-bios-high-levels-of-hcn-gas/ (accessed on 21 September 2021).
33. Oswald, F.; Zwick, M.; Omar, O.; Hotz, E.N.; Neumann, A. Growth and Product Formation of Clostridium Ljungdahlii in Presence of Cyanide. *Front. Microbiol.* **2018**, *9*, 1213. [CrossRef] [PubMed]

34. Kumagai, S.; Hosaka, T.; Kameda, T.; Yoshioka, T. Removal of Toxic HCN and Recovery of H2-Rich Syngas via Catalytic Reforming of Product Gas from Gasification of Polyimide over Ni/Mg/Al Catalysts. *J. Anal. Appl. Pyrolysis* **2017**, *123*, 330–339. [CrossRef]
35. Elgowainy, A.; Han, J.; Cai, H.; Wang, M.; Forman, G.S.; Divita, V.B. Energy efficiency and greenhouse gas emission intensity of petroleum products at US refineries. *Environ. Sci. Technol.* **2020**, *48*, 7612–7624. [CrossRef] [PubMed]
36. Mekonnen, M.M.; Romanelli, T.L.; Ray, C.; Hoekstra, A.Y.; Liska, A.J.; Neale, C.M.U. Water, Energy, and Carbon Footprints of Bioethanol from the U.S. and Brazil. *Environ. Sci. Technol.* **2018**, *52*, 14508–14518. [CrossRef]
37. Wei, L.; Pordesimo, L.O.; Igathinathane, C.; Batchelor, W.D. Process Engineering Evaluation of Ethanol Production from Wood through Bioprocessing and Chemical Catalysis. *Biomass Bioenergy* **2009**, *33*, 255–266. [CrossRef]

Article

Performance Evaluation of Pressurized Anaerobic Digestion (PDA) of Raw Compost Leachate

Alessio Siciliano *, Carlo Limonti and Giulia Maria Curcio

Department of Environmental Engineering, University of Calabria, 87036 Rende, CS, Italy; carlo.limonti@unical.it (C.L.); gmaria.curcio@unical.it (G.M.C.)
* Correspondence: alessio.siciliano@unical.it

Abstract: Anaerobic digestion (AD) represents an advantageous solution for the treatment and valorization of organic waste and wastewater. To be suitable for energy purposes, biogas generated in AD must be subjected to proper upgrading treatments aimed at the removal of carbon dioxide and other undesirable gases. Pressurized anaerobic digestion (PDA) has gained increasing interest in recent years, as it allows the generation of a high-quality biogas with a low CO_2 content. However, high pressures can cause some negative impacts on the AD process, which could be accentuated by feedstock characteristics. Until now, few studies have focused on the application of PAD to the treatment of real waste. The present work investigated, for the first time, the performance of the pressurized anaerobic digestion of raw compost leachate. The study was conducted in a lab-scale pressurized CSTR reactor, working in semi-continuous mode. Operating pressures from the atmospheric value to 4 bar were tested at organic loading rate (OLR) values of 20 and 30 kg_{COD}/m^3d. In response to the rise in operating pressure, for both OLR values tested, a decrease of CO_2 content in biogas was observed, whereas the CH_4 fraction increased to values around 75% at 4 bar. Despite this positive effect, the pressure growth caused a decline in COD removal from 88 to 62% in tests with OLR = 20 kg_{COD}/m^3d. At OLR = 30 kg_{COD}/m^3d, an overload condition was observed, which induced abatements of about 56%, regardless of the applied pressure. With both OLR values, biogas productions and specific methane yields decreased largely when the pressure was brought from atmospheric value to just 1 bar. The values went from 0.33 to 0.27 $L_{CH4}/g_{CODremoved}$ at 20 kg_{COD}/m^3d, and from 0.27 to 0.18 $L_{CH4}/g_{CODremoved}$ at 30 kg_{COD}/m^3d. Therefore, as the pressure increased, although there was an enhanced biogas quality, the overall amount of methane was lowered. The pressured conditions did not cause substantial modification in the characteristics of digestates.

Keywords: biogas; biomethane; compost leachate; pressurized anaerobic digestion

Citation: Siciliano, A.; Limonti, C.; Curcio, G.M. Performance Evaluation of Pressurized Anaerobic Digestion (PDA) of Raw Compost Leachate. *Fermentation* **2022**, *8*, 15. https://doi.org/10.3390/fermentation8010015

Academic Editor: Alessia Tropea

Received: 29 November 2021
Accepted: 28 December 2021
Published: 30 December 2021

1. Introduction

Energy consumption increases every year with technological and social development causing significant environmental impacts. In recent years, the exploitation of organic waste as a source to produce energy and to recover chemical compounds has gained increasing interest [1–6]. Among the different technologies, anaerobic digestion (AD) is widely used [1,2,7–9], as it represents a sustainable approach to obtain biofuel and bioproducts from the treatment of wet biomass [8–10]. AD evolves according to a series of biochemical reactions involving different groups of microorganisms [3]. A wet digestate and biogas are generated because of the organic matter degradation under anaerobic conditions. Digestate can be generally used for agronomic purposes, due to its high nutrient content [11–15]. Biogas is a mixture mainly composed of methane (CH_4) and carbon dioxide (CO_2), with a lower heating value (LHV) of about 21.5 MJ/m^3 [14]. It also contains traces of other non-condensable gases such as H_2S, H_2, N_2, NH_3, O_2, CO, volatile organic compounds (VOCs), and steam [13,14,16]. Biogas composition varies according to the type of feedstock and the process conditions. Generally, the content of CH_4 ranges between 55 and 65%, while the fraction of CO_2 is between 35 and 40%. However, CO_2 percentage can reach 50%, which leads to a reduction in the LHV value [17].

The use of biogas for energy purposes requires the application of effective upgrading treatments to remove CO_2 and other undesirable gases [4,13,15,18]. These treatments permit the production of biogas with a CH_4 content greater than 95% [4,13,15,19–21]. Upgrading technologies such as adsorption, high-pressure washing, high-pressure adsorption, cryogenic separation, and membrane separation are widely used in current practices [4,13,15,20,22]. However, these technologies are affected by high costs, clogging and foaming phenomena, water and energy consumption, and process complexity [4,13,15].

In particular, the high costs for biogas upgrading make biomethane less attractive than other biofuels [23]. In recent years, a lot of research has been carried out to reduce the drawbacks related to the purification of biogas. In this regard, pressurized anaerobic digestion (PAD) has attracted great attention [23–26]. In pressurized anaerobic digestion the pressure of the biogas is gradually autogenerated during fermentation. Therefore, PAD processes are carried out at pressures greater than atmospheric, which allows the obtainment of a biogas with a high methane fraction and a low carbon dioxide content [23,27]. This is due to the different properties of CH_4 and CO_2. Methane has a very low solubility in water and mainly remains in the gaseous phase, regardless of the pressure conditions in the digester. CO_2 is characterized by greater solubility that significantly grows with increasing pressure [28,29]. Consequently, the pressure rise causes a greater solubilization of carbon dioxide in the liquid phase, which results in a biogas with a higher CH_4 fraction and a greater LHV value than that generated under atmospheric conditions. PAD processes have so far been tested within the pressure range of 1–100 bar [30,31].

Merkel et al. [29] observed a CH_4 growth from 79.08% at 10 bar, to 90.45% at 50 bar. Lindeboom et al. [27] monitored CH_4 yields between 90% and 95% at pressures up to 90 bar. Bär et al. [32] found that the two-stage high-pressure digestion in biofilm reactors significantly improves the biogas quality at high operating pressure. In particular, methane fractions up to 85% were achieved at an operating pressure of 25 bar, whereas a CH_4 percentage of 93% was detected by feeding the methanogenesis reactor with permeate from a microfiltration pretreatment [32]. Chen et al. [25,33] detected an increase in the CH_4 fraction from 66.2 to 74.5% by raising the pressure from 1 to 9 bar, during the treatment of grass silage leachate in two-stage pressurized reactors. In addition to the generation of a biogas with a high CH_4 content, PAD could potentially improve the characteristics of the digestate. Indeed, high pressures in the reactors promote the solubilization of nutrient compounds in the liquid phase [19,33,34]. In agreement with this statement, Latif et al. [34] detected an enhancement in soluble phosphate concentration working at 6 bar. Despite these benefits, PAD may suffer from some negative aspects. In particular, the greater CO_2 solubilization could lead to an excessive reduction of pH, inhibiting the methanogens activity [25]. This is particularly relevant when wastes with a considerable content of easily degradable substrates are digested. On the other hand, some studies have postulated that high levels of total Kjeldahl nitrogen (TKN) in the feedstock could lead to the formation of ammonia sufficient to increase the buffer capacity, and to hinder the pH drop induced by the CO_2 solubilization [19]. Other works have stated that the accumulating pressure impacts on microorganism consortium cause the inhibition of the methane production yield [35]. These considerations make clear the need for further research to assess the actual performance of PAD in relation to the type and characteristics of the waste to be digested.

Until now, PAD has been mainly tested in the digestion of synthetic substrates [24,27], activated sludge [34], and silage waste [19,25,26,29,33]. In the present work, the pressurized anaerobic digestion of compost leachate, generated by the aerobic stabilization of the organic fraction of municipal solid waste, was investigated. This is a new contribution to the development of PAD as, to the best of our knowledge, no previous work has focused on the treatment of compost leachate in pressurized digesters. This type of wastewater is produced in large quantities worldwide, and has peculiar characteristics such as a high content of organic matter with considerable aliquots of volatile fatty acids (VFA), acidic pH, high salinity, etc. [1,36,37]. The experiments were conducted on a pilot plant, working in semi-continuous mode under mesophilic conditions. The study assessed the effects of pressure increase, at different organic load rate (OLR) values, on process performance. Biogas composition, specific biogas yield (SBY), specific methane yield (SMY) and the main process parameters such as pH, VFA/alkalinity ratio, nutrients concentrations, etc., were evaluated in response to the pressure change. In addition, the characteristics of the produced digestates were analyzed.

2. Materials and Methods

2.1. Materials

In this study, compost leachate from a tunnel composting facility located in Rende near Cosenza (Calabria Region, Italy) was used as feedstock. Activated sludge, collected from the recirculation line of the wastewater treatment plant of Lamezia Terme (Calabria Region, Italy) and maintained in anaerobic conditions for 15 days, was used as inoculum for the AD start-up. This operation mode was selected on the basis of our previous works, which proved the applicability of activated sludge for the inoculation of an anaerobic digestion process [1,5]. The samples were stored in 30 L tanks at 4 °C, to avoid any degradation.

2.2. Pressurized Pilot Plant

The PDA tests were performed using a laboratory pilot-plant designed and built in the Laboratory of Sanitary and Environmental Engineering of the University of Calabria. The pilot plant was composed of a completely stirred tank reactor (CSTR), and some auxiliary devices (biogas measurement system, heating device, connecting pipelines, and data acquisition system) (Figure 1a). The digester consisted of a 3 L cylindrical reactor, made of stainless steel 316 and wrapped in an insulated heating jacket, able to withstand high operating pressures (up to 40 bar). The overall unit was hermetically closed by a top flange provided with a nozzle pipe suitable for feeding the leachate and collecting the digestate (Figure 1b). The digester was equipped with a vertical steel mixer powered by a gear motor. Mesophilic conditions (37 °C) were maintained by a heating device connected to the reactor. The pressure inside the reactor was controlled by a system of adjustable valves.

The produced biogas was left inside the reactor until the set pressure value was reached. At the required pressure, the excess biogas was extracted from the reactor.

The biogas measurement system consisted of a vertical cylindrical PVC tank divided into two equal septa, communicating with each other by a pipe (Figure 1c). In the lower septum, two silicone pipes were connected to a two-way solenoid valve that allowed the inlet and outlet of the biogas (Figure 1c). The system operated as described below.

The produced biogas flowed from the digester in the lower septum, which was initially filled with water. The volume of biogas progressively displaced the water from the lower to the upper septum. Therefore, the lower septum was gradually filled with biogas and, simultaneously, the upper septum became filled with water. Once the maximum level was reached, the solenoid valve discharged the biogas from the lower septum and the system returned to the initial condition. The water level in the upper compartment was measured by an ultrasonic sensor (Microsonic® mic + 25/DIU/TC; Microsonic GmbH, Dortmund, Germany), equipped with integrated temperature compensation. The sensor sent an analogue signal to the PLC (Arduino® Mega 250; Arduino, Turin, Italy) that switched it into biogas volume.

2.3. PDA Experimental Set-Up

The effects of the pressure increase on the AD performance at high OLR values were investigated. In particular, five operating pressures, from atmospheric to 4 bar, and two different organic loading rate (OLR) values, 20 kg_{COD}/m^3d and 30 kg_{COD}/m^3d, were tested.

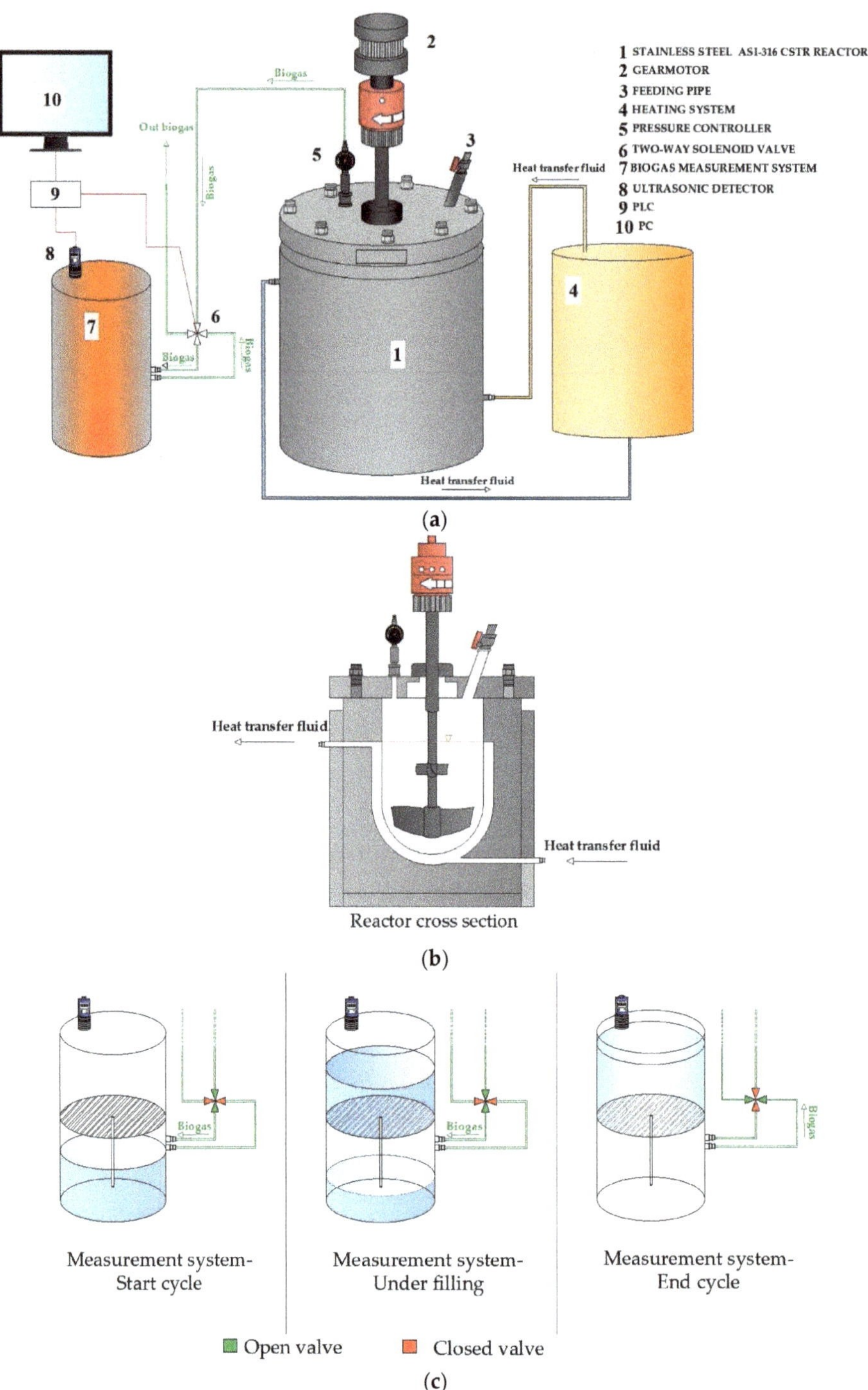

Figure 1. PAD laboratory pilot plant (**a**); reactor cross section (**b**); biogas measurement system (**c**).

The OLR were selected to investigate PAD in an optimal condition (20 kg_{COD}/m^3d) and in an unfavorable state (30 kg_{COD}/m^3d). In fact, our previous studies proved that, under atmospheric conditions, the digestion of compost leachate can efficiently evolve up to a high OLR of about 25 kg_{COD}/m^3d, beyond which the performance rapidly deteriorates [1].

Before performing the pressurized tests, the reactor was inoculated with activated sludge that was maintained in anaerobic conditions for 15 days. After the preliminary phase, the CSTR was started in semi-continuous mode keeping the working volume at 1.5 L. The organic loading rate was gradually increased up to 20 kg_{COD}/m^3d. Subsequently, holding the OLR of 20 kg_{COD}/m^3d, the pressure was increased stepwise from the atmospheric value up to 4 bar. At this pressure, the OLR was then brought to 30 kg_{COD}/m^3d. Finally, keeping the last OLR constant, decreasing pressures were applied until the atmospheric value was re-established. For each OLR, the values of pressure were maintained for two weeks. The average values of the parameters monitored during these two weeks were assumed and presented in the following sections.

The feeding of the reactor and the extraction of the digestate were carried out manually. PDA tests lasted approximately five months, during which no additional chemicals were added. For each operating condition, a chemical–physical characterization of digestate was carried out.

2.4. Analytical Methods

Conductivity and pH were measured through benchtop analyzers (Crison BASIC 30 EC, Crison BASIC 20 pH; Hach Lange, Barcelona, Spain). Total solids (TS) and volatile solids (VS) were measured by weight analysis after drying the samples at 105 and 550 °C [38]. Alkalinity was measured by the potentiometric method [38]. Total COD and soluble COD were measured after digestion with potassium dichromate ($K_2Cr_2O_7$ 0.5N) and volumetric titration with ammonium iron sulphate $(NH_4)_2Fe(SO_4)_2 \cdot 6H_2O$ [38]. Volatile fatty acids (VFA) were detected after distillation of the sample (VELP UDK 127 distillation unit; VELP Scientifica srl, Usmate, MB, Italy) and titration with sodium hydroxide (NaOH 0.01N) [38]. Ammonia nitrogen (N-NH_4^+), orthophosphates (P-PO_4^{3-}) and sulphates (SO_4^{2-}) were detected by spectrophotometric analysis with a UV-Vis (Thermo Spectronic Genesys 10uv; Thermo Fischer Scientific, Waltham, MA, USA) [38]. Metals were determined, after calcination of the sample at 550 °C, with atomic absorption spectrophotometry (GBC 933 PLUS; GBC Scientific Equipment, Braeside, VIC, Australia) [38]. Each analysis was carried out in triplicate. The biogas production was detected at standard conditions. Methane percentage was detected daily after acidic gases and carbon dioxide neutralization through sodium hydroxide, as described in our previous work [1].

3. Results and Discussion

3.1. Composting Leachate and Activated Sludge Characteristics

Composting leachate used in this study was characterized by a moderately acidic pH of 5.3, and by a notable conductivity value of around 5.6 mS/cm (Table 1).

A large amount of organic matter was detected with a COD concentration equal to 66.5 g/L and a soluble fraction of about 80%. Moreover, a very high content of volatile fatty acids (VFA), close to 15.2 $g_{CH3COOH}/L$, was observed. The total Kjeldahl nitrogen concentration (TKN) resulted of about 1.5 gN/L, and the ammoniacal form was found to be 0.66 gN/L. Remarkable quantities of orthophosphate (P-PO_4^{3-}) and sulphate (SO_4^{2-}) were also measured. Due to the low pH value, dissolved metals such as Zinc (Zn), Nickel (Ni), Iron (Fe), Lead (Pb), and Manganese (Mn) were detected.

The characteristics presented above are representative of a leachate with a low maturation degree, generated in a composting process that evolved in micro-aerobic conditions. In fact, the values of total and solubilized COD indicate that the organic matter was not yet degraded. Furthermore, the large amount of VFA and the acidic pH suggested that the biological transformations evolved with lack of oxygen, similarly to the typical acid-acetogenic phases in fermentation processes.

Table 1. Characteristics of raw compost leachate and activated sludge (d.l. detection limit).

Parameters	Measure Unit	Compost Leachate	Activated Sludge
pH	-	5.35 ± 0.2	6.87 ± 0.1
Conductivity	mS/cm	5.62 ± 0.1	1.19 ± 0.1
TS	g/L	61.89 ± 2.01	10.85 ± 0.08
VS	g/L	38.23 ± 2.11	8.91 ± 0.09
COD	g/L	66.50 ± 3.5	12.84 ± 0.33
COD_{sol}	g/L	54.28 ± 0.24	1.76 ± 0.11
Alkalinity	g_{CaCO3}/L	12.56 ± 0.77	0.51 ± 0.04
VFA	$g_{CH3COOH}/L$	15.23 ± 0.78	0.08 ± 0.003
TKN	g/L	1.52 ± 0.14	3.05 ± 0.32
$N\text{-}NH_4^+$	g/L	0.66 ± 0.05	1.40 ± 0.11
$P\text{-}PO_4^{3-}$	g/L	0.55 ± 0.03	0.039 ± 0.003
SO_4^{2-}	g/L	0.45 ± 0.028	0.088 ± 0.002
Ca^{2+}	g/L	3.55 ± 0.021	0.098 ± 0.002
Mg^{2+}	g/L	0.82 ± 0.036	0.039 ± 0.001
K^+	mg/L	0.61 ± 0.017	<d.l.
Fe^{2+}	mg/L	113.8 ± 4.1	0.31 ± 0.01
Pb^{2+}	mg/L	34.37 ± 1.1	<d.l.-
Mn^{2+}	mg/L	10.61 ± 0.21	0.10 ± 0.005
Zn^{2+}	mg/L	20.02 ± 0.4	<d.l.-
Ni^{2+}	mg/L	0.21 ± 0.01	<d.l.-

The properties of inoculum were in line with the values of a typical activated sludge taken from the recirculation line of a municipal wastewater treatment plant. In particular, a total solids (TS) content of 10.8 g/L and a volatile fraction of 82% were found.

3.2. *Performance of PAD*

3.2.1. COD Removal

The characterization of the feedstock used during the experiments showed a high content of organic matter, which could lead to a notable biogas production. However, some of the leachate properties have the potential to hinder the evolution of the digestion process. Indeed, the acidic pH and the high level of VFA are adverse factors that can inhibit the methanogens activity. These negative effects can be overcome if an adequate buffer capacity is reached in the digester [1]. However, as previously described, the consequences of low pH values and great amounts of VFA could be even more marked in pressurized anaerobic digestion. The experiments conducted permitted to verify whether high-pressure digestion is suitable for the treatment of compost leachate.

In this regard, PAD tests were carried out at 20 and 30 kg_{COD}/m^3d, varying the process pressure between the atmospheric value up to 4 bar. Under atmospheric conditions the removal yield was around 88 and 56% at 20 and 30 kg_{COD}/m^3d, respectively (Figure 2). These values, in agreement with our previous work [1], confirmed that at atmospheric pressure, the anaerobic digestion of compost leachate significantly deteriorates when the OLR exceeds values of about 20 kg_{COD}/m^3d. This deterioration of COD conversion is a clear consequence of a substrate overload condition. With the highest OLR applied, the organic matter abatement was independent of the operating pressure, and fluctuated between 50 and 60%. On the other hand, the detected results proved that the pressure growth causes a significant negative effect on COD degradation when the process operates under favorable OLR values. Indeed, at an OLR of 20 kg_{COD}/m^3d, the pressure increase led to a decrease in COD removal efficiency from 88% at atmospheric pressure, to 62% at 4 bar (Figure 2).

In particular, there was a marked reduction in COD yield in response to a slight pressure increase to just 1 bar. Above this value, a slower decreasing trend of the COD abatement was observed.

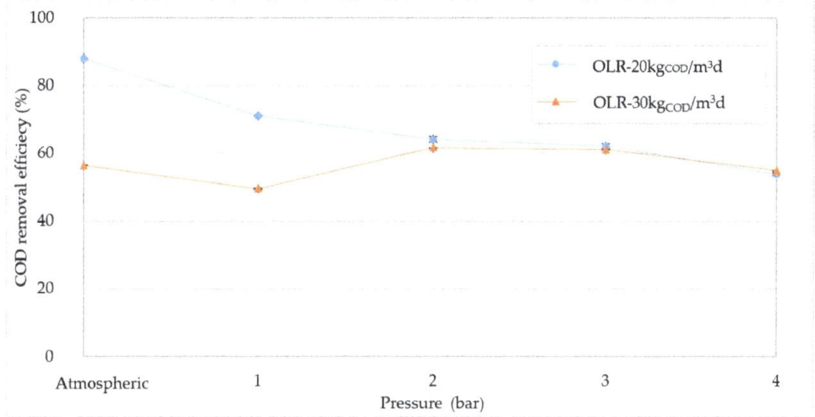

Figure 2. Chemical oxygen demand (COD) removal efficiency at increasing pressures.

The adverse impact of pressure on COD removal was reported in previous studies [25,34,36,39]. Latif et al. [34], in a single-stage reactor, achieved COD removal efficiencies about 35 and 30% at 4 bar and 6 bar, respectively. Chen et al. [25] raised the working pressure to 9 bar and observed a worsening of the process performance at high OLR values and short hydraulic retention times.

3.2.2. Biogas Production and Composition

At OLR = 20 kg_{COD}/m^3d, the trend of biogas production, as a function of process pressure (Figure 3), was generally consistent with the amounts of COD removed (Figure 2). Indeed, due to the substantial reduction in COD abatement with the pressure growth to 1 bar, the generated biogas showed a similar decrement. In particular, the biogas volume diminished by approximately 23% from 16.8 L, recorded at atmospheric pressure, to 12.8 L, at 1 bar. The production further decreased to about 10.4 L as the pressure increased to 4 bar (Figure 3).

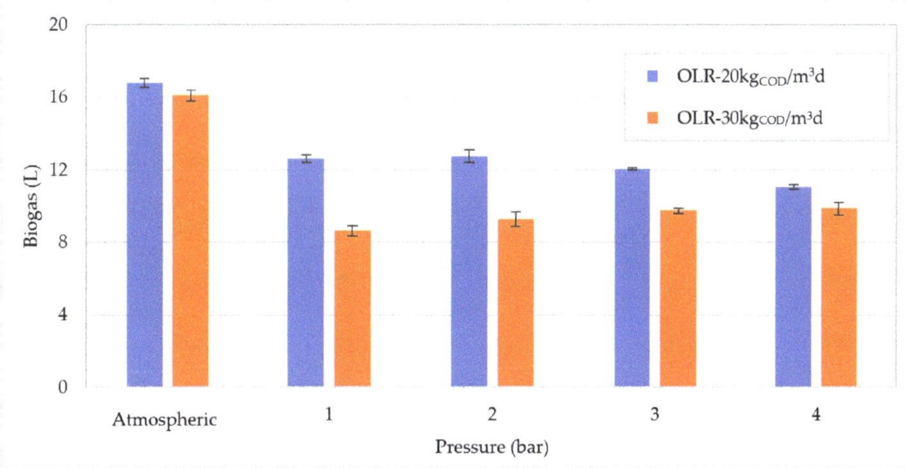

Figure 3. Biogas production at increasing pressures for organic load rate (OLR) of 20 kg_{COD}/m^3d and 30 kg_{COD}/m^3d.

At OLR = 30 kg_{COD}/m^3d, despite the lower yield in COD degradation, the volume of biogas produced by operating at atmospheric pressure (16 L) was quite similar to that detected at 20 kg_{COD}/m^3d (Figure 3). This can be explained by the fact that the overall amount of COD converted in atmospheric conditions was analogous with the two OLR values tested. With the growth in pressure there was a stabilization of the biogas production, and the values ranged between 9.28 and 9.87 L. The results presented above indicate that, as the pressure grows, the organic matter digestion and the biogas production deteriorate, and this effect is more marked at 20 kg_{COD}/m^3d. On the other hand, the quality of biogas was notably improved by the increase in pressure. Indeed, as shown in Figures 4 and 5, at increasing pressure a linear increase in CH_4 percentage was detected for both applied OLR values. In particular, the methane fraction grew by approximately 12–14% from the percentage of 62% at atmospheric pressure, to 74–76% at 4 bar (Figures 4 and 5). The enhancement in methane fraction can be attributed to the higher solubilization of CO_2 at increasing pressure generated in the reactor. This was confirmed by the progressive decrease in the CO_2 aliquot in the biogas. The improvement in biogas composition represents an undoubted advantage of PAD processes which was proven in several previous works. In particular, Lindeboom et al. [24] observed an increase in methane from 49 $\pm$ 2% up to a maximum of 73 $\pm$ 2% at 5 bar in the digestion of starch. The authors found that no further improvement in biogas composition occurred at pressures above 5 bar. Other works, however, reported an enhancement in biogas quality at pressures above 10 bar. Merkle et al. [29] observed an increase in CH_4 content to over 90% at 50 bar, by treating leachate of grass and maize silage.

In our experiments, despite the enhancement in biogas quality, the overall amount of methane decreased with the pressure rise. Indeed, the reduction in the volume of produced biogas was not compensated by the increase in the CH_4 fraction. Therefore, higher pressures in the reactor led to lower methane yields.

The worsening of the process performance was confirmed by the trends of the specific biogas yield (SBY) and of the specific methane yield (SMY) (Figures 6 and 7), defined as the volume produced per gram of COD removed.

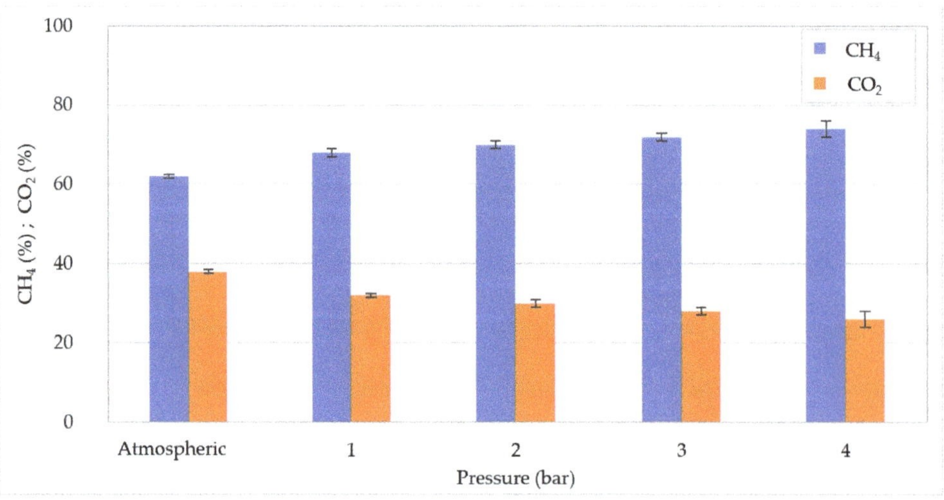

Figure 4. CH_4 and CO_2 percentages at increasing pressures for organic load rate (OLR) of 20 kg_{COD}/m^3d.

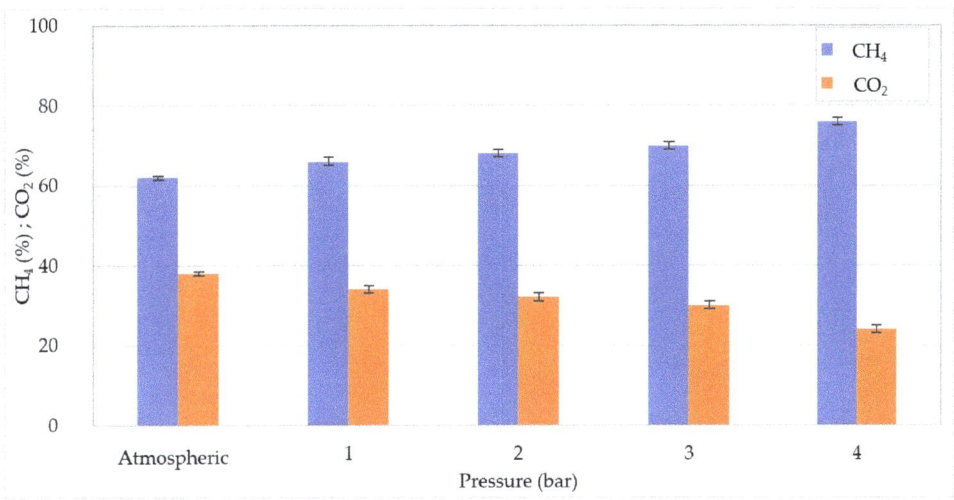

Figure 5. CH_4 and CO_2 percentages at increasing pressures for organic load rate (OLR) of 30 kg_{COD}/m^3d.

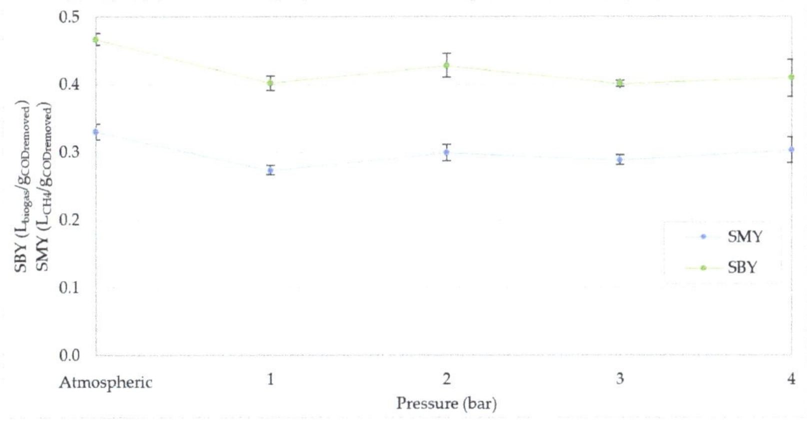

Figure 6. Specific biogas yield (SBY) and specific methane yield (SMY) at increasing pressures for organic load rate (OLR) of 20 kg_{COD}/m^3d.

In particular, at an OLR of 20 kg_{COD}/m^3d, in agreement with our previous work, the SMY was around 0.33 $L_{CH4}/g_{CODremoved}$ operating in atmospheric conditions (Figure 6). This value is quite close to the stoichiometric yield at 37 °C, equal to 0.4 $L_{CH4}/g_{CODremoved}$, which confirms the high efficiency obtainable on the digestion of compost leachate, even at such a high OLR value [1,3]. There was a consistent decrease in SMY and SBY, bringing the pressure to 1 bar while, beyond this value, the variations were not significant.

By working at OLR = 30 kg_{COD}/m^3d and atmospheric pressure, there was a worse conversion of the organic matter, and lower SMY and SBY were detected than those achieved at 20 kg_{COD}/m^3d (Figures 6 and 7). The yields underwent a reduction of about 35% with the pressure rise to 1 bar (Figure 7), beyond which SMY and SBY oscillated around 0.18 $L_{CH4}/g_{CODremoved}$ and 0.25 $L/g_{CODremoved}$, respectively. Consistent with our results, Chen et al. [25] found a notable attenuation of the specific methane yield between 1.5 bar to 9 bar, when the OLR was raised over to 15 kg_{COD}/m^3d. Li et al. [40] observed a production yield at a pressure of 3 bar, significantly higher than that detected at 10 bar.

Lemmer et al. [26] found a decrease in the SMY in the digestion of grass/maize-silage from 303.8 ± 47.2 mL/$g_{CODadded}$ to 258.0 ± 45.3 mL/$g_{CODadded}$, at increasing pressure from 1 bar to 9 bar. The same authors [39] reported a stable value of the specific yields between 10 and 30 bar. These statements suggest that the greatest negative effects on SMY occur at moderate increases in pressure, and then the specific productions tend to stabilize. In particular, our results showed that the reduction in SMY and SBY mainly occurred at 1 bar.

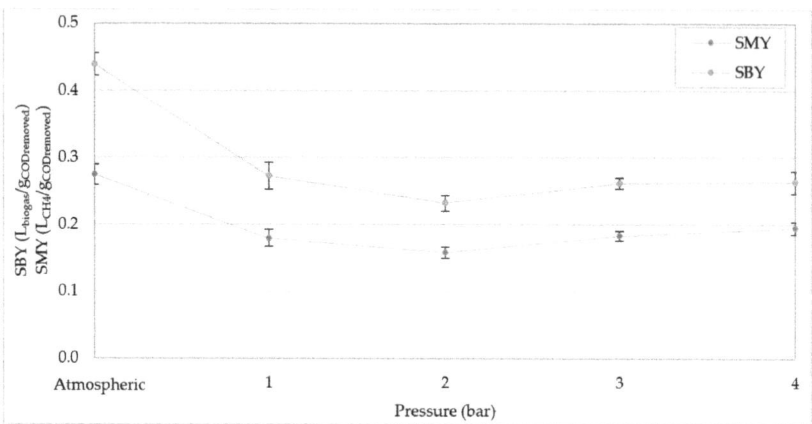

Figure 7. Specific biogas yield (SBY) and specific methane yield (SMY) at increasing pressures for organic load rate (OLR) of 30 kg_{COD}/m^3d.

3.2.3. pH and VFA/Alkalinity

Generally, the deterioration of digestion performance is attributed to the decrease in pH caused by the greater CO_2 solubilization with increasing pressure. Moreover, this phenomenon can favor the activity of acetogenic bacteria leading to an accumulation of VFA in the reactor [24], which further accentuates the acidification of digesting mixture. Clearly, such an effect could cause the inhibition of methanogens and, therefore, lower methane production. Our results confirmed the reduction in pH with the pressure growth, and decreasing linear trends were detected for both OLR values tested (Figure 8). At OLR = 30 kg_{COD}/m^3d the pH values were always lower than those measured at 20 kg_{COD}/m^3d (Figure 8), as a consequence of higher levels of VFA. In fact, as shown in Figure 9, at atmospheric pressure the VFA reached a value around 7.5 $g_{CH3COOH}/L$, while the concentration was below 6 $g_{CH3COOH}/L$ at 20 kg_{COD}/m^3d. The higher amount of VFA found at OLR = 30 kg_{COD}/m^3d corresponded to a greater value of VFA/Alkalinity that overcame 0.45 $g_{CH3COOH}/g_{CaCO3}$ (Figure 10). These values are representative of an overload condition (excessive OLR value) that justifies the lower digestion performance detected at atmospheric pressure. With increasing pressure, both the VFA concentrations and VFA/Alkalinity ratio increased. Therefore, under overload conditions (30 kg_{COD}/m^3d) the pressure increase aggravates the digestion upset by promoting the accumulation of volatile fatty acids. However, with an organic load of 20 kg_{COD}/m^3d, the pH values (Figure 8) were significantly higher compared with those monitored at 30 kg_{COD}/m^3d and, furthermore, the concentration of volatile fatty acids did not show a significant accumulation with the operating pressure (Figure 9). Consistent with the trend of VFA, the values of VFA/Alkalinity were always below 0.41 $g_{CH3COOH}/g_{CaCO3}$ (Figure 10). According to our previous study, these values are tolerable in the digestion of compost leachate in CSTR systems without the occurrence of inhibition effects [1]. Based on these considerations, at OLR = 20 kg_{COD}/m^3d the deterioration in biogas production with increasing pressure is not related to acidification conditions. Therefore, it can be affirmed that, contrarily to that hypothesized in some literature reports, the pressurized conditions in the digestion of waste with a high content of easily degradable substrates, such as compost leachate, do not cause detrimental acidification of mixture if suitable OLR values are applied.

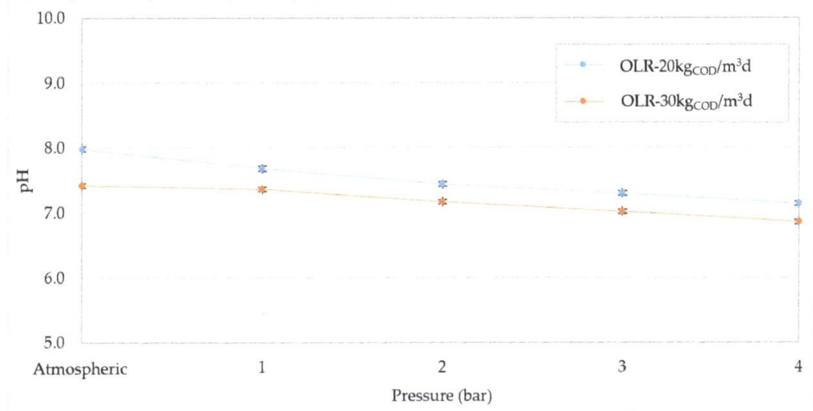

Figure 8. pH values at increasing pressures for organic load rate (OLR) of 20 kg_{COD}/m^3d and 30 kg_{COD}/m^3d.

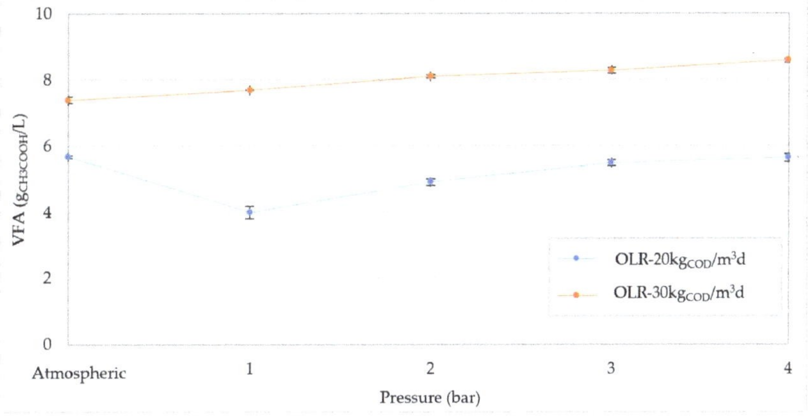

Figure 9. Volatile fatty acids (VFA) trends at increasing pressures for organic load rate (OLR) of 20 kg_{COD}/m^3d and 30 kg_{COD}/m^3d.

Other mechanisms probably play a role in the adverse effects of pressure on the evolution of digestion. Some authors argued that high-pressure conditions can reduce the hydrolytic capacity of the system, which would be consistent with the lack of VFA accumulation. Other research took into consideration the effects of pressure in microbial community and microbial activity. Under pressurized conditions, a lower diversity and richness in microbial species was observed. Abe and Horikoshi [41] reported that the growth rates of piezosensitive microbes drop with the pressure. At high pressures, Li et al. [40] found a low abundance of Archea able to use the hydrogenotrophic methanogenesis pathways. This condition can negatively impact the direct interspecies electron transfer (DIET) mechanism [42], unbalancing the syntrophic relationship between the microbial community and worsening the overall digestion performance.

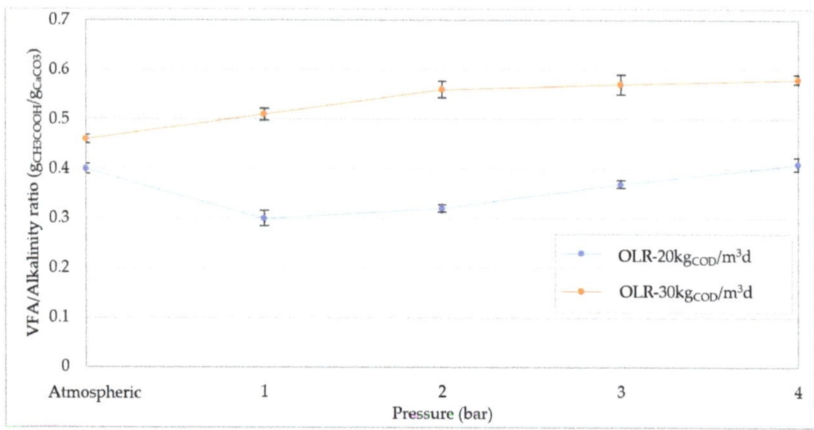

Figure 10. Volatile fatty acids/Alkalinity ratio at increasing pressures for organic load rate (OLR) of 20 kg_{COD}/m^3d and 30 kg_{COD}/m^3d.

3.2.4. Sulphate (SO_4^{2-})

In general, high levels of sulfate in the digester promote the development of sulfate-reducing bacteria (SRB), which oxidize the organic matter using SO_4^{2-} as an electron acceptor. The sulfate-reducing mechanism could induce positive effects in pressurized reactors, as the sulfate reduction generates alkalinity, which could counteract the digestate acidification. However, the development of SRB can upset the activity of methanogens, as the two groups of microorganisms compete with each other. Moreover, the reduction of sulfate generates sulfide, which can induce toxic effects on methane-producing Archea. These adverse effects are related to the COD/SO_4 ratio, and a total or partial inhibition of methanogenesis might occur for values below 4 or between 4–10 g_{COD}/g_{SO4}, respectively [1]. During our experiments, a sulphate amount able to cause the competition or inhibition phenomena was not reached. Indeed, the COD/SO_4 always remained above 10. Moreover, no significant changes in sulfate concentration were observed with increasing pressure (Figure 11). Our results suggest that pressure increase does not have any influence on the sulfate degradation and SRB development. Previous works have reported that pressured conditions do not adversely affect the growth of sulphate-reducing bacteria [43,44].

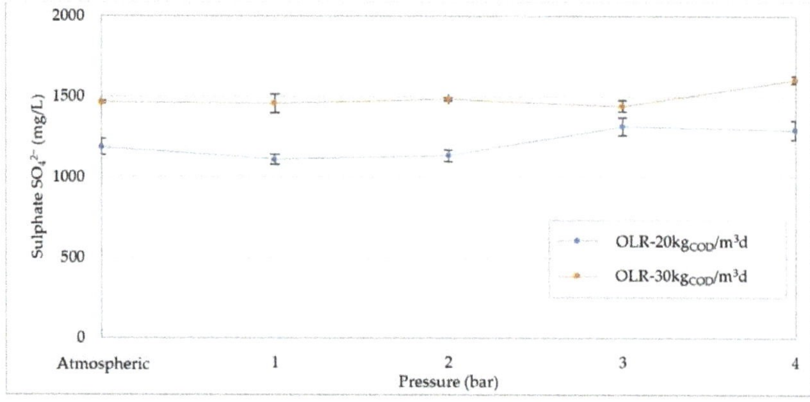

Figure 11. Sulphate (SO_4^{2-}) concentration at pressure increases for organic load rate (OLR) of 20 kg_{COD}/m^3d and 30 kg_{COD}/m^3d.

3.2.5. Ammonia Nitrogen ($N\text{-}NH_4^+$)

The presence of ammonia nitrogen is an important factor for a stable evolution of AD. Ammonia is generated by the breakdown of proteins and amino acids [3] and could reach harmful levels during the digestion of wastes with a high content of organic nitrogen. According to Hansen et al. [45], under thermophilic conditions, at pH 8 a free ammonia concentration of 1100 mg/L is toxic to anaerobic microorganisms. Khanal et al. [3] reported that an ammonium ion concentration above 3000 mg/L causes inhibitory effects, regardless of the operating conditions. On the other hand, some works considered a high ammonia nitrogen production beneficial in pressurized digesters, as it could increase the buffer capacity and hinder the acidification phenomena [26]. In our experiments, with an OLR of 20 kg_{COD}/m^3d, the ammonium concentration increased by about 15%, as the pressure rose from the atmospheric value to 1 bar (Figure 12). Over 1 bar, the $N\text{-}NH_4^+$ slightly grew, reaching a maximum value of around 2100 mg/L. At OLR = 30 Kg_{COD}/m^3d, higher $N\text{-}NH_4^+$ values were detected in atmospheric conditions as a clear consequence of the increase in organic load, whereas a lower increase was monitored in response to the pressure growth. For each applied condition, the values of ammonium were quite high, but remained below the threshold considered able to cause inhibition effects [3]. Therefore, no negative impacts on biogas production can be attributed to the ammonia production. On the other hand, the increase in $N\text{-}NH_4^+$, which occurred mainly at 1 bar, did not avoid the pH from falling at increasing pressure. Based on these results, it can be affirmed that the generation of ammonia does not have a significant effect on the pressurized anaerobic digestion of compost leachate.

3.2.6. Phosphate ($P\text{-}PO_4^{3-}$)

During the AD process, only small amounts of nutrients are used for cell synthesis and, thus, most of the initial amount in the feedstock is released into the digestate [3]. Phosphorus generally remains in the liquid phase as orthophosphate ions. However, depending on the presence of metallic elements, PO_4^{3-} could precipitate in the form of insoluble compounds such as $Ca_3(PO_4)_2$, $FePO_4$, $AlPO_4$ or struvite ($MgNH_4PO_4{\cdot}6H_2O$) [12,46]. This phenomenon tends to occur on the surface of the pipelines which connect the various units (digesters, thickeners, etc.) and could cause the ducts obstruction. The high pressure in anaerobic digestion may enhance the phosphate solubility due to the increased CO_2 concentration in the liquid phase. This helps to limit the adverse effect of the uncontrolled precipitation of phosphate. Moreover, after the pressurized digestion and the digestate thickening, the soluble PO_4^{3-} in the liquid phase could be recovered through controlled precipitation treatments, in the form of valuable compounds such as struvite, which are potentially reusable as slow-release fertilizers [12]. The results of our investigations confirmed the increase in dissolved PO_4^{3-} concentration in response to the pressure growth. As shown in Figure 13, no significant differences were found with the two applied OLR. In both cases, the phosphorus concentration linearly increased by about 35%, bringing the pressure from the atmospheric value to 4 bar. In agreement with these results, Latif et al. [34] found an enhancement in phosphate concentration from 51 to 73 mg/L, between 2 and 6 bar. Despite the increase in phosphorus solubility, it should be noted that the PO_4^{3-} concentrations were quite low compared with the amount of P in the feedstock (Table 1). Therefore, under pressurized conditions, only a small enhancement in phosphorus solubilization can be reached.

3.3. *Digestate Characteristics*

Table 2 shows the physical–chemical characteristics of the digestates obtained in the different operating conditions tested.

High contents of fertilizing elements such as $N\text{-}NH_4^+$, K^+, Ca^{2+}, Mg^{2+} and SO_4^{2-} were found for all samples. These elements make the digestate a valuable matrix potentially exploitable in agronomic practices. Furthermore, in compliance with the recommendations for agricultural applications, digestates were characterized by low ratios between organic matter content and nitrogen (Table 2). This parameter is of great importance as it affects the availability of nitrogen in soil [1]. The orthophosphate content was small compared with the amount of nitrogen. The low content of the soluble phosphorus, as previously discussed, is mainly attributable to the precipitation of phosphate salts [1,3,34]. Therefore, the use of compost leachate digestate for fertilizing purposes would be more suitable in soils with sufficient amounts of phosphorus. The presence of hazardous metal ions, such as Pb, Ni and Zn, was negligible. Their concentrations were lower than those required by current fertilizer regulations [47]. These low concentrations were due to the pH values which allowed the precipitation of most of the metallic species present in the mixture. Overall, for most of the monitored parameters, a clear effect of the pressure increase on their release into the digestate was not found.

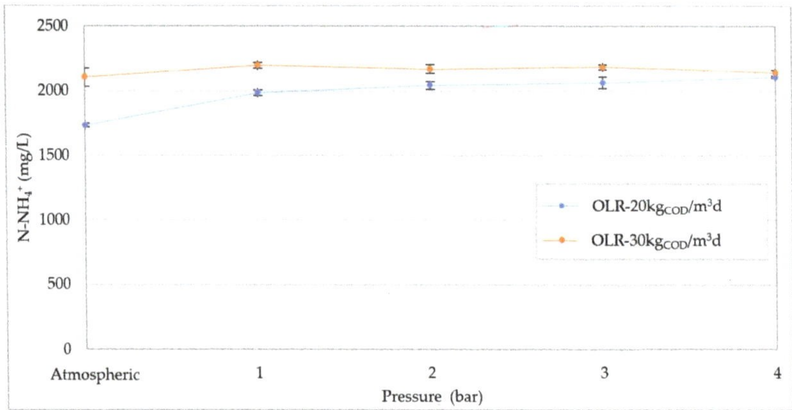

Figure 12. Ammoniacal nitrogen (N-NH_4^+) concentration at increasing pressures for organic load rate (OLR) of 20 kg_{COD}/m^3d and 30 kg_{COD}/m^3d.

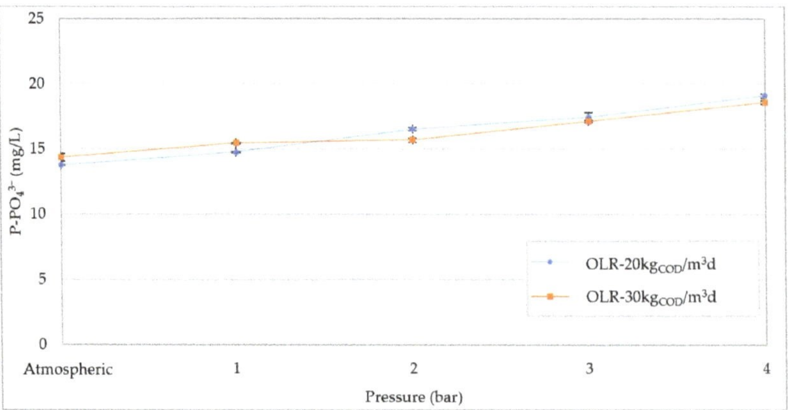

Figure 13. Phosphate (P-PO_4^{3-}) concentration at increasing pressures for organic load rate (OLR) of 20 kg_{COD}/m^3d and 30 kg_{COD}/m^3d.

Table 2. Digestate's chemical–physical characteristics (measure unit (M.U.); detection limit (d.l.)).

Parameters		OLR (Organic Load Rate)									
	M.U.	20 kg_{COD}/m^3d					30 kg_{COD}/m^3d				
Pressure	bar	Atmospheric	1	2	3	4	Atmospheric	1	2	3	4
pH	-	7.98	7.68	7.44	7.31	7.14	7.42	7.36	7.17	7.02	6.86
Conductivity	mS/cm	19.01	21.9	22.94	23.81	25.62	21.22	22.51	24.43	25.51	26.24
COD	g/L	4.82	8.61	10.64	11.53	12.92	18.81	15.53	15.91	15.94	17.32
$N\text{-}NH_4^+$	mg/L	1735.06	1986.31	2043.14	2060.03	2138.16	2106.09	2199.12	2169.34	2185.21	2145.11
$P\text{-}PO_4^{3-}$	mg/L	13.72	14.74	15.71	18.56	19.15	14.54	15.51	16.54	17.12	18.63
SO_4^{2-}	mg/L	1187.00	1110.15	1132.24	1317.09	1295.31	1466.19	1458.13	1484.26	1442.27	1607.39
Ca^{2+}	mg/L	249.31	348.12	299.33	314.74	364.71	302.44	409.31	376.11	342.93	415.64
Mg^{2+}	mg/L	139.03	145.12	137.19	132.03	135.14	142.23	153.41	139.17	112.09	129.13
K^+	mg/L	566.44	459.50	476.11	503.33	452.72	489.64	556.14	504.83	567.41	499.72
Fe^{2+}	mg/L	19.31	19.83	17.42	18.90	21.41	26.44	19.50	18.53	22.71	21.61
Pb^{2+}	mg/L	0.47	0.34	0.39	0.46	0.51	0.31	0.48	0.31	0.33	0.42
Mn^{2+}	mg/L	0.57	0.72	0.84	0.79	0.63	0.73	0.71	0.64	0.49	0.76
Zn^{2+}	mg/L	0.27	0.22	0.26	0.19	0.31	0.37	0.18	0.26	0.34	0.28
Ni^{2+}	mg/L	<d.l.	<d.l.	<d.l.	<d.l.	<d.l.	<d.l.	<d.l.	<d.l.	<d.l.	<d.l.

4. Conclusions

In this study, the pressurized anaerobic digestion of compost leachate was investigated. The experiments were carried out by testing operating pressures up to 4 bar, with organic load rates of 20 kg_{COD}/m^3d and 30 kg_{COD}/m^3d. The detected results confirmed that pressure growth leads to the production of biogas with an increase in the methane fraction and a decrease in CO_2 content. At a pressure of 4 bar, a percentage of methane of around 75% was reached for both OLR values tested. On the other hand, there was a general decline of the digestion performance under high pressure conditions. At 20 kg_{COD}/m^3d, the COD removal was close to 88% at atmospheric pressure, and then decreased by about 35% as the pressure increased. At 30 kg_{COD}/m^3d, the removal of organic load remained around a quite low value of 56%, which is representative of overload conditions.

Regardless of the applied organic load, biogas production and specific methane yield notably worsened as the pressure increased to 1 bar. In particular, the SMY values fell from 0.33 to 0.27 $L_{CH4}/g_{CODremoved}$, at 20 kg_{COD}/m^3d, and from 0.27 to 0.18 $L_{CH4}/g_{CODremoved}$, at 30 kg_{COD}/m^3d. As expected, the acidification of digestate was monitored at growing pressures. However, the magnitude of this phenomenon was not sufficient to justify the worsening of the digestion performance. Moreover, no adverse effects imputable to sulphates and ammonia compounds were identified. Therefore, the detected results suggested the occurrence of other effects such as the reduction of hydrolytic capability, or the alteration of microbial activity. An enhanced solubilization of nutrient compounds, such as orthophosphates, was observed with increasing pressures. However, the increase in PO_4^{3-} was quite low up to a pressure of 4 bar. More generally, the pressure values did not substantially affect the chemical composition of the digestates, which showed characteristics compatible for agronomic utilization.

In conclusion, the digestion of compost leachate in pressurized reactors does not appear to be advantageous compared with digestion in atmospheric conditions. However, further studies, under different operating conditions (OLR, pressure values), are necessary to confirm these statements. Furthermore, additional research should be conducted to better clarify the mechanisms that mainly affect the biogas production.

Author Contributions: Conceptualization, A.S.; methodology, A.S. and C.L.; formal analysis, C.L. and G.M.C.; investigation, C.L.; data curation, A.S., C.L. and G.M.C.; writing—original draft preparation, A.S. and C.L.; writing—review and editing, A.S. and C.L.; supervision, A.S. All authors have read and agreed to the published version of the manuscript.

Funding: This research received no external funding.

Data Availability Statement: The data presented in this study are available on request from the corresponding author.

Conflicts of Interest: The authors declare no conflict of interest.

References

1. Siciliano, A.; Limonti, C.; Curcio, G.M.; Calabrò, V. Biogas Generation through Anaerobic Digestion of Compost Leachate in Semi-Continuous Completely Stirred Tank Reactors. *Processes* **2019**, *7*, 635. [CrossRef]
2. Siciliano, A.; Stillitano, M.A.; Limonti, C. Energetic Valorization of Wet Olive Mill Wastes through a Suitable Integrated Treatment: H2O2 with Lime and Anaerobic Digestion. *Sustainability* **2016**, *8*, 1150. [CrossRef]
3. Khanal, S. *Anaerobic Biotechnology for Bioenergy Production: Principles and Applications*; Wiley-Blackwell: Ames, IA, USA, 2008.
4. Barbera, E.; Menegon, S.; Banzato, D.; D'Alpaos, C.; Bertucco, A. From biogas to biomethane: A process simulation-based techno-economic comparison of different upgrading technologies in the Italian context. *Renew. Energy* **2019**, *135*, 663–673. [CrossRef]
5. Siciliano, A.; Limonti, C.; Mehariya, S.; Molino, A.; Calabrò, V. Biofuel Production and Phosphorus Recovery through an Integrated Treatment of Agro-Industrial Waste. *Sustainability* **2018**, *11*, 52. [CrossRef]
6. Barros, M.V.; Salvador, R.; de Francisco, A.C.; Piekarski, C.M. Mapping of research lines on circular economy practices in agriculture: From waste to energy. *Renew. Sustain. Energy Rev.* **2020**, *131*, 109958. [CrossRef]
7. Ubando, A.T.; Felix, C.B.; Chen, W.-H. Biorefineries in circular bioeconomy: A comprehensive review. *Bioresour. Technol.* **2020**, *299*, 122585. [CrossRef] [PubMed]
8. Mehariya, S.; Patel, A.K.; Obulisamy, P.K.; Punniyakotti, E.; Wong, J.W. Co-digestion of food waste and sewage sludge for methane production: Current status and perspective. *Bioresour. Technol.* **2018**, *265*, 519–531. [CrossRef]
9. Ren, Y.; Yu, M.; Wu, C.; Wang, Q.; Gao, M.; Huang, Q.; Liu, Y. A comprehensive review on food waste anaerobic digestion: Research updates and tendencies. *Bioresour. Technol.* **2018**, *247*, 1069–1076. [CrossRef]
10. Siciliano, A.; Stillitano, M.; De Rosa, S. Increase of the anaerobic biodegradability of olive mill wastewaters through a pre-treatment with hydrogen peroxide in alkaline conditions. *Desalin. Water Treat.* **2014**, *55*, 1735–1746. [CrossRef]
11. Monfet, E.; Aubry, G.; Ramirez, A.A. Nutrient removal and recovery from digestate: A review of the technology. *Biofuels* **2018**, *9*, 247–262. [CrossRef]
12. Siciliano, A.; Limonti, C.; Curcio, G.M.; Molinari, R. Advances in Struvite Precipitation Technologies for Nutrients Removal and Recovery from Aqueous Waste and Wastewater. *Sustainability* **2020**, *12*, 7538. [CrossRef]
13. Sun, Q.; Li, H.; Yan, J.; Liu, L.; Yu, Z.; Yu, X. Selection of appropriate biogas upgrading technology-a review of biogas cleaning, upgrading and utilisation. *Renew. Sustain. Energy Rev.* **2015**, *51*, 521–532. [CrossRef]
14. Liu, J.; Zhong, J.; Wang, Y.; Liu, Q.; Qian, G.; Zhong, L.; Guo, R.; Zhang, P.; Xu, Z.P. Effective bio-treatment of fresh leachate from pretreated municipal solid waste in an expanded granular sludge bed bioreactor. *Bioresour. Technol.* **2010**, *101*, 1447–1452. [CrossRef] [PubMed]
15. Chen, X.Y.; Vinh-Thang, H.; Ramirez, A.A.; Rodrigue, D.; Kaliaguine, S. Membrane gas separation technologies for biogas upgrading. *RSC Adv.* **2015**, *5*, 24399–24448. [CrossRef]
16. Toledo-Cervantes, A.; Estrada, J.M.; Lebrero, R.; Muñoz, R. A comparative analysis of biogas upgrading technologies: Photosynthetic vs physical/chemical processes. *Algal Res.* **2017**, *25*, 237–243. [CrossRef]
17. Demirbaş, A. Biogas Potential of Manure and Straw Mixtures. *Energy Sources* **2006**, *28*, 71–78. [CrossRef]
18. Pellegrini, L.A.; De Guido, G.; Consonni, S.; Bortoluzzi, G.; Gatti, M. From biogas to biomethane: How the biogas source influences the purification costs. *Chem. Eng. Trans.* **2015**, *43*, 409–414.
19. Lemmer, A.; Chen, Y.; Wonneberger, A.-M.; Graf, F.; Reimert, R. Integration of a Water Scrubbing Technique and Two-Stage Pressurized Anaerobic Digestion in One Process. *Energies* **2015**, *8*, 2048–2065. [CrossRef]
20. Niesner, J.; Jecha, D.; Stehlik, P. Biogas upgrading technologies: State of art review in European region. *Chem. Eng. Trans.* **2013**, *35*, 517–522.
21. Köppel, W.; Götz, M.; Graf, F. Biogas upgrading for injection into the gas grid. *GWF Gas Erdgas* **2009**, *150*, 26–35.
22. Leonzio, G. Upgrading of biogas to bio-methane with chemical absorption process: Simulation and environmental impact. *J. Clean. Prod.* **2016**, *131*, 364–375. [CrossRef]
23. Scamardella, D.; De Crescenzo, C.; Marzocchella, A.; Molino, A.; Chianese, S.; Savastano, V.; Tralice, R.; Karatza, D.; Musmarra, D. Simulation and Optimization of Pressurized Anaerobic Digestion and Biogas Upgrading Using Aspen Plus. *Chem. Eng. Trans.* **2019**, *74*, 55–60.
24. Lindeboom, R.E.; Ding, L.; Weijma, J.; Plugge, C.M.; van Lier, J.B. Starch hydrolysis in autogenerative high pressure digestion: Gelatinisation and saccharification as rate limiting steps. *Biomass Bioenergy* **2014**, *71*, 256–265. [CrossRef]
25. Chen, Y.; Rößler, B.; Zielonka, S.; Wonneberger, A.-M.; Lemmer, A. Effects of Organic Loading Rate on the Performance of a Pressurized Anaerobic Filter in Two-Phase Anaerobic Digestion. *Energies* **2014**, *7*, 736–750. [CrossRef]
26. Lemmer, A.; Chen, Y.; Lindner, J.; Wonneberger, A.; Zielonka, S.; Oechsner, H.; Jungbluth, T. Influence of different substrates on the performance of a two-stage high pressure anaerobic digestion system. *Bioresour. Technol.* **2015**, *178*, 313–318. [CrossRef] [PubMed]

27. Lindeboom, R.E.F.; Fermoso, F.; Weijma, J.; Zagt, K.; Van Lier, J.B. Autogenerative high pressure digestion: Anaerobic digestion and biogas upgrading in a single step reactor system. *Water Sci. Technol.* **2011**, *64*, 647–653. [CrossRef] [PubMed]
28. Wang, L.-K.; Chen, G.-J.; Han, G.-H.; Guo, X.-Q.; Guo, T.-M. Experimental study on the solubility of natural gas components in water with or without hydrate inhibitor. *Fluid Phase Equilibria* **2003**, *207*, 143–154. [CrossRef]
29. Merkle, W.; Baer, K.; Lindner, J.; Zielonka, S.; Ortloff, F.; Graf, F.; Kolb, T.; Jungbluth, T.; Lemmer, A. Influence of pressures up to 50 bar on two-stage anaerobic digestion. *Bioresour. Technol.* **2017**, *232*, 72–78. [CrossRef]
30. Merkle, W.; Baer, K.; Haag, N.L.; Zielonka, S.; Ortloff, F.; Graf, F.; Lemmer, A. High-pressure anaerobic digestion up to 100 bar: Influence of initial pressure on production kinetics and specific methane yields. *Environ. Technol.* **2016**, *38*, 337–344. [CrossRef] [PubMed]
31. Sarker, S.; Lamb, J.J.; Hjelme, D.R.; Lien, K.M. Overview of recent progress towards in-situ biogas upgradation techniques. *Fuel* **2018**, *226*, 686–697. [CrossRef]
32. Bär, K.; Merkle, W.; Tuczinski, M.; Saravia, F.; Horn, H.; Ortloff, F.; Graf, F.; Lemmer, A.; Kolb, T. Development of an innovative two-stage fermentation process for high-calorific biogas at elevated pressure. *Biomass Bioenergy* **2018**, *115*, 186–194. [CrossRef]
33. Chen, Y.; Rößler, B.; Zielonka, S.; Lemmer, A.; Wonneberger, A.-M.; Jungbluth, T. The pressure effects on two-phase anaerobic digestion. *Appl. Energy* **2014**, *116*, 409–415. [CrossRef]
34. Latif, M.A.; Mehta, C.M.; Batstone, D.J. Enhancing soluble phosphate concentration in sludge liquor by pressurised anaerobic digestion. *Water Res.* **2018**, *145*, 660–666. [CrossRef] [PubMed]
35. Lindeboom, R.E.F.; Shin, S.G.; Weijma, J.; van Lier, J.B.; Plugge, C.M. Piezo-tolerant natural gas-producing microbes under accumulating pCO_2. *Biotechnol. Biofuels* **2016**, *9*, 236. [CrossRef] [PubMed]
36. Mokhtarani, N.; Bayatfard, A.; Mokhtarani, B. Full scale performance of compost's leachate treatment by biological anaerobic reactors. *Waste Manag. Res.* **2012**, *30*, 524–529. [CrossRef]
37. Roy, D.; Azaïs, A.; Benkaraache, S.; Drogui, P.; Tyagi, R.D. Composting leachate: Characterization, treatment, and future perspectives. *Rev. Environ. Sci. Biotechnol.* **2018**, *17*, 323–349. [CrossRef]
38. APHA. *Standard Methods for the Examination of Water and Wastewater*, 20th ed.; American Public Health Association and Water Environment Federation: Washington, DC, USA, 1998.
39. Lemmer, A.; Merkle, W.; Baer, K.; Graf, F. Effects of high-pressure anaerobic digestion up to 30 bar on pH-value, production kinetics and specific methane yield. *Energy* **2017**, *138*, 659–667. [CrossRef]
40. Li, Y.; Liu, H.; Yan, F.; Su, D.; Wang, Y.; Zhou, H. High-calorific biogas production from anaerobic digestion of food waste using a two-phase pressurized biofilm (TPPB) system. *Bioresour. Technol.* **2017**, *224*, 56–62. [CrossRef]
41. Abe, F.; Horikoshi, K. The biotechnological potential of piezophiles. *Trends Biotechnol.* **2001**, *19*, 102–108. [CrossRef]
42. Calabrò, P.; Fazzino, F.; Limonti, C.; Siciliano, A. Enhancement of Anaerobic Digestion of Waste-Activated Sludge by Conductive Materials under High Volatile Fatty Acids-to-Alkalinity Ratios. *Water* **2021**, *13*, 391. [CrossRef]
43. Kallmeyer, J.; Boetius, A. Effects of Temperature and Pressure on Sulfate Reduction and Anaerobic Oxidation of Methane in Hydrothermal Sediments of Guaymas Basin. *Appl. Environ. Microbiol.* **2004**, *70*, 1231–1233. [CrossRef]
44. Weber, A.; Jørgensen, B.B. Bacterial sulfate reduction in hydrothermal sediments of the Guaymas Basin, Gulf of California, Mexico. *Deep Sea Res. Part I Oceanogr. Res. Pap.* **2002**, *49*, 827–841. [CrossRef]
45. Hansen, K.H.; Angelidaki, I.; Ahring, B.K. Anaerobic digestion of swine manure: Inhibition by ammonia. *Water Res.* **1998**, *32*, 5–12. [CrossRef]
46. Siciliano, A.; Stillitano, M.A.; Limonti, C.; Marchio, F. Ammonium Removal from Landfill Leachate by Means of Multiple Recycling of Struvite Residues Obtained through Acid Decomposition. *Appl. Sci.* **2016**, *6*, 375. [CrossRef]
47. European Parliament. *Regulation of the European Parliament and of the Council Laying Down Rules on the Making Available on the Market of CE Marked Fertilising Products and Amending Regulations (EC) No 1069/2009 and (EC) No 1107/2009 (COM(2016)0157-C8-0123/2016-2016/0084(COD))*; European Parliament: Strasbourg, France, 2019.

MDPI
St. Alban-Anlage 66
4052 Basel
Switzerland
Tel. +41 61 683 77 34
Fax +41 61 302 89 18
www.mdpi.com

Fermentation Editorial Office
E-mail: fermentation@mdpi.com
www.mdpi.com/journal/fermentation